AIP Handbook of Modern Sensors

Modern Sensors

Physics, Designs and Applications

■ Modern Instrumentation and Measurements in Physics & Engineering

Series Editor-in-Chief
Ray Radebaugh
National Institute of Standards and Technology
Boulder, Colorado

Modern Instrumentation and Measurements
in Physics & Engineering

AIP Handbook of Modern Sensors

Physics, Designs and Applications

Jacob Fraden

American Institute of Physics   **New York**

American Institute of Physics
335 East 45th Street
New York, NY 10017-3483

Library of Congress Cataloging-in-Publication Data

Fraden, Jacob
AIP Handbook of modern sensors: physics, designs and applications/
     Jacob Fraden.
     p. cm--(Modern instrumentation and measurements in physics &
     engineering)
     Includes bibliographical references and index.
     ISBN 1-56396-108-3
     1. Detectors. 2. Interface circuits. I. American Institute of
Physics. II. Title. III Series.
     TA165.F72 1993   93-84
     681'.2--dc20   CIP

Preface

The invention of the microprocessor has brought highly sophisticated instruments into our everyday life. Numerous computerized appliances, of which microprocessors are integral parts, wash clothes and prepare coffee, play music, guard homes, and control room temperature. What are these microprocessors? Microprocessors are digital devices that manipulate binary codes generally represented by electric signals. Yet we live in an analog world where these devices function among objects that are mostly not digital. Moreover, this world is generally not electrical (apart from the atomic level). Digital systems, however complex and intelligent they might be, must receive information from the outside world. Sensors are interface devices between various physical values and electronic circuits that "understand" only a language of moving electrical charges. In other words, sensors are eyes, ears, and noses of silicon chips.

Some sensors are relatively simple, for instance a photodiode, while others are complex devices which may comprise combinations of resistive, magnetic, optical, and other parts. Sensors operate on fundamental scientific principles that are taught in college undergraduate courses. The understanding of these devices generally requires an interdisciplinary background in such fields as physics, electronics, chemistry, etc.

In the course of my work, I often felt a strong need for a book that would combine practical information on diversified subjects related to the most important physical principles, design, and use of various sensors. Surely, I could find almost all I had to know in texts on physics, electronics, technical magazines, and manufacturers' catalogs. However, the information is scattered over many publications, and almost every question I was pondering required substantial research work and numerous trips to the library. Little by little, I have been gathering practical information on everything that in any way was related to various sensors and their applications to scientific and engineering measurements. Soon, I realized that the information I collected might be quite useful to more than one person. This idea prompted me to write this book.

This volume covers numerous modern sensors and detectors. Its goal is to examine basic physical principles which form a foundation for their operations and, thus, to stimulate a creative reader to choose alternative ways of design and to apply nontrivial solutions to trivial problems. It is clear that one book can not embrace the whole variety of sensors and their applications, even if it is called something like "The Encyclopedia of Sensors." This is a different book, and the author's task was much less ambitious. Here, an attempt has been made to generate a reference text that could be used by students, researchers interested in modern instrumentation (applied physicists and engineers), sensor designers, application engineers, and technicians whose jobs are to understand, select, and/or design sensors for practical systems. Obviously, to make an intelligent decision, one must well understand the physical principles behind the sensor operation. In addition, it is imperative to learn what is available in today's marketplace and what would be the best choice from two standpoints: performance and cost.

While this volume covers a broad range of sensors and detectors, an emphasis is on those devices that are less known, whose technology is still on the rise, and whose use permits the measurement of variables which were previously inaccessible. It is the author's intention to present a comprehensive and up-to-date account of the theory (physical principles), design, and practical implementations of various (especially the newest) sensors for scientific, industrial, and consumer applications.

This book contains 17 chapters; however, it is actually divided into three major parts. The first part consists of three chapters (1–3) to provide a general framework of the overall characteristics and physical principles of sensors and effects which form a foundation for their practical designs. Chapter 4 describes some useful electronic interfaces between sensors and peripheral processing devices. Here, the emphasis is made on those circuits that are suitable for integration with sensors. Certainly, the "meat" of this book is contained in the third part (chapters 5–17), which is organized by the type of variables that are measured. Numerous sensors are described in these chapters. Many of them are presently being manufactured by various companies, while some devices have only been patented and still remain just bright ideas on paper. Some sensors were presented by their designers during symposia and conferences, or published in books and professional journals. Certainly, the topics included in this part of the book reflect the author's own preferences and interpretations. Some may find a description of a particular sensor either too detailed, too broad, or too brief. In most cases, the author tried to make an attempt to strike a balance between a detailed description and a simplicity of coverage.

Quite often, the best sensor is the simplest one. It is, therefore, appropriate to quote the great American inventor Charles F. Kettering: "Inventing is a combination of brains and materials. The more brains you use, the less materials you need."

Jacob Fraden
San Diego, CA
January 1993

Table of Contents

To my children Roman and Julia

1
Data Acquisition

1.1 SENSORS, SIGNALS, AND SYSTEMS

A sensor is often defined as a *"device that receives and responds to a signal or stimulus"*. This definition is broad. In fact, it is so broad that it covers almost everything from a human eye to a trigger in a pistol. Consider the level control system shown in Fig. 1-1 [1]. The operator controls the level of fluid in the tank by manipulating its valve. Variations in the inlet flow rate, temperature changes (these would alter the fluids viscosity and consequently the flow rate through the valve) and similar disturbances must be compensated for by the operator. Without control, the tank is likely to flood, or run dry. To act appropriately, the operator must obtain information about the level of fluid in the tank on a timely basis. In this example, the information is perceived by the sensor which consists of two main parts: the sight tube on the tank and the operator's eye which generates an electric response in the optic nerve. The sight tube by itself is not a sensor and, in this particular control system, the eye is not a sensor either. Only the combination of these two components makes a narrow purpose sensor (detector) which is *selectively* sensitive to fluid level. If a sight tube is designed properly, it will very quickly reflect variations in the level and, it is said, that the sensor has a fast speed response. If internal diameter of the tube is too small for a given fluid viscosity, the level in the tube may lag behind the level in the tank. Then, we have to consider a phase characteristic of such a sensor. In some cases, the lag may be quite acceptable, while in other cases, a better sight tube design would be required. Hence, the sensor's performance must be assessed only as a part of a data acquisition system.

This World is divided into natural and man-made objects. The natural sensors, like those found in living organisms, usually respond with signals, having electrochemical character. That is, their physical nature is based on ion transport, like in the nerve fibers (such as an optic nerve in the fluid tank operator). In man-made devices, information is also transmitted and processed in electrical form, however, through the transport of electrons. Sensors which are used in artificial systems must speak the same language as the devices with which they are interfaced. This language is electrical in its nature and a man-made sensor should be capable of responding with signals where information is carried by displacement of electrons, rather than ions[1]. Thus, it should be possible to connect a sensor to an electronic system through electrical wires, rather than through an electrochemical solution or a nerve fiber. Hence, in this book, we use a somewhat narrower definition of sensors, which may be phrased as *"a sensor is a device that receives a signal or stimulus and responds with an electrical signal"*. The term stimulus is used throughout this book and needs to be clearly understood. The stimulus is the quantity, property, or condition that is sensed and converted into electrical signal. Some texts (for instance, [2]) use a different term measurand which has the same meaning, however with the stress on quantitative characteristic of sensing.

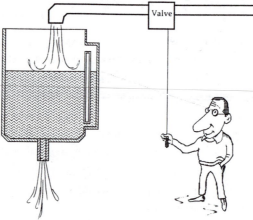

Fig. 1-1 Level Control System
A sight tube and operator's eye form a sensor:
a device which converts information into electrical signal.

The purpose of a sensor is to respond to some kind of an input physical property (stimulus) and to convert it into an electrical signal which is compatible with electronic circuits. We may say that a sensor is a translator of a generally nonelectrical value into an electrical value. When we say "electrical" we mean a signal, which can be channeled, amplified, and modified by electronic devices. The sensor's output signal may be in a form of voltage, current, or charge. These may be

---

1   There is a very exciting field of optical computing and communications where information is processed by a transport of photons. That field is beyond the scope of this book.

further described in terms of amplitude, frequency, and phase. This set of characteristics is called the output signal format. Therefore, a sensor has input properties (of any kind) and electrical output properties.

A sensor does not function by itself - it is always a part of a larger system which may incorporate many other detectors, signal conditioners, signal processors, memory devices, data recorders, and actuators. The sensor's place in a device is either intrinsic or extrinsic. It may be positioned at the input of a device to perceive the outside effects and to signal the system about variations in the outside stimuli. Also, it may be an internal part of a device which monitors the devices' own state to cause the appropriate performance. A sensor is always a part of some kind of a data acquisition system. Often such a system may be a part of a larger control system which includes various feedback mechanisms. To select an appropriate sensor, a system designer must address the question: "What is the *simplest* way to sense the stimulus without degradation of the overall system performance?".

All sensors may be of two kinds: *passive* and *active*. The passive sensors directly generate an electric signal in response to an external stimulus. That is, the input stimulus energy is converted by the sensor into output energy without the need for an additional power source. The examples are a thermocouple, a pyroelectric detector, and a piezoelectric sensor. The active sensors require external power for their operation, which is called an *excitation signal*. That signal is modified by the sensor to produce the output signal. The active sensors sometimes are called *parametric* because their own properties change in response to an external effect and these properties can be subsequently converted into electric signals. For example, a thermistor is a temperature sensitive resistor. It does not generate any signal, but by passing an electric current through it (excitation signal), its resistance can be measured by detecting variations in current and/or voltage across the thermistor. These variations (presented in ohms) directly relate to temperature.

To illustrate a place of sensors in a larger system, Fig. 1-2 shows a block diagram of a data acquisition and control device. An object can be anything: car, space ship, animal or human, liquid or gas. Any material object may become a subject of some kind of a measurement. Data are collected from an object by a number of sensors. Some of them (2, 3, and 4) are positioned directly on or inside the object. Sensor 1 perceives the object without a physical contact and, therefore, is called a *noncontact* sensor. Examples of such a sensor is a radiation detector and a TV camera. Sensor 5 serves a different purpose. It monitors internal conditions of a data acquisition system itself. Some sensors (1 and 3) can not be directly connected to standard electronic circuits because of inappropriate output signal formats. They require the use of interface devices (signal conditioners). Sensors 1, 2, 3, and 5 are passive. They generate electric signals without energy consumption from the electronic circuits. Sensor 4 is active. It requires an operating signal which is provided by an excitation circuit. An example of an active sensor is a thermistor which is a temperature sensitive resistor. It may operate with a constant current source which is an excitation circuit. Depending on the complexity of the system, the total number of sensors may vary from as little as one (a home thermostat) to many thousands (a space shuttle).

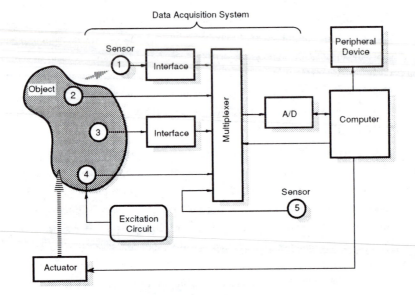

Fig. 1-2 Positions of sensors in a data acquisition system
Sensor 1 is noncontact, sensors 2 and 3 are passive, sensor 4 is active, and sensor 5 is internal to a data acquisition system.

Electrical signals from the sensors are fed into a multiplexer (MUX), which is a switch or a gate. Its function is to connect sensors one at a time to an analog-to-digital (A/D) converter or directly to a computer (if a sensor produces signals in a digital format). The computer controls a multiplexer and an A/D converter for the appropriate timing. Also, it may send control signals to the actuator which acts on the object. Examples of the actuators are an electric motor, a solenoid, a relay, and a pneumatic valve. The system contains some peripheral devices (for instance, a data recorder, a display, an alarm, etc.) and a number of components which are not shown in the block diagram. These may be filters, sample-and-hold circuits, amplifiers, etc.

To illustrate how such a system works, let us consider a simple car door monitoring arrangement. Every door in a car is supplied with a sensor which detects the door position (open or closed). In most cars, the sensor is a simple electric switch. Signals from all door sensors go to the car internal microprocessor (no need for an A/D converter as all door signals are in a digital format: ones or zeros). The microprocessor identifies which door is open and sends an indicating signal to the peripheral devices (a dash board display and an audible alarm). A car driver (the actuator) gets the message and acts on the object (closes the door).

An example of a more complex device is an anesthetic vapor delivery system. It employs several active and passive sensors. Vapor concentration of anesthetic agents (Halothane, Isoflurane, or Enflurane) is selectively monitored by an active piezoelectric sensor, installed into a ventilation tube. Molecules of anesthetic vapors add mass to the oscillating crystal in the sensor and change its natural frequency which is a measure of vapor concentration. Several other sensors monitor the concentration of CO_2, to distinguish exhale from inhale, and temperature and pressure, to compensate for additional variables. All these data are multiplexed, digitized, and fed into the microprocessor which calculates the actual vapor concentration. An anesthesiologist presets a desired delivery level and the processor adjusts the actuator (the valves) to maintain anesthetics at the correct concentration.

In the following chapters we concentrate on methods of sensing, physical principles of sensors operations, practical designs, and interface electronic circuits. Generally, sensor's input signals (stimuli) may have almost any conceivable physical or chemical nature. For instance: light flux, temperature, pressure, vibration, displacement, position, velocity, ion concentration, etc. The sensor's design may be of a general purpose. A special packaging and housing should be built to adapt it for a particular application. For instance, a micromachined piezoresistive pressure sensor may be housed into a water-tight enclosure for the invasive measurement of aortic blood pressure through a catheter. The same sensor will be given an entirely different enclose when it is intended for measuring blood pressure by a noninvasive oscillometric method. Some sensors are specifically designed to be very selective in a particular range of input stimulus and be quite immune to signals outside the desirable limits. For instance, a motion detector for a security system should be sensitive to movement of humans and not responsive to the movement of smaller animals, like dogs and cats.

1.2   SENSOR CLASSIFICATION

Sensor classification schemes range from very simple to the complex. One good way to look at a sensor is to consider all of its properties, such as what it measures (stimulus), what its specifications are, what physical phenomenon it is sensitive to, what conversion mechanism is employed, what material it is fabricated from and what is its field of application. Six tables bellow, adapted from [3], represent such a classification scheme which is pretty much broad and representative. If we take, for the illustration a surface acoustic-wave oscillator accelerometer, the table entries might be as follows:

Stimulus   acceleration
Specifications: sensitivity in frequency shift per *g* of
                                 acceleration, short and long terms stability in
                                 Hz per unit time, etc.
Detection means: mechanical
Conversion phenomenon: elastoelectric
Material: inorganic insulator
Field: automotive, marine, space and
                                 scientific measurement

Table 1-1 Specifications

Sensitivity	Stimulus range (span)
Stability (short and long term)	Resolution
Accuracy	Selectivity
Speed of response	Environmental conditions
Overload characteristics	Linearity
Hysteresis	Dead band
Operating life	Output format
Cost, size, weight	Other

Table 1-2 Sensor material

Inorganic	Organic
Conductor	Insulator
Semiconductor	Liquid gas or plasma
Biological substance	Other

Table 1-3 Detection means used in sensors

Biological
Chemical
Electric, magnetic or electromagnetic wave
Heat, temperature
Mechanical displacement or wave
Radioactivity, radiation
Other

Table 1-4 Conversion phenomena

Physical		Biological	
	Thermoelectric		Biochemical transformation, Physical transformation
	Photoelectric		
	Photomagnetic		Effect on test organism Spectroscopy
	Magnetoelectric		Other
	Electromagnetic		
	Thermoelastic		
	Electroelastic		
	Thermomagnetic		
	Thermooptic		
	Photoelastic		
	Other		
Chemical	Chemical transformation		
	Physical transformation Electrochemical process Spectroscopy		
	Other		

Table 1-5 Field of applications

Agriculture	Automotive
Civil engineering, construction	Domestic, appliances
Distribution, commerce, finance	Environment, meteorology, security
Energy, power	Information, telecommunication
Health, medicine	Marine
Manufacturing	Recreation, toys
Military	Space
Scientific measurement	Other
Transportation (excluding automotive)	

Table 1-6 Stimulus

STIMULUS		STIMULUS	
Acoustic	Wave amplitude, phase, polarization,	Mechanical	Position (linear, angular)
			Acceleration
	Spectrum		Force
	Wave velocity		Stress, pressure
	Other		
Biological	Biomass (types, concentration, states)		Strain
	Other		Mass, density
			Moment, torque
Chemical	Components (identities, concentration, states)		Speed of flow, rate of mass transport
	Other		Shape, roughness, orientation
			Stiffness, compliance
Electric	Charge, current		Viscosity
	Potential, voltage		Crystallinity, structural integrity
	Electric field (amplitude, phase, polarization, spectrum)		Other
	Conductivity	Radiation	Type
	Permitivity		Energy
	Other		Intensity
			Other
Magnetic	Magnetic field (amplitude, phase, polarization, spectrum)		
	Magnetic flux	Thermal	Temperature
	Permeability		Flux
	Other		Specific heat
			Thermal conductivity
Optical	Wave amplitude, phase, polarization, spectrum		Other
	Wave velocity		
	Refractive index		
	Emissivity, reflectivity, absorption		
	Other		

1.3 SENSOR SELECTION

When an engineer realizes that a certain variable has to be monitored, he/she faces a dilemma: what would be the best sensor for the job? Any sensor operation is based on a simple concept — a physical property of a sensor must be altered by an external stimulus to cause that property either to produce an electric signal or to modulate (to modify) an external electric signal. Quite often, the same stimulus may be measured by using quite different physical phenomena, and, subsequently, by different sensors. It is, therefore, a matter of an engineering choice to select the best sensor for the particular application. Selection criteria depend on many factors, such as availability, cost, power consumption, environmental conditions, etc. The best choice can be done only after all variables are considered. It is, therefore, not a correct question to ask, "What sensor should I use to measure X?". The proper question must include a wide spectrum of conditions to narrow the choice to one or two options.

1.4 UNITS OF MEASUREMENTS

In this book, we use base units which have been established in The 14th General Conference on Weights and Measures (1971). The base measurement system is known as SI which stands for French "Le Systéme International d'Unités" (Table 1-7). All other physical quantities are derivatives of these base units. Some of them are listed in Table 1-9 [4].

Table 1-7 SI basic units

Quantity	Name	Symbol	Defined by... (year established)
Length	meter	m	...the length of the path traveled by light in vacuum in 1/299,792,458 of a second... (1983)
Mass	kilogram	kg	...after a platinum-iridium prototype (1889)
Time	second	s	...the duration of 9,192,631,770 periods of the radiation corresponding to the transition between the two hyperfine levels of the ground state of the cesium-133 atom (1967)
Electric current	ampere	A	force equal to $2 \cdot 10^{-7}$ newton per meter of length exerted on two parallel conductors in vacuum when they carry the current (1946)
Thermodynamic temperature	kelvin	K	The fraction 1/273.16 of the thermodynamic temperature of the triple point of water (1967)
Amount of substance	mole	mol	...the amount of substance which contains as many elementary entities as there are atoms in 0.012 kg of carbon 12 (1971)
Luminous intensity	candela	cd	...intensity in the perpendicular direction of a surface of 1/600,000 m^2 of a blackbody at temperature of freezing Pt under pressure of 101,325 newton per m^2 (1967)
Plane angle	radian	rad	(supplemental unit)
Solid angle	steradian	sr	(supplemental unit)

Often it is not convenient to use base or derivative units directly - in practice quantities may be either too large or too small. For convenience in the engineering work, multiples and submultiples of the units are generally employed. They can be obtained by multiplying a unit by a factor from Table 1-8. When pronounced, in all cases the first syllable is accented. For example, 1 ampere (A) may be multiplied by factor of 10^{-3} to obtain a smaller unit: 1 milliampere (mA) which is one thousandth of an ampere.

Sometimes, two other systems of units are used. They are the Gaussian System and the British System, which in the U.S.A. its modification is called the US customary system. The United States is the only developed country where SI still is not in common use. However, with the end of Communism and the increase of World integration, international cooperation gains strong momentum. Hence, it is unavoidable that America will convert to SI[1] in the very near future. In this book, we will generally use SI, however, for the convenience of the reader, the US customary system units will be used in places where US manufacturers employ them for sensor specifications. For the conversion to SI from other systems[2] the reader may use Tables 1-10. To make a conversion, a nonSI value should be multiplied by a number given in the table. For instance, to convert acceleration of 55 ft/s² to SI, it must to be multiplied by 0.3048:

$$55 \text{ ft/s}^2 \times 0.3048 = 16.764 \text{ m/s}^2$$

Similarly, to convert electric charge of 1.7 faraday, it must be multiplied by $9.65 \cdot 10^{19}$:

$$1.7 \text{ faraday} \times 9.65 \cdot 10^{19} = 1.64 \cdot 10^{20} \text{ C}$$

Table 1-8 SI multiples

Factor	Prefix	Symbol	Factor	Prefix	Symbol
10^{18}	exa	E	10^{-1}	deci	d
10^{15}	peta	P	10^{-2}	centi	c
10^{12}	tera	T	10^{-3}	milli	m
10^{9}	giga	G	10^{-6}	micro	μ
10^{6}	mega	M	10^{-9}	nano	n
10^{3}	kilo	k	10^{-12}	pico	p
10^{2}	hecto	h	10^{-15}	femto	f
10^{1}	deka	da	10^{-18}	atto	a

[1] SI is often called the modernized metric system
[2] Nomenclature, abbreviations and spelling in the conversion tables are in accordance with *"Standard practice for use of the International System of units (SI) (the Modernized Metric System)"*. Standard E380-91a. ©1991 ASTM, 1916 Race St., Philadelphia, PA 19103.

Table 1-9 Derivative SI units

Quantity	Name of unit	Expression in terms of basic units
Area	square meter	m^2
Volume	cubic meter	m^3
Frequency	hertz (Hz)	s^{-1}
Density (concentration)	kilogram per cubic meter	kg/m^3
Velocity	meter per second	m/s
Angular velocity	radian per second	rad/s
Acceleration	meter per second squared	m/s^2
Angular acceleration	radian per second squared	rad/s^2
Volumetric flow rate	cubic meter per second	m^3/s
Force	newton (N)	$kg \cdot m/s^2$
Pressure	newton per square meter (N/m^2) or pascal (Pa)	$kg/m \cdot s^2$
Work energy heat torque	joule (J), newton-meter (N·m) or watt-second (W·s)	$kg \cdot m^2/s^2$
Power heat flux	watt (W) Joule per second (J/s)	$kg \cdot m^2/s^3$
Heat flux density	watt per square meter (W/m^2)	kg/s^3
Specific heat	joule per kilogram degree (J/kg·deg)	$m^2/s^2 \cdot deg$
Thermal conductivity	watt per meter degree (W/m·deg) or (J·m/s·m^2·deg)	$kg \cdot m/s^3 \cdot deg$
Mass flow rate (mass flux)	kilogram per second	kg/s
Mass flux density	kilogram per square meter-second	$kg/m^2 \cdot s$
Electric charge	coulomb (C)	$A \cdot s$
Electromotive force	volt (V) or (W/A)	$kg \cdot m^2/A \cdot s^3$
Electric resistance	ohm (Ω) or (V/A)	$kg \cdot m^2/A^2 \cdot s^3$
Electric conductivity	ampere per volt-meter (A/V·m)	$A^2 \cdot s^3/kg \cdot m^3$
Electric capacitance	farad (F) or (A·s/V)	$A^3 \cdot s^4/kg \cdot m^2$
Magnetic flux	weber (Wb) or (V·s)	$kg \cdot m^2/A \cdot s^2$
Inductance	henry (H) or (V·s/A)	$kg \cdot m^2/A^2 \cdot s^2$
Magnetic permeability	henry per meter (H/m)	$kg \cdot m/A^2 \cdot s^2$
Magnetic flux density	tesla (T) or weber per square meter (Wb/m^2)	$kg/A \cdot s^2$
Magnetic field strength	ampere per meter	A/m
Magnetomotive force	ampere	A
Luminous flux	lumen (lm)	$cd \cdot sr$
Luminance	candela per square meter	cd/m^2
Illumination	lux (lx) or lumen per square meter (lm/m^2)	$cd \cdot sr/m^2$

Tables 1-10 SI conversion multiples

Acceleration: (m/s^2)

ft/s^2	0.3048	gal	0.01
free fall (g)	9.80665	in/s^2	0.0254

Angle: radian (rad)

degree	0.01745329	second	$4.848137 \cdot 10^{-6}$
minute	$2.908882 \cdot 10^{-4}$	grade	$1.570796 \cdot 10^{-2}$

Area: (m^2)

acre	4046.873	hectare	$1 \cdot 10^4$
are	100.00	mi^2 (US statute)	$2.589998 \cdot 10^6$
ft^2	$9.290304 \cdot 10^{-2}$	yd^2	0.8361274

Bending Moment or torque: $(N \cdot m)$

dyne·cm	$1 \cdot 10^{-7}$	lbf·in	0.1129848
kgf·m	9.806650	lbf·ft	1.355818
ozf·in	$7.061552 \cdot 10^{-3}$		

Electricity and Magnetism[1]

ampere hour	3600 coulomb (C)	EMU of inductance	$8.987 \cdot 10^{11}$ henry (H)
EMU of capacitance	10^9 farad (F)	EMU of resistance	$8.987 \cdot 10^{11}$ (Ω)
EMU of current	10 ampere (A)	faraday	$9.65 \cdot 10^{19}$ coulomb (C)
EMU of elec. potential	10^{-8} volt (V)	gamma	10^{-9} tesla (T)
EMU of inductance	10^{-9} henry (H)	gauss	10^{-4} tesla (T)
EMU of resistance	10^{-9} ohm (Ω)	gilbert	0.7957 ampere (A)
ESU of capacitance	$1.112 \cdot 10^{-12}$ farad (F)	maxwell	10^{-8} weber (Wb)
ESU of current	$3.336 \cdot 10^{-10}$ ampere (A)	mho	1.0 siemens (S)
EMU of elec. potential	299.79 volt (V)	ohm centimetre	0.01 ohm metre ($\Omega \cdot m$)

---

[1] ESU means electrostatic cgs unit; EMU means electromagnetic cgs unit

Tables 1-10 (cont.)

Energy (work): joule (J)

British thermal unit (Btu)	1055	kilocalorie	4187
calorie	4.18	kW·h	$3.6 \cdot 10^6$
calorie (kilogram)	4184	ton (nuclear equiv. TNT)	$4.184 \cdot 10^9$
electronvolt	$1.60219 \cdot 10^{-19}$	therm	$1.055 \cdot 10^8$
erg	10^{-7}	W·h	3600
ft·lbf	1.355818	W·s	1.0
ft-poundal	0.04214		

Force: newton (N)

dyne	10^{-5}	ounce-force	0.278
kilogram-force	9.806	pound-force (lbf)	4.448
kilopond (kp)	9.806	poundal	0.1382
kip (1000 lbf)	4448	ton-force (2000 lbf)	8896

Heat

Btu·ft/(h·ft^2·°F) (thermal conductivity)	1.7307 W/(m·K)	cal/cm^2	$4.18 \cdot 104$ J/m^2
Btu/lb	2324 J/kg	cal/(cm^2·min)	697.3 W/m^2
Btu/(lb·°F) (heat capacity)	4186 J/(kg·K)	cal/s	4.184 W
Btu/ft^3	$3.725 \cdot 10^4$ J/m^3	°F·h·ft^2/Btu (thermal resistance)	0.176 K·m^2/W
cal/(cm·s·°C)	418.4 W/(m·K)	ft2/h (thermal diffusivity)	$2.58 \cdot 10^{-5}$ m^2/s

Length: metre (m)

angstrom	10^{-10}	microinch	$2.54 \cdot 10^{-8}$
astronomical unit	$1.495979 \cdot 10^{11}$	micrometre (micron)	10^{-6}
chain	20.11	mil	$2.54 \cdot 10^{-5}$
fermi (femtometre)	10^{-15}	mile (nautical)	1852.000
foot	0.3048	mile (international)	1609.344
inch	0.0254	pica (printer's)	$4.217 \cdot 10^{-3}$
light year	$9.46055 \cdot 10^{15}$	yard	0.9144

Tables 1-10 (cont.)

Light

cd/in^2	1550 cd/m^2	lambert	$3.183 \cdot 10^3$ cd/m^2
footcandle	10.76 lx (lux)	lm/ft^2	10.76 lm/m^2
footlambert	3.426 cd/m^2		

Mass: kilogram (kg)

carat (metric)	$2 \cdot 10^{-4}$	ounce (troy or apothecary)	$3.110348 \cdot 10^{-2}$
grain	$6.479891 \cdot 10^{-5}$	pennyweight	$1.555 \cdot 10^{-3}$
gram	0.001	pound (lb avoirdupois)	0.4535924
hundredweight (long)	50.802	pound (troy or apothecary)	0.3732
hundredweight (short)	45.359	slug	14.5939
kgf·s^2/m	9.806650	ton (long, 2240 lb)	907.184
ounce (avoirdupois)	$2.834952 \cdot 10^{-2}$	ton (metric)	1000

Mass: per unit time (includes Flow)

perm (0°C)	$5.721 \cdot 10^{-11}$ kg/(Pa·s·m^2)	lb/(hp·h) SPC - specific fuel consumption	$1.689659 \cdot 10^{-7}$ kg/J
lb/h	$1.2599 \cdot 10^{-4}$ kg/s	ton (short)/h	0.25199 kg/s
lb/s	0.4535924 kg/s		

Mass per unit volume (includes Density and Capacity): (kg/m^3)

oz (avoirdupois)/gal (UK liquid)	6.236	oz (avoirdupois)/gal (US liquid)	7.489
oz (avoirdupois)/in^3	1729.99	slug/ft^3	515.3788
lb/gal (US liquid)	11.9826 kg/m^3	ton (long)/yd^3	1328.939

Power: watt (W)

Btu (International)/s	1055.056	horsepower (electric)	746
cal/s	4.184	horsepower (metric)	735.499
erg/s	10^{-7}	horsepower (UK)	745.7
horsepower (550 ft·lbf/s)	745.6999	ton of refrigeration (12 000 Btu/h)	3517

Tables 1-10 (cont.)

Pressure or Stress: pascal (Pa)

atmosphere, standard	$1.01325 \cdot 10^5$	dyne/cm^2	0.1
atmosphere, technical	$9.80665 \cdot 10^4$	foot of water (39.2°F)	2988.98
bar	10^5	poundal/ft2	1.488164
centimeter of mercury (0°C)	1333.22	psi (lbf/in^2)	6894.757
centimeter of water (4°C)	98.0638	torr (mm Hg, 0°C)	133.322

Radiation units

curie	$3.7 \cdot 10^{10}$ becquerel (Bq)	rem	0.01 sievert (Sv)
rad	0.01 gray (Gy)	roentgen	$2.58 \cdot 10^{-4}$ C/kg

Temperature

°Celsius	$T_K = t_C + 273.15$ K	°Fahrenheit	$T_C = (t_F - 32)/1.8$ °C
°Fahrenheit	$T_K = (t_F + 459.67)/1.8$ K	°Rankine	$T_K = T_R/1.8$

Velocity (includes Speed): (m/s)

ft/s	0.3048	mi/h (International)	0.44704
in/s	$2.54 \cdot 10^{-2}$	rpm (r/min)	0.1047 rad/s
knot (International)	0.51444		

Viscosity: (Pa·s)

centipose (dynamic viscosity)	10^{-3}	lbf·s/in^2	6894.757
centistokes (kinematic viscosity)	10^{-6}	rhe	10 1/(Pa·s)
poise	0.1	slug/(ft·s)	47.88026
poundal·s/ft^2	1.488164	stokes	10^{-4} m^2/s
lb/(ft·s)	1.488164		

Volume (includes Capacity): (m^3)

acre-foot	1233.489	gill (U.S.)	$1.182941 \cdot 10^{-4}$
barrel (oil, 42 gal)	0.1589873	in^3	$1.638706 \cdot 10^{-5}$
bushel (U.S.)	$3.5239 \cdot 10^{-2}$	litre	10^{-3}
cup	$2.36588 \cdot 10^{-4}$	ounce (U.S. fluid)	$2.957353 \cdot 10^{-5}$
ounce (U.S. fluid)	$2.95735 \cdot 10^{-5}$	pint (U.S. dry)	$5.506105 \cdot 10^{-4}$
ft^3	$2.83168 \cdot 10^{-2}$	pint (U.S. liquid)	$4.731765 \cdot 10^{-4}$
gallon (Canadian, U.K. liquid)	$4.54609 \cdot 10^{-3}$	tablespoon	$1.478 \cdot 10^{-5}$
gallon (U.S. liquid)	$3.7854 \cdot 10^{-3}$	ton (register)	2.831658
gallon (U.S. dry)	$4.40488 \cdot 10^{-3}$	yd^3	0.76455

References

1. Thompson, S. Control systems: engineering & design, Longman Scientific & Technical, Essex, England, 1989

2. Norton, H. N. Handbook of transducers, Prentice Hall, Englewood Cliffs, NJ, 1989.

3. White, R. W. A sensor classification scheme, Microsensors, IEEE Press, New York, pp: 3-5, 1991

4. Halliday, D., Resnick, R. Fundamentals of Physics, John Wiley & Sons, New York, 2nd ed., 1986

2

Sensor Characteristics

From the input to the output, a sensor may have several conversion steps before it produces an electrical signal. For instance, pressure inflicted on the fiber-optic sensor, first results in strain in the fiber, which, in turn, causes deflection in its refractive index, which, in turn, results in an overall change in optical transmission and modulation of photon density. Finally, photon flux is detected and converted into electric current. Here, we discuss the overall sensor characteristics, regardless of its physical nature or steps which are required to make a conversion. We regard a sensor as a "black box" where we concern only with relationships between its output and input signals.

2.1 TRANSFER FUNCTION

An *ideal* or *theoretical* output-stimulus relationship exists for every sensor. If the sensor is ideally designed and fabricated with ideal materials by ideal workers, the output of such a sensor would always represent the true value of the stimulus. The ideal function may be stated in the form of a table of values, a graph, or a mathematical equation. An ideal (theoretical) output-stimulus relationship is characterized by the so-called *transfer function*. This function establishes dependence between the electrical signal, S, produced by the sensor, and the stimulus, s. That function may be a simple linear connection or a nonlinear dependence, for instance logarithmic, exponential, or power function. A linear relationship is represented by equation

$$S = a + bs \ ,$$ (2.1)

where a is the intercept, that is, the output signal at zero input signal, and b is the slope, which is sometimes called *sensitivity*. S is one of the characteristics of the output electric signal. It may be amplitude, frequency or phase, depending on the sensor properties.

A logarithmic function is

$$S = a + b \ln s , \tag{2.2}$$

Exponential function is

$$S = a e^{ks} , \tag{2.3}$$

Power function is

$$S = a_o + a_1 s^k , \tag{2.4}$$

where k is a constant number.

A sensor may have such a transfer function that none of the above approximations fit sufficiently well. In that case, a higher order polynomial approximation is often employed.

For a nonlinear transfer function, sensitivity, b, is not a fixed number as for the linear relationship (Eq. 2.1). At any particular input value, s_o, it can be defined as

$$b = \frac{dS(s_o)}{ds} . \tag{2.5}$$

In many cases, a nonlinear sensor may be considered linear over a limited range. Over the extended range, a nonlinear transfer function may be modeled by several straight lines. This is called a piece-wise approximation. To determine whether a function can be represented by a linear model, incremental variables are introduced for the input while observing the output. A difference between the actual response and a liner model is compared with the specified accuracy limits (see below).

2.2 SPAN (INPUT)

A dynamic range of stimuli which may be converted by a sensor is called a *span* or an *input full scale* (FS). It represents the highest possible input value which can be applied to the sensor without causing unacceptably large inaccuracy. For the sensors with a very broad and nonlinear response characteristic, a dynamic range of the input stimuli is often expressed in decibels, which is a logarithmic measure of ratios of either power, or force (voltage). It should be emphasized that decibels do not measure absolute values, but a ratio of values only. A decibel scale represents signal magnitudes by much smaller numbers, which in many cases is far more convenient. Being a nonlinear scale, it may represent low level signals with high resolution while compressing the high level numbers. In other words, the logarithmic scale for small objects works as a microscope and for the large objects as a telescope. By definition, decibels are equal to ten times the log of the ratio of powers (Table 2.1)

$$1 dB = 10 \log \frac{P_2}{P_1} . \tag{2.6}$$

In a similar manner, decibels are equal to 20 times the log of the force, or current, or voltage

$$1 dB = 20 \log \frac{S_2}{S_1} . \tag{2.7}$$

Table 2-1 Relationship between power, force (voltage, current), and decibels

Power ratio	1.023	1.26	10.0	100	10^3	10^4	10^5	10^6	10^7	10^8	10^9	10^{10}
Force ratio	1.012	1.12	3.16	10.0	31.6	100	316	10^3	316^2	10^4	$3 \cdot 10^4$	10^5
Decibels	0.1	1.0	10.0	20.0	30.0	40.0	50.0	60.0	70.0	80.0	90.0	100.0

2.3 FULL SCALE OUTPUT

Full scale output (FSO) is the algebraic difference between the electrical output signals measured with maximum input stimulus and the lowest input stimulus applied. This must include all deviations from the ideal transfer function. For instance, the FSO output in Fig. 2-1A is represented by S_{FS}.

2.4 ACCURACY

A very important characteristic of a sensor is *accuracy* which really means *inaccuracy*. Inaccuracy is measured as a ratio of the highest deviation of a value represented by the sensor to the ideal value. That deviation can be described as a difference between the ideal input value and a value which was converted by the sensor into voltage and than, without any error, converted back. For example, a linear displacement sensor ideally should generate 1 mV per 1 mm displacement. However, in the experiment, a displacement of 10 mm produced an output of 10.5 mV. Converting this number back (1 mm per 1 mV), we would expect that the displacement was

10.5 mm, that is 0.5 mm more than the actual. This extra 0.5 mm is a deviation and, therefore, in a 10 mm range the sensor's inaccuracy is 0.5 mm/10 mm x 100%= 5%.

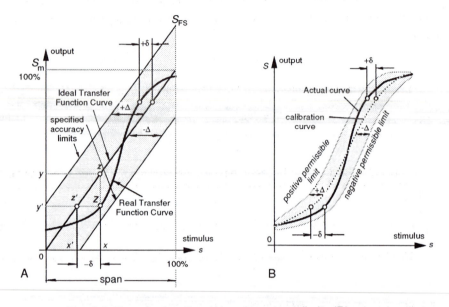

Fig. 2-1 Transfer function (A) and accuracy limits (B)
Error is specified in terms of input value

Fig. 2-1A shows an ideal or theoretical transfer function. In the real world, any sensor performs with some kind of imperfection. A possible *real* transfer function is represented by a thick line, which generally may be neither linear nor monotonic. A real function rarely coincides with the ideal. Because of material variations, workmanship, design errors, manufacturing tolerances, and other limitations, it is possible to have a large family of real transfer functions, even when sensors are tested under the identical conditions. However, all runs of the real transfer functions must fall within the limits of a specified accuracy. These permissive limits differ from the ideal transfer function line by $\pm\Delta$. The real functions deviate from the ideal by $\pm\delta$, where $\delta<=\Delta$. For example, let us consider a stimulus having value x. Ideally, we would expect this value to correspond to point z on the transfer function, resulting in the output value y. Instead, the real function will respond at point Z producing output value y'. This output value corresponds to point z' on the ideal transfer function, which, in turn, relates to a "would-be" input stimulus x' whose value is smaller than x. Thus, in this example, imperfection in the sensor's transfer function leads to a measurement error $-\delta$.

The accuracy rating includes a combined effect of part-to-part variations, a hysteresis, a dead band, calibration and repeatability errors (see below). The specified accuracy limits generally are used in the worst case analysis to determine the worst possible performance of the system. Fig. 2-1B shows that $\pm\Delta$ may more closely follow the real transfer function, meaning better tolerances of the sensor's accuracy. This can be accomplished by a multiple-point calibration. Thus, the specified accuracy limits are established not around the theoretical (ideal) transfer function, but around the calibration curve which is determined during the actual calibration procedure. Then, the permissive limits become narrower as they do not embrace part-to-part variations between the sensors and are geared specifically to the calibrated unit. Clearly, this method allows more accurate sensing, however in some applications it may be prohibitive because of a higher cost.

Inaccuracy rating may be represented in a number of forms:

1. Directly in terms of measured value (Δ);

2. In % of input span (full scale);

3. In terms of output signal.

For example, a piezoresistive pressure sensor has a 100 kPa input full scale and 10 Ω full scale output. Its inaccuracy may be specified as $\pm0.5\%$, or ±500 Pa, or ±0.05 Ω.

2.5 CALIBRATION ERROR

Calibration error is inaccuracy permitted by a manufacturer when a sensor is calibrated in the factory. This error is of a systematic nature, meaning that it is added to all possible real transfer functions. It shifts the accuracy of transduction for each stimulus point by a constant. This error is not necessarily uniform over the range and may change depending on the type of error in calibration. For example, let us consider a two-point calibration of a real linear transfer function (thick line in Fig. 2-1B). To determine the slope and the intercept of the function, two stimuli, s_1 and s_2, are applied to the sensor. The sensor responds with two corresponding output signals A_1 and A_2. The first response was measured absolutely accurately, however, the higher signal was measured with error $-\Delta$. This results in errors in the slope and intercept calculation. A new intercept, a_1 will differ from the real intercept a by

$$\delta_a = a_1 - a = \frac{\Delta}{s_2 - s_1},$$
(2.8)

and the slope will be calculated with error:

$$\delta_b = -\frac{\Delta}{s_2 - s_1}, \qquad (2.9)$$

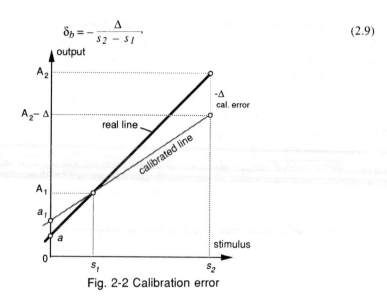

Fig. 2-2 Calibration error

2.6 HYSTERESIS

A *hysteresis error* is a deviation of the sensor's output at a specified point of input signal when it is approached from opposite directions (Fig. 2-3). For example, at 50°C, a thermometer shows 49° when the object is warmed up and the same thermometer at 50° shows 51° when the object is cooled down. In this case, the hysteresis is specified as 2°C (51°- 49°) or ±1°C from the ideal transfer function.

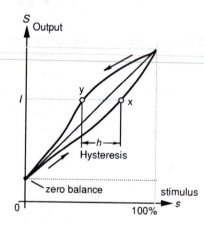

Fig. 2-3 Transfer function with hysteresis

2.7 NONLINEARITY

Nonlinearity error is specified for sensors whose transfer function may be approximated by a straight line (Eq. 2.1). A nonlinearity is a maximum deviation (L) of a real transfer function from the approximation straight line. The term "linearity" actually means "nonlinearity". When more than one calibration run is made, the worst linearity seen during any one calibration cycle should be stated. Usually, it is specified either in % of span or in terms of measured value, for instance, in kPa or °C. "Linearity", when not accompanied by a statement explaining what sort of straight line it is referring to, is meaningless. There are several ways to specify a nonlinearity, depending how the line is superimposed on the transfer function. One way is to use *terminal* points (Fig. 2-4A), that is, to determine output values at the smallest and highest stimulus values and to draw a straight line through these two points (line 1). Here, near the terminal points, the nonlinearity error is the smallest and it is higher somewhere in between.

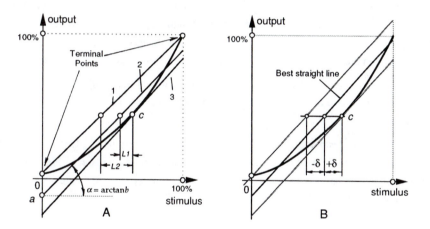

Fig. 2-4 Linear approximations of a nonlinear transfer function (A);
and independent linearity (B)

Another way to define the approximation line is to use a method of *least squares* (line 2 in Fig. 2-4A). This can be done in the following manner. Measure several (*n*) output values *S* at input values *s* over a substantially broad range, preferably over an entire full scale. Use the following formulas for linear regression to determine intercept *a* and slope *b* of the best fit straight line:

$$a = \frac{\Sigma S \Sigma s^2 - \Sigma s \Sigma s S}{n \Sigma s^2 - (\Sigma s)^2} \quad ,$$

(2.10)

$$b = \frac{n \Sigma s S - \Sigma s \Sigma S}{n \Sigma s^2 - (\Sigma s)^2} \quad ,$$

where Σ is the summation of n numbers.

In some applications, higher accuracy may be desirable in a particular narrower section of the input range. For instance, a medical thermometer should have the best accuracy in a fever definition region which is between 37 and 38°C. It may have a somewhat lower accuracy beyond these limits. Usually, such a sensor is calibrated in the region where the highest accuracy is desirable. Then, the approximation line may be drawn through the calibration point c (line 3 in Fig. 2-4A). As a result, nonlinearity has the smallest value near the calibration point and it increases toward the ends of the span. In this method, the line is often determined as tangent to the transfer function in point c. If the actual transfer function is known, the slope of the line can be found from Eq. (2.5).

Independent linearity is referred to the so-called "best straight line" (Fig. 2-4B), which is a line midway between two parallel straight lines closest together and enveloping all output values on a real transfer function.

Depending on the specification method, approximation lines may have different intercepts and slopes. Therefore, nonlinearity measures may differ quite substantially from one another. A user should be aware that manufacturers often publish the smallest possible number to specify nonlinearity, without defining what method was used.

2.8 SATURATION

Almost any sensor has its operating limits. Even if it is considered linear, at some levels of the input stimuli, its output signal no longer will be responsive. Further increase in stimulus does not produce a desirable output. It is said that the sensor exhibits a span-end nonlinearity or saturation (Fig. 2-5).

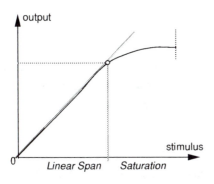

Fig. 2-5 Transfer function with saturation

2.9 REPEATABILITY

Repeatability (reproducibility) error is caused by the inability of a sensor to represent the same value under identical conditions. It is expressed as the maximum difference between output readings as determined by two calibrating cycles (Fig. 2-6A), unless otherwise specified. It is usually represented as % of FS:

$$\delta_r = \frac{\Delta}{FS} \ 100\% \ .$$ (2.11)

Possible sources of the repeatability error may be thermal noise, build up charge, material plasticity, etc.

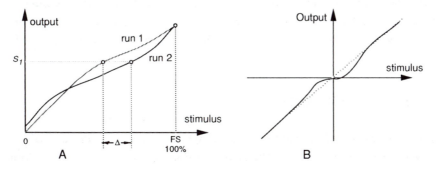

Fig. 2-6 A: Repeatability error
The same output signal S_1 corresponds to two different input signals
B: Dead-band zone in a transfer function

2.10 DEAD BAND

Dead band is the insensitivity of a sensor in a specific range of input signals (Fig. 2-6B). In that range, the output may remain near a certain value (often zero) over an entire dead band zone.

2.11 RESOLUTION

Resolution describes smallest increments of stimulus which can be sensed. When a stimulus continuously varies over the range, the output signals of some sensors will not be perfectly smooth, even under the no-noise conditions. The output may change in small steps. This is typical for potentiometric transducers, for occupancy infrared detectors with grid masks, and other sensors where the output signal change is enabled only upon a certain degree of stimulus variation. The magnitude of the input variation which results in the output smallest step is specified as resolution under specified conditions (if any). For instance, for the occupancy detector the resolution may be specified as follows: "resolution – minimum equidistant displacement of the object for 20 cm at 5 m distance". For wire-wound potentiometric angular sensors, resolution may be specified as "a minimum angle of 0.5°". Sometimes, it may be specified as percents of full scale (FS). For instance, for the angular sensor having 270° FS, the 0.5° resolution may be specified as 0.181% of FS. It should be noted, that the step size may vary over the range, hence, the resolution may be specified as typical, average, or "worst". The resolution of digital output format sensors is given by the number of bits in the data word. For instance, the resolution may be specified as "8-bit resolution". When there are no measurable steps in the output signal, it is said that the sensor has *continuous* or *infinitesimal* resolution (sometimes erroneously referred to as "infinite resolution").

2.12 SPECIAL PROPERTIES

Special input properties may be needed to specify for some sensors. For instance, light detectors are sensitive within a limited optical bandwidth. Therefore, it is appropriate to specify for them a spectral response.

2.13 OUTPUT IMPEDANCE

Output impedance, Z_{out}, is important to know to better interface a sensor with the electronic circuit. This impedance is connected either in parallel with the input impedance, Z_{in}, of the circuit (voltage connection) or in series (current connection). Fig. 2-10 shows two connections. Output and input impedances generally should be represented in a complex form, as they may include active and reactive components. To minimize output signal distortions, the sensor generating current (B) should have output impedance as high as possible and the circuit's input impedance should be low. For the voltage connection (A), a sensor is preferable with lower Z_{out} and the circuit should have Z_{in} as high as practical.

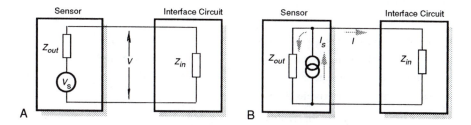

Fig. 2-7 Sensor connection to an interface circuit
A: sensor has voltage output
B : sensor has current output

2.14 EXCITATION

Excitation is the electrical signal needed for the active transducer operation. Excitation is specified as a range of voltage and/or current. For some transducers, the frequency of the excitation signal, and its stability must also be specified. Variations in the excitation may alter the transducer's transfer function and cause output errors.

An example of excitation signal specification is

Maximum current through a thermistor		
	in still air	50 µA
	in water	200 µA

2.15 DYNAMIC CHARACTERISTICS

Under static conditions a sensor is fully described by its transfer function, span, calibration, etc. However, when an input stimulus varies, a sensor response generally does not follow with perfect fidelity. The reason it that both the sensor and its coupling with the source of stimulus can not always respond instantly. In other words, a sensor may be characterized with a time dependent characteristic, which is called a *dynamic characteristic*. If a sensor does not respond instantly, it may indicate values of stimuli which are somewhat different from the real, that is, the sensor responds with a *dynamic error*. A difference between static and dynamic errors is that the latter is always time dependent. If a sensor is a part of a control system which has its own dynamic characteristics, the combination may cause oscillations.

Warm-up time is the time between applying to the sensor power or excitation signal and the moment when the sensor can operate within its specified accuracy. Many sensors have a negligibly short warm-up time. However, some detectors, especially those that operate in a thermally controlled environment (a thermostat) may require seconds and minutes of warm-up time before they are fully operational within the specified accuracy limits.

Frequency response is an important dynamic characteristic of a detector as it specifies how fast the sensor can react to a change in the input stimulus. The frequency response is expressed in Hz or rad/sec to specify the relative reduction in the output signal at certain frequency (Fig. 2-8A). A commonly used reduction number (frequency limit) is -3 dB. It shows at what frequency the output voltage (or current) drops by about 30%. Frequency response limit, f_u, is often called the upper cutoff frequency, as it is considered the highest frequency which a sensor can process.

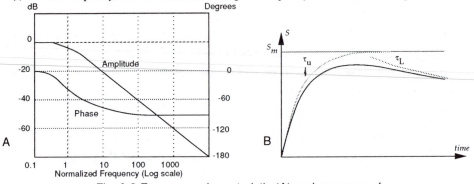

Fig. 2-8 Frequency characteristic (A) and response of
a first order sensor (B) with limited upper and lower cutoff frequencies
τ_U and τ_L are corresponding time constants

The frequency response directly relates to a *speed response*, which is defined in units of input stimulus per unit of time. Which response, frequency or speed, to specify in any particular case, depends on the sensor type, its application, and a preference of a designer.

Another way to specify speed response is by time which is required by the sensor to reach 90% of a steady-state or maximum level upon exposure to a step stimulus. For the first-order response, it is very convenient to use a so-called *time constant*. Time constant τ is a measure of the sensor's inertia. In electrical terms, it is equal to a product of electrical capacitance and resistance: $\tau = CR$. In thermal terms, thermal capacity and thermal resistances should be used instead. Practically, time constant can be easily measured. A first order system response is

$$S = S_m(1 - e^{-t/\tau}) \ , \tag{2.12}$$

where S_m is steady-state output, t is time, and e is base of natural logarithm.

Substituting $t = \tau$, we get

$$\frac{S}{S_m} = 1 - \frac{1}{e} = 0.6321 \ . \tag{2.13}$$

In other words, after an elapse of time equal to one time constant, the response reaches about 63% of its steady-state level. Similarly, it can be shown that after two time constants, the height will be 86.5% and after three time constants it will be 95%.

Lower cutoff frequency shows what is the lowest frequency of stimulus the sensor can process. There is a lot of similarities between definitions of the upper and the lower cutoff frequencies. They are defined in the same terms and the time constants have the same meanings. It should be emphasized that while the upper cutoff frequency shows how fast the sensor reacts, the lower cutoff frequency shows how slowly changing stimuli the sensor can process. Fig. 2-8B depicts the sensor's response when both upper and lower cutoff frequencies are limited. Eventually, the response never reaches its would-be steady state level S_m. For the first order response, it can be expressed as a product of two exponential processes:

$$S = S_m(1 - e^{-\frac{t}{\tau_u}})e^{-\frac{t}{\tau_L}} \ . \tag{2.14}$$

As a rule of thumb, a simple formula can be used to establish a connection between the cutoff frequency f_c (either upper and lower) and time constant in a first-order sensor:

$$f_c \approx \frac{0.159}{\tau} \tag{2.15}$$

Clearly, for a relatively narrow bandwidth sensor (when the upper and lower cutoff frequencies are close to one another), using of time constants becomes inappropriate, because it is almost impossible to separate two exponential slopes in measurements. However, for a broad-bandwidth sensor (when the upper cutoff frequency is much higher, say 50 times), both time constants can be measured quite accurately.

There is a large class of sensors which may respond to constant stimuli. Such sensors, is said, have a dc response, therefore $\tau_L = \infty$ and $f_L = 0$. Fig. 2-9 shows typical responses of sensors which are the result of various combinations of cutoff frequencies.

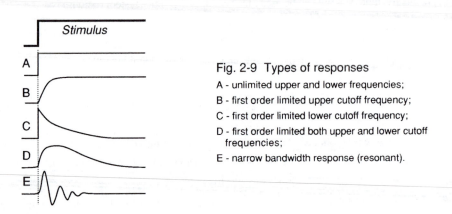

Fig. 2-9 Types of responses

A - unlimited upper and lower frequencies;

B - first order limited upper cutoff frequency;

C - first order limited lower cutoff frequency;

D - first order limited both upper and lower cutoff frequencies;

E - narrow bandwidth response (resonant).

Phase shift at a specific frequency defines how the output signal lags behind in representing the stimulus change (Fig. 2-8A). The shift is measured in angular degrees or rads. If a sensor is a part of a feedback control system, it is very important to know it's phase characteristic. Phase lag reduces the phase margin of the system and may result in the overall instability.

Resonant (natural) frequency is a number expressed in Hz or rad/sec which shows where the sensor's output signal increases considerably. Many sensors behave as linear, first-order systems which do not resonate. However, if a dynamic transducer's output conforms to the standard curve of a second-order response, the manufacturer will state the natural frequency and the damping ratio of the transducer. The resonant frequency may be related to mechanical, thermal or electrical properties of the detector. Generally, the operating frequency range for the sensor should be selected well below (at least 60%) or above the resonant frequency. However, in some sensors, the resonant frequency is the operating point. For instance, in glass breakage detectors (used in security systems) the resonant makes the sensor selectively sensitive to a narrow bandwidth which is specific for the acoustic spectrum produced by shuttered glass.

Damping is the progressive reduction or suppression of the oscillation in the sensor having higher than the first order response. When the sensor's response is as fast as possible without overshoot, the response is said to be critically damped (Fig. 2-10). Underdamped response is when the overshoot occurs and the overdamped response is slower than the critical. The damping ratio is a number expressing the quotient of the actual damping of a second-order linear transducer by its critical damping. The second order transfer function must include a quadratic factor: $s^2+2z\omega_n s+\omega_n^2$, where ω_n is natural frequency (rad/sec), s is the complex variable, and z is *damping ratio*. For a critically damped detector $z=1$. The damping factor is defined as:

$$z = \frac{\sigma}{\omega_n} = \frac{\sigma}{\sqrt{\sigma^2 + \omega^2}}, \tag{2.16}$$

where σ is real part of a complex variable. For an oscillating response, as shown in Fig. 2-9, a *damping factor* is a measure of damping, expressed (without sign) as the quotient of the greater by the lesser of pair of consecutive swings in opposite directions of the output signal, about an ultimately steady-state value. Hence, the damping factor can be measured as:

$$\text{damping factor} = \frac{F}{A} = \frac{A}{B} = \frac{B}{C} = \text{etc.} \tag{2.17}$$

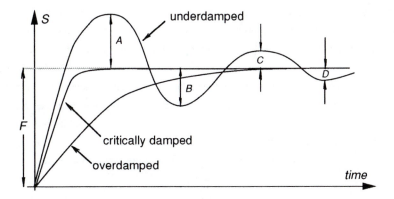

Fig. 2-10 Responses of sensors with different damping characteristics

2.16 ENVIRONMENTAL FACTORS

Storage conditions are nonoperating environmental limits to which a sensor may be subjected during a specified period without permanently altering its performance under normal operating conditions. Usually, storage conditions include the highest and the lowest storage temperatures and maximum relative humidities at these temperatures. Words "noncondensing" may be added to the relative humidity number. Depending on the sensor's nature, some specific limitation for the storage may need to be considered. For instance, maximum pressure, presence of some gases or contaminating fumes, etc.

Short and long term stabilities (drift) are parts of the accuracy specification. The short term stability is manifested as changes in the sensor's performance within minutes, hours or even days. Eventually, it is another way to express repeatability (see above) as drift may be bidirectional. That is, the sensor's output signal may increase or decrease, which is, in other terms, may be described as ultralow frequency noise. The long term stability may be related to aging of the sensor materials, which is an irreversible change in the material's electrical, mechanical, chemical, or thermal properties. That is, the long term drift is usually unidirectional. It happens over a relatively long time span, such as months and years. Long term stability is one of the most important for the sensors that are used for precision measurements. Aging greatly depends on environmental storage and operating conditions, how well the sensor components are isolated from the environment and what materials are used for their fabrication. For instance, glass coated metal-oxide thermistors exhibit much greater long term stability as compared with epoxy coated. A powerful way to improve long term stability is to pre-age the component at extreme conditions. The extreme conditions may be cycled from the lowest to the highest. For instance, a sensor may be periodically swung from freezing to hot temperatures. Such accelerated aging not only enhances stability of the sensor's characteristics, but also improves the reliability (see below), as the pre-aging process reveals many hidden defects. For instance, epoxy-coated thermistors may be substantially improved if they are maintained at +150° for 1 month before they are calibrated and installed into a product.

Environmental conditions to which a sensor is subjected do not include variables which the sensor measures. For instance, an air pressure sensor usually is subjected not just to air pressure, but to other influences as well, such as temperatures of air and surrounding components, humidity, vibration, ionizing radiation, electromagnetic fields, gravitational forces, etc. All these factors may and usually do affect the sensor's performance. Both static and dynamic variations in these conditions should be considered. Some environmental conditions are of a multiplicative nature, that is they alter a transfer function of the sensor, for instance changing its gain. One example is resistive strain gauge whose sensitivity increases with temperature.

Environmental stability is quite broad and usually a very important requirement. Both the sensor designer and the application engineer should consider all possible external factors which may affect the sensor's performance. For instance, a pyroelectric sensor for detecting movement of people may generate spurious signals if affected by a sudden change in ambient temperature, electrostatic discharge, formation of electrical charges (triboelectric effect), wind, loud noise, vibration of supporting structures, electromagnetic interferences (EMI), etc. Even if a manufacturer does not specify such effects, an application engineer should simulate them during the prototype phase of the design process. If, indeed, the environmental factors degrade the sensor's performance, additional corrective measures may be required (see Chapter 4). For instance, placing the sensor in a protective box, electrical shielding, using a thermal insulation, or a thermostat.

Temperature factors are very important for sensor performance, they must be known and accounted for. The operating temperature range is the span of ambient temperatures given by their upper and lower extremes (e.g., "-20 to +100°C") within which the sensor maintains its specified accuracy. Many sensors change with temperature and their transfer functions may shift significantly. Special compensating elements are often incorporated either directly into the sensor or into signal conditioning circuits, to compensate for temperature errors. The simplest way of specifying tolerances of thermal effects is provided by the error-band concept, which is simply the error band that is applicable over the operating temperature band. A temperature band may be divided into sections while the error band is separately specified for each section. For example, a sensor may be specified to have an accuracy of ±1% in the range from 0 to 50°C, ±2% from -20 to 0°C, and from +50 to 100°C, and ±3% beyond these ranges with operating limits which are specified from -40 to +150°C.

Temperatures will also affect dynamic characteristics, particularly when they employ viscous damping. A relatively fast temperature change may cause the sensor to generate a spurious output signal. For instance, a dual pyroelectric sensor in a motion detector is insensitive to slow varying ambient temperature. However, when the temperature changes fast, the sensor will generates electric current which may be recognized by a processing circuit as a valid response to a stimulus, thus causing a false positive detection.

A *self-heating error* may be specified when an excitation signal is absorbed by a sensor and changes its temperature by such a degree that it may affect its accuracy. For instance, a thermistor temperature sensor requires passage of electric current, causing heat dissipation within the sensor's body. Depending on its coupling with the environment, the sensors' temperature may increase due to a self-heating effect. This will result in errors in temperature measurement. The coupling depends on the media where the sensor operates – a dry contact, liquid, air, etc. A worst coupling may be through still air. For thermistors, manufacturers often specify self-heating errors in air, stirred liquid, or other media.

Sensor's temperature increase above its surroundings may be found from formula:

$$\Delta T^\circ = \frac{V^2}{(\xi vc + \alpha)R} , \qquad (2.19)$$

where ξ is the sensor's mass density, c is specific heat, v is the volume of the sensor, α is the coefficient of thermal coupling between the sensor and the outside (thermal conductivity), R is the electrical resistance, and V is the effective voltage across the resistance. If a self-heating results in an error, formula (2.19) may be used as a design guidance. For instance, to increase α, a thermistor detector should be well coupled to the object by increasing the contact area, applying thermally conductive grease or using thermally conductive adhesives. Also, high resistance sensors and low measurement voltages are preferable.

2.17 RELIABILITY

Reliability is the ability of a sensor to perform a required function under stated conditions for a stated period. It is expressed in statistical terms as a probability that the device will function without failure over a specified time or a number of uses. It should be noted, that reliability is not a characteristic of drift or noise stability. It specifies a failure, that is, temporary or permanent, exceeding the limits of a sensor's performance under normal operating conditions.

Reliability is an important requirement, however, it is rarely specified by the sensor manufacturers. Probably, the reason for that is the absence of a commonly accepted measure for the term. In the U.S.A., for many electronic devices, the procedure for predicting in-service reliability is the MTBF (mean-time-between-failure) calculation described in MIL-HDBK-217 standard. Its basic approach is to arrive at a MTBF rate for a device by calculating the individual failure rates of the individual components used and by factoring in the kind of operation the device will see: its temperature, stress, environmental, and screening level (measure of quality). Unfortunately, MTBF reflects reliability only indirectly and it is often hardly applicable to everyday use of the device. The qualification tests on sensors are performed at combinations of the worst possible conditions. One approach (suggested by MIL-STD-883) is 1,000 hours, loaded at maximum temperature. This test does not qualify for such important impacts as fast temperature changes. The most appropriate method of testing would be accelerated life qualification. It is a procedure that emulates the sensor's operation, providing real-world stresses, but compressing years into weeks. Three goals are behind the test: to establish MTBF; to identify first failure points that can then be strengthened by design changes; and to identify the overall system practical life time.

One possible way to compress time is to use the same profile as actual operating cycle, including maximum loading and power-on, power-off cycles, but expanded environmental highest and lowest ranges (temperature, humidity, and pressure). The highest and lowest limits should be substantially broader than normal operating conditions. Performance characteristics may be outside specifications, but must return to those when the device is brought back to the specified operating range. For example, if a sensor is specified to operate up to 50°C at the highest relative humidity (RH) of 85% at maximum supply voltage of +15V, it may be cycled up to 100°C at 99% RH and at +18V power supply. To estimate number of test cycles (n), the following empirical formula (developed by Sandstrand Aerospace, Rockford, IL and Interpoint Corp., Redmond, WA) [1] may be useful:

$$n = N \left(\frac{\Delta T_{max}}{\Delta T_{test}} \right)^{2.5}, \qquad (2.18)$$

where N is the estimated number of cycles per lifetime, ΔT_{max} is the maximum specified temperature fluctuation and ΔT_{test} maximum cycled temperature fluctuation during the test. For instance, if the normal temperature is 25°C, the maximum specified temperature is 50°C, cycling was up to 100°C, and over the life time (say, 10 years) the sensor was estimated will be subjected to 20,000 cycles, then the number of test cycles is calculated as:

$$n = 20,000 \cdot \left(\frac{50\text{-}25}{100\text{-}25} \right)^{2.5} = 1283 \ .$$

As a result, the accelerated life test requires about 1,300 cycles instead of 20,000. It should be noted, however, that the 2.5 factor was derived from a solder fatigue multiple, since that element is heavily influenced by cycling. Some sensors have no solder connections at all, and some might have even more sensitive to cycling substances than solder, for instance, electrically conductive epoxy. Then, the factor should be selected somewhat smaller. As a result of the accelerated life test, the reliability may be expressed as a probability of failure. For instance, if 2 out of 100 sensors (with an estimated life time of 10 years) failed the accelerated life test, the reliability is specified as 98% over 10 years.

A sensor, depending on its application, may be subjected to some other environmental effects which potentially can alter its performance or uncover hidden defects. Among such additional tests are:

- High temperature/high humidity while being fully electrically powered. For instance, a sensor may be subjected to its maximum allowable temperature at 85-90% relative humidity (RH) and kept under these conditions during 500 hours. This test is very useful for detecting contaminations and evaluation of packaging integrity. Life of sensors, operating at normal room temperatures, is often accelerated at 85°C and 85%RH, that sometimes is called an "85-85 test".

- Mechanical shocks and vibrations may be used to simulate adverse environmental conditions, especially in evaluation wire bonds, adhesion of epoxy, etc. A sensor may be dropped to generate high level accelerations (up to 3,000 g's of force). The drops should be made on different axes. Harmonic vibrations should be applied to the sensor over the range which includes its natural frequency. In the US, military standard #750, methods 2016 and 2056 are often used for mechanical tests.

- Extreme storage conditions may be simulated, for instance at +100 and -40°C while maintaining a sensor for at least 1000 hours under these conditions. This test simulates storage and shipping conditions and usually is performed on nonoperating devices. The upper and lower temperature limits must be consistent with the sensor's physical nature. For example, a TGS pyroelectric sensors manufactured in the past by Philips are characterized by a Curie temperature of +60°C. Approaching and surpassing this temperature results in a permanent destruction of sensitivity. Hence, the temperature of such sensors should never exceed +50°C, which must be clearly specified and marked on its packaging material.

- Thermal shock or temperature cycling (TC) is subjecting a sensor to alternate extreme conditions. For example, it may be dwelled for 30 minutes at -40°C, then rapidly moved to +100°C for 30 minutes, and then back to cold. The method must specify total number of cycling, like 100 or 1000. This test helps to uncover die bond, wire bond, epoxy connections and packaging integrity.

- To simulate sea conditions, sensors may be subjected to a salt spray atmosphere for a specified time, for example 24 hours. This helps to uncover its resistance to corrosion and structural defects.

2.18 APPLICATION CHARACTERISTICS

Design, weight, and overall *dimensions* are geared to specific areas of applications.

Price may be a secondary issue when the sensor's reliability and accuracy are of paramount importance. If a sensor is intended for life support equipment, weapons or spacecraft, a high price tag may be well justified to assure high accuracy and reliability. On the other hand, for a very broad range of consumer applications, the price of a sensor often becomes a corner stone of a design. For instance, human body temperatures preferably should be taken from the ear canal by thermal radiation thermometers. During the first several years after introduction of these instruments in 1986, the applications of ear thermometry were limited only to hospitals. The reason was that thermopile detectors (which are relatively expensive) were employed in the ear thermometers for sensing thermal radiation. As a result, the price tag for such an instrument was in the range of $500 to $700. In 1991, a home model of a medical infrared thermometer HM-1 was introduced for 1/5 of that price. It was made possible only because a thermopile was replaced by a pyroelectric detector - a much less expensive sensor.

2.19 STATISTICAL ANALYSIS

When information from a sensor is received and analyzed, a question has to be answered - how closely does it represent the actual stimulus? As was noted before - there is a great variety of distortions which may alter the representation and lead to error in measurement. Some distortions are *systematic*, that is they are consistent and predictable. Examples are nonlinearity, hysteresis, dead band, miscalibration, etc. Many other distortions are *random* and have characteristics of noise. Therefore, under the noisy conditions, the sensors and data acquisition systems are tested and analyzed by statistical methods. Benjamin Disraeli, the British Prime Minister, once said, "There are three kinds of lies — a simple lie, a damn lie and the statistics". This funny observation is the result of an erroneous transposition of statistical values onto any particular event. It must be clearly understood that methods of statistical analysis produce values which are descriptive only of a large number of measurements. They represent a crowd rather than an individual, a forest – not a tree. Statistical numbers are definitely not a lie when applied to a large number or to an imaginary "average" event which may never happen in the reality in that exact form. A statistical description gives numbers which tell about the device's expected performance under uncertain conditions. The word "statistics" was first applied to affairs of the state, to data that government finds necessary for the effective planning, ruling, and tax collecting. Collectors and analyzers of this information were once called "statists" reflecting that their business was directly related to the facts of the state.

There are two kinds of statistics which are useful for the sensors evaluation. The first is a *descriptive* statistics. It classifies data — performance histograms that correspond to frequency distribution that result after the data are classified, the representation of data by other sorts of graphs, such as line graphs, bar graphs, pictograms, the computation of sample means, medians, or modes; the computation of variance, means absolute deviations, and ranges - all these activities deal with descriptive statistics, whose principles have been largely established in the nineteenth century and the early part of this century.

The second important kind of statistics are known as *inferential statistics*. It is described as the science of making a decision in the face of uncertainty; that is, making the best decision on the basis of incomplete and presumably distorted information. On the basis of numerous testing of a device and measuring its output signal over a representative range of the input stimuli, we *infer* things about its fidelity under the influence of disturbances. Inferential statistics relies on a limited number of data (sample) to make decision about an overall performance of the device.

When taking individual measurements (samples) under the noisy conditions, we expect that stimulus s is represented by the sensor as having a somewhat different value s'. Thus, the error in measurement is represented as

$$\Delta = s' - s \; . \tag{2.20}$$

Assuming that the stimulus is constant, randomness in error leads to multiple stimulus representation when the measurement is repeated. After many measurements of the same stimulus, its true value can be estimated through the *mean value* (expected value). This expected value is not exactly equal to the actual stimulus s however, it is the best guess under the circumstances. The mean value may be expressed as

$$\bar{s} = \frac{s_1 + s_2 + \dots + s_i + \dots + s_n}{n} \; , \tag{2.21}$$

where n is a number of measurements. A convenient measure of deviation from the mean value is the mean absolute deviation. The term "absolute deviation" means the numerical (i.e. positive) value of deviation. It is simply the arithmetic mean of the absolute deviation and is denoted M.A.D.:

$$\text{M.A.D.} = \frac{\sum\limits_{i=1}^{n} |s_i - \bar{s}|}{n} \; . \tag{2.22}$$

The M.A.D. is an easy measure of dispersion, however, it does not yield any further elegant mathematical statistical results, as does the variance (see below), because the absolute values are rather unusable for mathematical analysis, especially comparing data when stimuli vary. A more convenient measure of dispersion is variance which is defined as an average of squared deviations from the mean value:

$$v^2 = \frac{\sum\limits_{i=1}^{n}(s_i - \bar{s})^2}{n - 1} \qquad . \tag{2.23}$$

If a stimulus is measured, for instance in meters, the variance will be defined in units of area (m^2). It is often desirable to measure dispersion in the same units as the original stimulus, then a square root is taken from formula (2.23):

$$v = \sqrt{\frac{\sum\limits_{i=1}^{n}(s_i - \bar{s})^2}{n - 1}} \qquad . \tag{2.24}$$

To simplify calculations, an alternative formula may be used:

$$v = \sqrt{\frac{1}{n - 1}\left(\sum\limits_{i=1}^{n}s_i^2 - n\bar{s}^2\right)} \qquad . \tag{2.25}$$

The above formulas represent the so-called *sample standard deviation*, which is a convenient measure of dispersion in the sampled data. Sample standard deviation for many practical purposes when n is large may be considered equal to true standard deviation σ: $v = \sigma$. For example, lets take measurements of temperatures from an object by a sensor whose readings are as follows (in °C)

    60.1   60.6   59.8   58.7   60.5   59.9   60.0   61.1   60.2   60.2

The total number of readings was $n = 10$. From formula (2.21) we can find that the mean value is equal to 60.11°C and the standard deviation from formula (2.25) is 0.63°C. It should be emphasized that the mean value may be quite different from the actual stimulus because of systematic errors either in the sensor, or in other parts of the data acquisition system. For example, if the sensor is miscalibrated by -1°C, all the above readings will have a fixed error of −1°C. This contemplates that the mean value should be corrected by the same number: −1°C. However, if the sensor is linear (or quasi-linear near the mean value), the standard deviation is not affected by a miscalibration or any other systematic error.

Deviations from a mean value while being random, have something in common, i.e., the number of positive and negative deviations are usually close and the sum of all deviations is near zero. In many cases, the larger is a deviation, the less is the probability of its occurrence. For instance, in the above set, deviations larger than ±1°C occur only once, while deviations between 0 and ±0.2° occur 7 times. The frequency of occurrence is called a distribution. The *Gaussian* (or *normal*) distribution, is probably the most extensively used probability distribution in engineering practice. Apart from the ease of use, another justification for its popularity is provided by the *central limit theorem*. This theorem states that a random variable that is formed by summing a very large number of independent random variables takes Gaussian distribution in the limit. Since many engineering phenomena are the results of numerous independent causes, the assumption of normal distribution is justified in many cases. The Gaussian distribution can be plotted according to probability density function which is given by

$$f(s) = \frac{1}{\sqrt{2\pi}\,\sigma}\, e^{-\frac{(s-\bar{s})^2}{2\sigma^2}}.$$

$$(2.26)$$

A ratio $\dfrac{s-\bar{s}}{\sigma}$ is called standard normal value. Note that only two parameters, mean, s, and standard deviation σ are necessary to determine a normal distribution. The density curve calculated according to equation (2.26) is shown in Fig. 2-11. The curve is symmetrical around the mean value which in this graph is equal to 0. The shape of the curve and its value does not change when the mean is not equal to zero. The total area under the curve is equal to 1, however, within the limits between -3σ and +3σ, the area is equal to 0.997 which means that almost all samples will deviate from the mean for no more than ±3σ. It can be stated that approximately

68% of values will fall within ±σ

95% of values will fall within ±2σ

99.7% of values will fall within ±3σ .

When comparing several independent sets of the same kind of measurements, we may expect that data in each set will be different. However, if the number of measurements within each set is large, the means and standard deviations will be close but different for different sets of data. For the infinitely large number of data, we could calculate the so-called *theoretical means* μ. In reality, the number of experimental measurements is limited to n which allows us to calculate a mean value $\bar{s}$. This number is our best estimate of the actual stimulus s, which we can not measure precisely. We have no reason to believe, however, that the mean $\bar{s}$ is exactly equal to μ, because we operate with a limited number of data. We merely hope that it is very close. To say "very close" is often no enough for practical purposes. We need a more definite measure of closeness, or, what is even more convenient, instead of a single number representing mean value, we may state that the mean is situated within certain limits. For instance, after 10 measurements of the object's temperature, we calculated its mean value as 75.0°C. But perhaps the actual (theoretical) mean is somewhere between 74.5° and 75.5°C. Such estimated limits are known as interval limits.

A statement such as the above is not very meaningful until we know how likely it is that the theoretical mean lies in a certain interval. Therefore, a number known as a *confidence coefficient* or a *degree of confidence* is given along with the interval, which is called a *confidence interval*. The upper and lower intervals are denoted μ_U and μ_L. For the normal distribution, their absolute values are equal. Without going too deeply into a theoretical discussion we may just state that for many practical purposes when degree of confidence over 90% is required, the absolute value of one half of a confidence interval $|\Delta\mu|$ may be found from the following formula

$$|\Delta\mu| \approx \frac{1.65\sigma}{\sqrt{n}}.$$
(2.27)

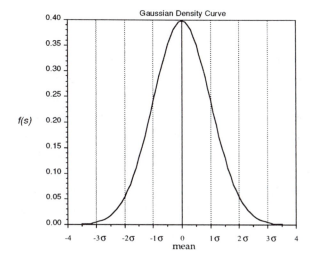

Fig. 2-11 Normal distribution

For example, let us assume that the temperature was measured from an object 10 times and has yielded a mean value 75.0°C and the standard deviation of 0.25°C. Applying these numbers to formula (2.27), we calculate that the theoretical mean value may be found somewhere between 75.0 - $\Delta\mu$ = 74.87°C and 75.0 + $\Delta\mu$ = 75.13°C. It should be noted that the confidence interval becomes narrower when the number of available data grows. However, since number n is under the square root, the growth is not proportional: doubling the number of data will narrow the confidence interval by about 1.4 times.

Table 2-2 Chauvenet's criterion for data rejection

Number of readings, n	ratio, d_{max}/σ
2	1.15
3	1.38
4	1.54
5	1.65
6	1.73
7	1.80
10	1.96
15	2.13
25	2.33
50	2.57
100	2.81
300	3.14
500	3.29
1000	3.48

Under some circumstances, a sensor may produce quite different output signals while being subjected several times to the same stimulus. That is, it has a reproducibility error. Similarly, several identical sensors exposed to the same stimulus may sometimes disagree with one another quite strongly. Here we are not talking about relatively small variations – they are always there. However, either a rare intrinsic noise spike or transmitted noise may cause a rare error. The user is therefore faced with the task of deciding whether these data should be seriously considered or they may be neglected as a result of some kind of gross blunder. On the other hand, the unusual response may represent a new physical phenomenon that is peculiar to certain, may be not fully understood at the time, conditions. The engineer should not throw away data that do not fit the expectations. Unless, of course, there is a some predetermined criteria for the elimination. Consider that the same stimulus is measured n times and that number n is large enough so we may expect a normal (Gaussian) distribution of the errors. This distribution may be used to compute the probability that a given reading will deviate a certain amount from the mean. We would not expect the probability much smaller than $1/n$ because this would be quite unlikely. Hence, we may set a criteria for elimination of erroneous data which appear with probabilities smaller than $1/n$. In practice, a more restrictive test is usually applied toward eliminating suspected errors from the data. It is known as *Chauvenet's criterion* and specifies that a reading may be rejected if the probability of obtaining the particular deviation from the mean is less than $1/2n$. Table 2-2 lists values of the ratio of residual d_{max} to standard deviation σ for various values of n according to Chauvenet's criterion.

To use Chauvenet's criterion, mean and standard deviation of data first should be calculated. Then, residuals of individual results are compared with information from Table 2-2 and the dubious points are eliminated. This method can be used for a selection of a threshold - a popular method of data testing in sensor technologies. Let us, for example, consider a motion detector used in security systems. The detector produces an analog output which is representative of moving activities within its field of coverage. There are two events of interest to the use of the device: there is no intrusion and there is an intrusion. Under no-intrusion conditions, the threshold comparator must produce a logic output Ø. Output signal, 1, should be generated by the threshold comparator only when someone enters into the protected area. An analog signal from the detector is compared with a predetermined threshold which is a decision level - any signal above that level must be considered an intrusion and logic number 1 must be produced. If the threshold is set too low, false positive detections may often occur. The threshold selection can be done with use of Chauvenet's criterion. Initially, the detectors' analog output is monitored every minute for some extended time, say, 5 hours under no-intrusion conditions. Within each minute, maximum analog output e_{max} is detected (for instance, by using a peak detector). After a 5-hour experiment, there will be 300 points, whose mean is 0 and the standard deviation was calculated $\sigma=265$ mV. From Table 2-2 we find that for $n = 300$, the ratio is $d_{max}/\sigma=3.14$ and any a deviation from the mean by $265 \cdot 3.14 = 832$ mV are considered abnormal for the nonintrusion conditions. As a result, the threshold may be set near that voltage and any signal crossing the threshold should be considered related to different conditions (intrusion) rather than to tested stand-by conditions (nonintrusion).

REFERENCE

1. Better Reliability Via System Tests. *Electronic Engineering Times*, CMP Publication, pp: 40-41, Aug. 19, 1991

3
Physical Principles of Sensing

3.1 ELECTRIC CHARGES, FIELDS AND POTENTIALS

There is a well-known phenomenon to those who live in dry climates – the possibility of the generation of sparks by friction involved in walking across the carpet. This is a result of the so-called *triboelectric effect*[1] which is a process of an electric charge separation due to object movements, friction of clothing fibers, air turbulence, atmosphere electricity, etc. There are two kinds of charges. Like charges repel and the unlike charges attract each other. Benjamin Franklin (1706-1790), among his other achievements, was the first American physicist. He named one charge *negative* and the other *positive*. These names have remained to this day. He made an elegant experiment with a kite flying in a thunderstorm to prove that the atmospheric electricity is of the same kind as produced by friction. In doing the experiment, Franklin was extremely lucky, as several Europeans who were trying to repeat his test were severely injured by the lightning and one was killed.

A triboelectric effect is a result of a mechanical charge redistribution. For instance, rubbing a glass rod with silk strips off electrons from the surface of the rod, thus leaving an abundance of positive charges, i.e., giving the rod a positive charge. It should be noted that the electric charge is conserved - it is neither created nor destroyed. Electric charges can be only moved from one place to another. Giving negative charge means taking electrons from one object and placing them onto another (charging it negatively). The object which loses some amount of electrons is said gets a positive charge.

A triboelectric effect influences an extremely small number of electrons as compared with the total electronic charge in an object. Actual amount of charges in any object is very large. To illustrate this let us consider total number if electrons in a US copper penny [1]. The coin weighs 3.1g, therefore, it can be shown that the total number of atoms in it is about 2.9×10^{22}. A copper atom has a positive nuclear charge of 4.6×10^{-18}C and, respectively, the same electronic charge of the

---

[1] prefix *tribo-* means "pertinent to friction".

opposite polarity. A combined charge of all electrons in a penny is $q=(4.6 \times 10^{-18}$ C/atom$)(2.9 \cdot 10^{22}$ atoms$)=1.3 \times 10^{5}$C. This electronic charge from a single copper penny may generate sufficient current of 0.91 A to operate a 100 W light bulb for 40 hours.

❖ With respect to electric charges, there are three kinds of materials: conductors, isolators, and semiconductors. In conductors, electric charges (electrons) are free to move through the material, whereas in isolators they are not. Although there is no perfect isolator, the isolating ability of fused quartz is about 10^{25} times as great as that of copper, so that for practical purposes many materials are considered perfect isolators. The semiconductors are intermediate between conductors and isolators in their ability to conduct electricity. Among the elements, silicon and germanium are well known examples. In semiconductors, the electrical conductivity may be greatly increased by adding small amounts of other elements: traces of arsenic or boron are often added to silicon for this purpose.

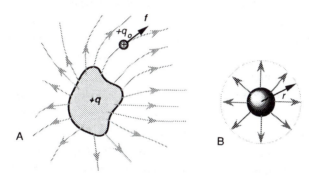

Fig. 3-1.1 Positive test charge in vicinity of a charged object (A)
and electric field of a spherical object (B)

❖ Fig. 3-1.1A shows an object which caries a positive electric charge q. If a small *positive* electric test charge, q_o, is positioned in the vicinity of a charged object, it will be subjected to a repelling electric force. If we place a negative charge on the object, it will attract the test charge. In a vector form, the repelling (or attracting) force is shown as *f*. The bold face indicates a vector notation. A fact that the test charge is subjected to force without a physical contact between charges means, that the volume of space which is occupied by the test charge, may be characterized by a so-called *electric field*.

The electric field in each point is defined through the force as

$$E = \frac{f}{q_o} \cdot \qquad\qquad (3.1.1)$$

Here E is vector of the same direction as f because q_o is scalar. Formula (3.1.1) expresses an electric field as a force divided by a property of a test charge. The test charge must be very small not to disturb the electric field. Ideally, it should be infinitely small, however, since the charge is quantized, we cannot contemplate a free test charge whose magnitude is smaller than the electronic charge: $e = 1.602 \cdot 10^{-19}$ C.

The field is indicated in Fig. 3-1.1A by the *field lines* which in every point of space are tangent to the vector of force. By definition, the field lines start on the positive plate and end on the negative. The density of field lines indicates the magnitude of electric field E in any particular volume of space.

For a physicist, any field is a physical quantity which can be specified simultaneously for all points within a given region of interest. Examples are pressure field, temperature fields, electric fields, and magnetic fields. A field variable may be a scalar (for instance, temperature field) or a vector (for instance, a gravitational field around the earth). The field variable may or may not change with time. A vector field may be characterized by a distribution of vectors which form the so-called flux (symbol Φ). Flux is a convenient description of many fields, such as electric, magnetic, thermal, etc. The word flux is derived from the Latin word *fluere* (to flow). A familiar analogy of flux is a stationary, uniform field of fluid flow (water) characterized by a constant flow vector v, the constant velocity of the fluid at any given point. In case of electric field, nothing flows in a formal sense. If we replace v by E (vector representing electric field) the field lines form flux. If we imagine a hypothetical closed surface (Gaussian surface) S, a connection between the charge, q, and flux can be established as

$$\varepsilon_0 \Phi_E = q \ , \qquad\qquad (3.1.2)$$

where $\varepsilon_0 = 8.8542 \cdot 10^{-12}$ C^2/Nm^2 is the permitivity constant, or by integrating flux over the surface

$$\varepsilon_0 \oint E dS = q \ , \qquad\qquad (3.1.3)$$

where the integral is equal to Φ_E. In the above equations, known as Gauss' law, charge q is the net charge surrounded by the Gaussian surface. If a surface encloses equal and opposite charges, the net flux Φ_E is zero. The charge outside the surface makes no contribution to the value of q, nor does the exact location of the inside charges affect this value. Gauss' law can be used to make an important prediction, namely: *An exact charge on an insulated conductor is in equilibrium, entirely on its outer surface.* This hypothesis was shown to be true even before either Gauss' law or Coulomb law was advanced. The Coulomb law itself can be derived from the Gauss' law. It states that the force acting on a test charge is inversely proportional to a squared distance from the charge

$$f = \frac{1}{4\pi\varepsilon_0} \cdot \frac{qq_0}{r^2} \; .$$

(3.1.4)

❖❖ Another result of Gauss' law is that the electric field outside any spherically symmetrical distribution of charge (Fig. 3-1.1B) is directed radially and has magnitude

$$E = \frac{1}{4\pi\varepsilon_0} \cdot \frac{q}{r^2} \; ,$$

(3.1.5)

where r is the distance from the sphere's center.

❖❖ Similarly, the electric field inside a uniform sphere of charge q is directed radially and has magnitude

$$E = \frac{1}{4\pi\varepsilon_0} \cdot \frac{qr}{R^3} \; ,$$

(3.1.6)

where R is the sphere's radius and r is the distance from the sphere's center. It should be noted that the electric field in the center of the sphere ($r=0$) is equal to zero.

❖❖ If the electric charge is distributed along an infinite (or, for the practical purposes, long) line (Fig. 3-1.2A), the electric field is directed perpendicularly to the line and has the magnitude

$$E = \frac{\lambda}{2\pi\varepsilon_0 r} \; ,$$

(3.1.7)

where r is the distance from the line and λ is the linear charge density (charge per unit length).

❖❖ The electric field due to an infinite sheet of charge (Fig. 3-1.2B) is perpendicular to the plane of the sheet and has magnitude

$$E = \frac{\sigma}{2\varepsilon_0} \; ,$$

(3.1.8)

where σ is the surface charge density (charge per unit area). However, for an isolated conductive object, the electric field is two times stronger

$$E = \frac{\sigma}{\varepsilon_0} \; .$$

(3.1.9)

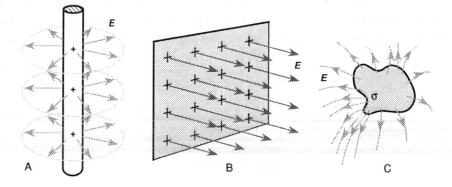

Fig. 3-1.2 Electric field around an infinite line (A) and near an infinite sheet (B).
A pointed conductor concentrates an electric field (C).

The apparent difference between electric fields is a result of different geometries – the former is an infinite sheet and the latter is an object of an arbitrary shape. A very important consequence of Gauss' law is that electric charges are distributed only on the outside surface. This is a result of repelling forces between charges of the same sign - all charges try to move as far as possible from one another. The only way to do this is to move to the foremost distant place in the material, which is the outer surface. Of all places on the outer surface the most preferable places are the areas with the highest curvatures. This is why pointed conductors are the best concentrators of the electric field (Fig. 3-1.2C). A very useful scientific and engineering tool is a Faraday cage - a room entirely covered by either grounded conductive sheets or a metal net. No matter how strong the external electric field, it will be essentially zero inside the cage. This makes cars and metal ships the best protectors during thunderstorms, because they act as virtual Faraday cages. It should be remembered however, that the Faraday cage, while being a perfect shield against electric fields, is of little use to protect against magnetic fields, unless it is made of a thick ferromagnetic material.

❖ An *electric dipole* is a combination of two opposite charges which are placed at a distance $2a$ apart (Fig. 3-1.3A). Each charge will act on a test charge with force which defines electric fields E_1 and E_2 produced by individual charges. A combined electric field of a dipole E is a vector sum of two fields. The magnitude of the field is

$$E = \frac{1}{4\pi\varepsilon_0} \cdot \frac{qa}{r^3} \quad , \tag{3.1.10}$$

where r is the distance from the center of the dipole. The essential properties of the charge distribution are magnitude of the charge, q, and the separation $2a$. In formula (3.1.10) charge and distance are entered only as a product. This means that, if we measure E at various distances from the electric dipole (assuming that distance is much longer than a), we can never deduce q and $2a$ separately, but only the product $2qa$. For instance, if q is doubled and a is cut in half, the electric field will not change. The product $2qa$ is called the electric dipole moment, p. Thus, equation (3.1.10) can be rewritten as

$$E = \frac{1}{4\pi\varepsilon_0} \cdot \frac{p}{r^3} \ . \tag{3.1.11}$$

The spatial position of a dipole may be specified by its moment in a vector form: $\boldsymbol{p}$. Not all materials have a dipole moment: gases such as methane, acetylene, ethylene, carbon dioxide, and many others have no dipole moment. On the other hand, carbon monoxide has a weak dipole moment ($0.37 \cdot 10^{-30}$C·m) and water has a strong dipole moment ($6.17 \cdot 10^{-30}$C·m).

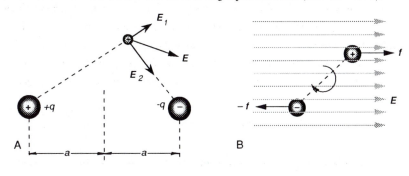

Fig. 3-1.3 Electric dipole (A);
an electric dipole in an electric field is subjected to a rotating force (B)

Dipoles are found in crystalline materials and form a foundation for such sensors as piezo and pyroelectric detectors. When a dipole is placed in an electric field, it becomes subjected to a rotation force (Fig. 3-1.3B). Usually, a dipole is a part of a crystal which defines its initial orientation. An electric field, if strong enough, will align the dipole along its lines. Torque which acts on a dipole in a vector form is

$$\tau = \boldsymbol{p}\boldsymbol{E} \ . \tag{3.1.12}$$

Work must be done by an external agent to change the orientation of an electric dipole in an external electric field. This work is stored as potential energy U in the system consisting of the dipole and the arrangement used to set up the external field. In a vector form this potential energy is

$$U = -pE .$$
(3.1.13)

A process of dipole orientation is called *poling*. The aligning electric filed must be strong enough to overcome a retaining force in the crystalline stricture of the material. To ease this process, the material during the poling is heated to increase mobility of its molecular structure. The poling is used in fabrication of piezo- and pyroelectric crystals.

❖ The electric field around the charged object can be described not only by the vector E, but by a scalar quantity, the *electric potential V* as well. Both quantities are intimately related and usually it is a matter of convenience which one to use in practice. A potential is rarely used as a description of an electric field in a specific point of space. A potential difference (*voltage*) between two points is the most common quantity in electrical engineering practice. To find the voltage between two arbitrary points, we may use the same technique as above - a small positive test charge q_o. If the electric charge is positioned in point A where its stays in equilibrium being under influence of force $q_o E$. It may remain there theoretically infinitely long. Now, if we try to move it to another point B, we have to work against the electric field. Work $-W_{AB}$ which is done against the field (that is why it has negative sign) to move the charge from A to B defines voltage between these two points

$$V_B - V_A = -\frac{W_{AB}}{q_o} .$$
(3.1.14)

Correspondingly, the electrical potential at point B is smaller than at point A. The SI unit for voltage is 1 volt = 1 joule/coulomb. For convenience, point A is chosen to be very far away from all charges (theoretically at an infinite distance) and the electric potential at that point is considered to be zero. This allows us to define electric potential at any other point as

$$V = -\frac{W}{q_o} .$$
(3.1.15)

This equation tells us that the potential near the positive charge is positive, because moving the positive test charge from infinity to the point in a field, must be made against a repelling force. This will cancel the negative sign in formula (3.1.15). It should be noted that the potential difference between two points is independent on a path at which the test charge is moving. It is strictly a description of the electric field difference between the two points. If we travel through the electric field along a straight line and measure V as we go, the rate of change of V with distance, l, that we observe is the components of E in that direction

$$E_l = -\frac{dV}{dl} \; .$$

(3.1.16)

The minus sign tells us that E points in the direction of decreasing V. Therefore, the appropriate units for electric field is volts/meter (V/m).

3.2 CAPACITANCE

Let us take two isolated conductive objects of arbitrary shape (plates) and connect them to the opposite poles of a battery (Fig. 3-2.1A). The plates will receive equal amounts of opposite charges. That is, a negatively charged plate will receive additional electrons while there will be a deficiency of electrons in the positively charged plate. Now, let us disconnect the battery. If the plates are totally isolated and exist in a vacuum, they will remain charged theoretically infinitely long. A combination of plates which can hold an electric charge is called a *capacitor*. If a small *positive* electric test charge q_o is positioned between the charged objects, it will be subjected to an electric force from the positive plate to the negative. The positive plate will repel the test charge and the negative will attract it, thus resulting in a combined push-pull force. Depending on the position of the test charge between the oppositely charged objects, the force will have a specific magnitude and direction which is characterized by vector f.

The capacitor may be characterized by q, the magnitude of charge on either conductor, shown in Fig. 3-2.1A, and by V, the positive potential difference between the conductors. It should be noted that q is not a net charge on the capacitor, which is zero. Further, V is not the potential of either plate, but the potential difference between them. The ratio of charge to voltage is constant for each capacitor:

$$\frac{q}{V} = C \; .$$

(3.2.1)

This fixed ratio C is called the *capacitance* of the capacitor. Its value depends on the shapes and relative position of the plates. C also depends on the medium in which the plates are immersed. Note, that C is always positive since we use the same sign for both q and V. The SI unit for capacitance is 1 farad = 1 coulomb/volt which is represented by the abbreviation F. A farad is a very large capacitance, hence, in practice submultiples of the farad are generally used

1 picofarad (pF)	$= 10^{-12}$F
1 nanofarad (nF)	$= 10^{-9}$F
1 microfarad (μF)	$= 10^{-6}$F

When connected into an electronic circuit, capacitance may be represented as a "complex resistance"

$$\frac{V}{i} = -\frac{1}{j\omega C} \quad ,$$

(3.2.2)

where $j = \sqrt{-1}$ and i is the sinusoidal current having a frequency of ω, meaning that the complex resistance of a capacitor drops at higher frequencies. This is called Ohm's law for the capacitor. The minus sign and complex argument indicate that the voltage across the capacitor lags by 90° behind the current.

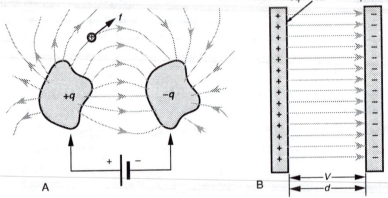

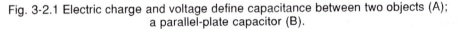

A   B

Fig. 3-2.1 Electric charge and voltage define capacitance between two objects (A);
a parallel-plate capacitor (B).

Capacitance is a very useful physical phenomenon in a sensor designer's toolbox. It can be successfully applied to measure distance, area, volume, pressure, force, etc. The following background establishes fundamental properties of the capacitor and gives some useful equations. Fig. 3-2.1B shows a parallel-plate capacitor in which the conductors take the form of two plane parallel plates of area A separated by a distance d. If d is much smaller that the plate dimensions, the electric field between the plates will be uniform, which means that the field lines (lines of force f) will be parallel and evenly spaced. The laws of electromagnetism requires that there be some "fringing" of the lines at the edges of the plates, but for small enough d we can neglect it for our present purpose.

3.2.1 Capacitor

To calculate the capacitance we must relate V, the potential difference between the plates, to q, the capacitor charge (3.2.1)

$$C = \frac{q}{V} \ . \tag{3.2.3}$$

Alternatively, the capacitance of a flat capacitor can be found from

$$C = \frac{\varepsilon_0 A}{d} \ . \tag{3.2.4}$$

Formula (3.2.4) is very important for the capacitive sensors design. It establishes a relationship between the plate area and distance between the plates. Varying either of them will linearly change the capacitor's value which can be measured quite accurately by an appropriate circuit. It should be noted that the above equations hold only for capacitors of the parallel type. A change in geometry will require modified formulas. A ratio A/d may be called a geometry factor for a parallel-plate capacitor.

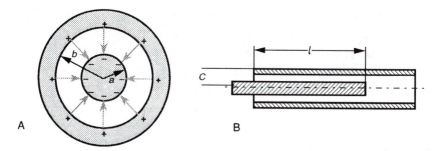

A   B

Fig. 3-2.2 Cylindrical capacitor (A); capacitive displacement sensor (B)

A cylindrical capacitor which is shown in Fig. 3-2.2A consists of two coaxial cylinders of radii a and b, and length l. For the case when $l >> b$ we can ignore fringing effects and calculate capacitance from the following formula

$$C = \frac{2\pi\varepsilon_0 l}{\ln\frac{b}{a}} \ . \tag{3.2.5}$$

In this formula l has a meaning of length of the overlapping conductors (Fig. 3-2.2B) and $2\pi l (\ln b/a)^{-1}$ may be called a geometry factor for a coaxial capacitor. A useful displacement sensor can be built with such a capacitor if the inner conductor can be moved in and out of the outer conductor. According to Eq. (3.2.5), the capacitance of such a sensor is in a linear relationship with the displacement l.

3.2.2. Dielectric constant

Equation (3.2.4) holds for a parallel-plate capacitor with its plates in vacuum (or air, for most practical purposes). In 1837, Michael Faraday first investigated the effect of completely filling the space between the plates with a dielectric. He had found that the effect of the filling is to increase the capacitance of the device by a factor of κ, which is known as the dielectric constant of the material.

The increase in capacitance due to the dielectric presence is a result of molecular polarization. In some dielectrics (for instance, in water), molecules have a permanent dipole moment, while in other dielectrics, molecules become polarized only when an external electric field is applied. Such polarization is called induced. In both cases, either permanent electric dipoles or acquired by induction, tend to align molecules with an external electric field. This process is called dielectric polarization. It is illustrated by Fig. 3–2.3A which shows permanent dipoles before and Fig. 3-2.3B after an external electric field is applied to the capacitor. In the former case, there is no voltage between the capacitor plates and all dipoles are randomly oriented. After the capacitor is charged, the dipoles will align with the electric field lines, however, thermal agitation will prevent a complete alignment. Each dipole forms its own electric field which is predominantly is oppositely directed with the external electric field E_o. Due to a combined effect of a large number of dipoles (E'), the electric filed in the capacitor becomes weaker ($E = E_o + E'$) when the field E_o would be in the capacitor without the dielectric.

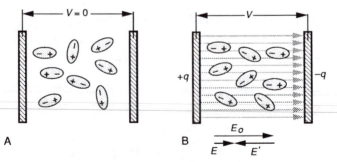

Fig. 3-2.3 Polarization of dielectric
A: Dipoles are randomly oriented without an external electric field;
B: Dipoles align with an electric field

Reduced electric field leads to a smaller voltage across the capacitor: $V = V_o/\kappa$. Substituting it into formula (3.2.3) we get an expression for the capacitor with dielectric

$$C = \kappa \frac{q}{V_o} = \kappa C_o \quad . \tag{3.2.6}$$

For the parallel plate capacitor we thus have

$$C = \frac{\kappa \varepsilon_0 A}{d} \quad . \tag{3.2.7}$$

Table 3-1 Dielectric constants of some materials at room temperature (25°C)

Material	κ	Frequency (Hz)	Material	κ	Frequency (Hz)
Air	1.00054	0	Plexiglas	3.12	10^3
Asphalt	2.68	10^6	Polyesters	3.22-4.3	10^3
Beeswax	2.9	10^6	Polyethylene	2.26	10^3-10^8
Benzene	2.28	0	Polyvinyl chloride	4.55	10^3
Carbon tetrachloride	2.23	0	Porcelain	6.5	0
Cellulose nitrate	8.4	10^3	Pyrex glass (7070)	4.0	10^6
Ceramic (titanium dioxide)	14-110	10^6	Pyrex glass (7760)	4.5	0
Ceramic(alumina)	4.5-8.4	10^6	Rubber (neoprene)	6.6	10^3
Compound for thick film capacitors	300-5000	0	Rubber (silicone)	3.2	10^3
Diamond	5.5	10^8	Rutile $\perp$ optic axis	86	10^8
Epoxy resins	3.65	10^3	Rutile $\parallel$ optic axis	170	10^8
Ferrous oxide	14.2	10^8	Silicone resins	3.85	10^3
Lead nitrate	37.7	$6 \cdot 10^7$	Tallium chloride	46.9	10^8
Methanol	32.63	0	Teflon	2.04	10^3-10^8
Nylon	3.50	10^3	Transformer oil	4.5	0
Paper	3.5	0	Vacuum	1	—
Paraffin	2.0-2.5	10^6	Water	78.5	0

In a more general form, the capacitance between two objects may be expressed through a geometry factor G

$$C = \varepsilon_0 \kappa G \quad , \tag{3.2.8}$$

G depends on the shape of the objects (plates) and their separation. Table 3-1 gives dielectric constants κ for various materials.

Dielectric constants must be specified for test frequency and temperature. Some dielectrics have a very uniform dielectric constant over a broad frequency range (for instance, polyethylene), while others display strong negative frequency dependence, that is, a dielectric constant decreases with frequency. Temperature dependence is also negative. Fig. 3-2.4 illustrates κ for water as a function of temperature.

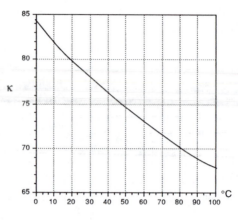

Fig. 3-2.4 Dielectric constant of water as a function of temperature

Dielectrics significantly increase capacitance and, therefore, are quite useful for applications in sensors. For example, let us consider a capacitive water level sensor (Fig. 3-2.5A). The sensor is fabricated in a form of a coaxial capacitor where the surface of each conductor is coated with a thin isolating layer to prevent an electric short circuit through water (the isolator is a dielectric which we disregard in the following analysis because it does not change in the process of measurement). The sensor is immersed in a water tank. When the level increases, water fills more and more space between the sensor's coaxial conductors, thus changing the sensor's capacitance.

Total capacitance of the coaxial sensor is

$$C = C_1 + C_2 = \varepsilon_0 G_1 + \varepsilon_0 \kappa G_2 \ , \qquad (3.2.9)$$

where C_1 is the capacitance of the water-free portion of the sensor and C_2 is the capacitance of the water filled portion. The corresponding geometry factors are designated G_1 and G_2.

From formulas (3.2.5) and (3.2.9), total sensor capacitance can be found as

$$C_h = \frac{2\pi\varepsilon_0}{\ln\frac{b}{a}} \left[H - h(1 - \kappa) \right] \ , \qquad (3.2.10)$$

where h is height of the water-filled portion of the sensor. If the water is at or below the level h_o, the capacitance remains constant

$$C_o = \frac{2\pi\varepsilon_0}{\ln\frac{b}{a}} H \ . \tag{3.2.11}$$

Fig. 3-2.5B shows a water level-capacitance dependence[1]. It is a straight line from level h_o.

The slope of the line depends on the liquid. For instance, if instead of water the sensor measures level of transformer oil, it will be 22 times less sensitive (See Table 3-1).

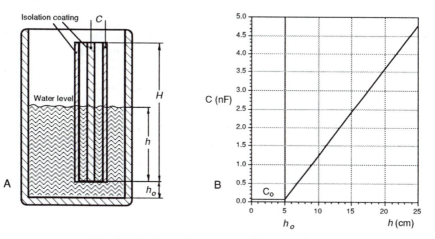

Fig. 3-2.5 Capacitive water level sensor (A);
capacitance as a function of the water level (B)

3.3 MAGNETISM

Magnetic properties were discovered in prehistoric times in certain specimens of an iron ore mineral known as magnetite (Fe_3O_4). It was also discovered that pieces of soft iron that rubbed against a magnetic material acquired the same property of acting as a magnet, i.e., attracting other magnets and pieces of iron. The first comprehensive study of magnetism was made by William Gilbert. His greatest contribution was his conclusion that the earth acts as a huge magnet. The word magnetism comes from the district of Magnesia in Asia Minor, which is one of the places at which the magnetic stones were found.

There is a strong similarity between electricity and magnetism. One manifestation of this is that two electrically charged rods have like and unlike ends, very much in the same way as two

---

[1]Sensor's dimensions are as follows: a=10 mm, b = 12 mm, H = 200 mm, liquid — water.

magnets have opposite ends. In magnets, these ends are called S (south) and N (north) poles. The like poles repeal and the unlike attract. Contrary to electric charges, the magnetic poles always come in pairs. This is proven by breaking magnets into any number of parts. Each part, no matter how small, will have a north pole and a south pole. This suggests that the cause of magnetism is associated with atoms or their arrangements or, more probably, with both.

If we place a magnetic pole in a certain space, that space about the pole appears to have been altered from what it was before. To demonstrate this, bring into that space a piece of iron. Now, it will experience a force that it will not experience if the magnet is removed. This altered space is called a magnetic field. The field is considered to exert a force on any magnetic body brought into the field. If that magnetic body is a small bar magnet or a magnetic needle, the magnetic field will be found to have direction. By definition, the direction of this field at any point is given by the direction of the force exerted on a small unit north pole. Directions of field lines are by definition from north to south pole. Fig. 3-3.1A shows the direction of the field by arrows. A tiny test magnet is attracted in the direction of the force vector *F*. Naturally, about the same force but of opposite direction is exerted on the south pole of the test magnet.

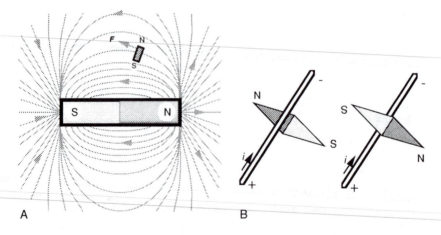

Fig. 3-3.1 Test magnet in a magnetic field (A);
compass needle rotates in accordance with the direction of the electric current (B)

The above description of the magnetic field was made for a permanent magnet. However, the magnetic field does not change its nature if it is produced by a different device - electric current passing through a conductor. It was Hans Christian Oersted, a Danish professor of physics, who in 1820 discovered that a magnetic field could exist where there were no magnets at all. In a series of experiments in which he was using an unusually large Voltaic pile (battery) so as to produce a large current, he happened to note that a compass in the near vicinity was behaving oddly. Further investigation showed that the compass needle always oriented itself at right angles to the current carrying wire, and that it reversed its direction if either current was reversed, or the compass was changed from a position below the wire to the one above (Fig. 3-3.1B). Stationary electric charges make no effect on a magnetic compass (in this experiment, a compass needle is used as a tiny test magnet). It was clear that the moving electric charges were the cause of the magnetic field. It can be shown that magnetic field lines around a wire are circular and their direction depends on the direction of electric current, i.e., moving electrons (Fig. 3-3.2). Above and below the wire, magnetic field lines are pointed in the opposite direction. That's why the compass needle turns around when it is placed below the wire.

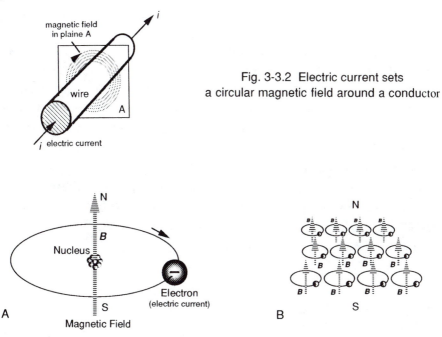

Fig. 3-3.2 Electric current sets
a circular magnetic field around a conductor

Fig. 3-3.3 Moving electron sets a magnetic field (A);
superposition of field vectors results in a combined magnetic field of a magnet (B)

A fundamental property of magnetism is that moving electric charges (electric current) essentially produce a magnetic field. Knowing this, we can explain the nature of a permanent magnet. A simplified model of a magnetic field origination process is shown in Fig. 3-3.3A. An electron continuously spins in an eddy motion around the atom. The electron movement constitutes a circular electric current around the atomic nucleus. That current is a cause for a small magnetic field. In other words, a spinning electron forms a permanent magnet of atomic dimensions. Now, let us imagine that many of such atomic magnets are aligned in an organized fashion (Fig. 3-3.3B), so that their magnetic fields add up. The process of magnetization then becomes quite obvious - nothing is added or removed from the material - only orientation of atoms is made. The atomic magnets may be kept in the aligned position in some materials which have an appropriate chemical composition and a crystalline structure. Such materials are called ferromagnetics.

3.3.1 Faraday law

Michael Faraday pondered the question, "If an electric current is capable of producing magnetism, is it possible that magnetism can be used to produce electricity?" It took him nine or ten years to discover how. If an electric charge is moved across a magnetic field, a deflecting force is acting on that charge. It must be emphasized that it is not important what actually moves - either the charge or the source of the magnetic field. What matters is a relative displacement of those. A discovery that a moving electric charge can be deflected as a result of its interaction with the magnetic field is a fundamental in electromagnetic theory. Deflected electric charges result in an electric field generation, which, in turn, leads to a voltage difference in a conducting material, thus producing an electric current.

The intensity of a magnetic field at any particular point is defined by vector B which is tangent to a magnetic field line at that point. For the better visual representation, the number of field lines per unit cross-sectional area (perpendicular to the lines) is proportional to the magnitude of B. Where the lines are close together, B is large and where they are far apart, B is small.

The flux of magnetic field can be defined as

$$\Phi_B = \oint B ds \quad , \tag{3.3.1}$$

where the integral is taken over the surface for which F_B is defined.

To define the magnetic field vector B we use a laboratory procedure where a positive electric charge q_o is used as a test object. The charge is projected through the magnetic field with velocity v. A sideways deflecting force F_B acts on the charge (Fig. 3-3.4A). By "sideways" we mean that F_B is at a right angle to v. It is interesting to note that vector v changes its direction while moving through the magnetic field. This results in a spiral rather than parabolic motion of the charge (Fig. 3-3.4B). The spiral movement is a cause for a magnetoresistive effect which forms a foundation for the magnetoresistive sensors (Section 15.1.3). Deflecting force F_B is proportional to charge, velocity, and magnetic field

$$F_B = q_o vB \ . \tag{3.3.2}$$

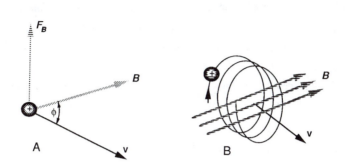

Fig. 3-3.4 Positive charge projected through a magnetic field
is subjected to a sideways force (A);
spiral movement of an electric charge in a magnetic field (B)

Vector F_B is always at right angles to the plane formed by v and B and thus is always at right angles to v and to B, that is why it is called a sideways force. The magnitude of magnetic deflecting force according to the rules for vector products is

$$F_B = q_o vB\sin\phi \ , \tag{3.3.3}$$

where ϕ is the angle between vectors v and B. The magnetic force vanishes if v is parallel to B. The above equation (3.3.3) is used for the definition of the magnetic field in terms of deflected charge, its velocity, and deflecting force. Therefore, the units of B is (Newton/coulomb)/(meter/second)$^{-1}$. In the system SI it is given name *tesla* (abbreviated T). Since coulomb/second is an ampere, we have 1T=1 newton/(ampere·meter). An older unit for B still is sometimes in use. It is the gauss:

$$1 \text{ tesla} = 10^4 \text{ gauss}$$

3.3.2 Biot-Savart law

The calculation of a magnetic field produced by an electric current can be done in the following way. Fig. 3-3.5A shows a wire of arbitrary shape. Electric current i is produced in a wire by an external source of an electric field, for instance, a battery (not shown). This current is a source of the magnetic field all around the wire. We are going to find the field produced in point P. The magnetic field in this point is a result of a superposition of the so-called elementary magnetic vectors produced by elementary pieces of a current carrying wire. The word elementary means a very small piece of wire having length dl and positioned at a point $\oint \mathbf{B} dl = \mu_0 i$. The piece is considered straight and directional (note vector notation for dl). The direction of the elementary wire piece is tangent to the wire (is indicated by a dashed line) and pointing toward the direction of current flow. We call the current element a product of current and elementary piece: idl. An entire current carrying wire may be considered as a vector sum of elementary current elements. Since our goal is to find the magnetic field at point P, we extend the vector from z to P to indicate the distance and direction to that point. The magnitude of elementary magnetic field set up at P is given by

$$dB = \frac{\mu_0}{4\pi} \frac{idl \, \sin\Theta}{r^2} \, .$$ (3.3.4)

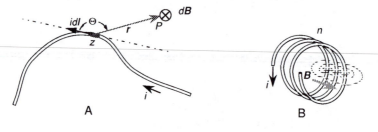

Fig. 3-3.5A Magnetic field produced by electric current
depends on current direction and its magnitude (A);
formation of magnetic field in a coil (B)

Equation (3.3.4) is called the Biot-Savart law and is analogous to the Coulomb law for the electric field. The law was named after two French scientists who experimentally established it within a short time from Oersted's discovery. The direction of dB is defined by a product of vectors dl and r, and is into the plane of Fig. 3-3.5A. This is denoted by a cross in the circle (a mnemonic representation of a tale of the magnetic vector arrow). To find a resultant magnetic field B we have to integrate all elementary magnetic fields set up by the elementary current pieces in the wire. The constant μ_0 which appears in equation (3.3.4) is called the permeability constant. In SI, its assigned value is

$$\mu_0 = 4\pi \cdot 10^{-7} \text{T·m/A} \, .$$ (3.3.5)

If we use the Biot-Savart law to calculate the magnetic field near a long straight wire, the result will be

$$B = \frac{\mu_0 i}{2\pi r} \quad , \tag{3.3.6}$$

where r is the perpendicular distance from the wire. If we have n thin parallel wires closely tied together and each carries equal currents i in the same direction, then the resultant magnetic field at distance, r, (which is much larger that the wire bundle diameter) is

$$B \approx n \frac{\mu_0 i}{2\pi r} \quad . \tag{3.3.7}$$

3.3.3 Ampere's law

Many electrostatic problems can be solved using Gauss' law. Similarly, in the theory of magnetism there is an equally efficient method which is based on *Ampere's* law. In a general form the law can be expressed as

$$\oint B dl = \mu_0 i \quad , \tag{3.3.8}$$

where l is the distance and the integral is over the closed loop drawn in the region where the magnetic field and current distribution exist. The integral on the left depends on the way B varies in magnitude and direction as we traverse the loop. The quantity of i on the right is the net current that pierces the loop. We are not going further to discuss the Ampere's law. For details the reader may be referred to many texts on the fundamentals of physics, specifically on electromagnetism. Here, we provide some useful results derived from Ampere's law which may be applicable to the design of sensors employing the magnetic components. There is a great variety of applications where a magnetic field is artificially produced by electric current, for instance – electric motors and galvanometers.

3.3.4 Solenoid

A practical device to produce a magnetic field is called a *solenoid*. It is a long wire wound in a close-packed helix and carrying a current i. In the following discussion we assume that the helix is very long as compared with its diameter. The solenoid magnetic field is the vector sum of the fields setup by all the turns that make up the solenoid.

If a coil (solenoid) has widely spaced turns (Fig. 3-3.5B), the fields tends to cancel between the wires. At points inside the solenoid and reasonably far from the wires, B is parallel to the solenoid axis. In the limiting case of adjacent very tightly packed wires (Fig. 3-3.6A), the solenoid becomes essentially a cylindrical current sheet. If we apply Ampere's law to that current sheet, the magnitude of magnetic field inside the solenoid becomes

$$B = \mu_0 i_o n ,$$ (3.3.9)

where n is the number of turns per unit length and i_o is the current through the solenoid wire. Although, this formula was derived for an infinitely long solenoid, it holds quite well for actual solenoids for internal points near the center of the solenoid. It should be noted that B does not depend on the diameter or the length of the solenoid and that B is constant over the solenoid cross section. Since the solenoid's diameter is not a part of the equation, multiple layers of winding can be used to produce a magnetic field of higher strength. It should be noted that magnetic field outside of a solenoid is weaker than that of the inside.

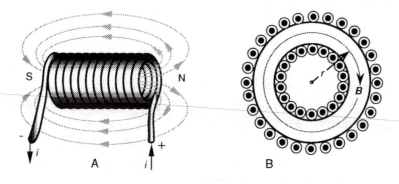

Fig. 3-3.6 Solenoid (A) and toroid (B)

3.3.5 Toroid

Another useful device that can produce a magnetic field is a toroid (Fig. 3-3.6B), which we can describe as a solenoid bent into the shape of a doughnut. A calculation of the magnetic field inside the toroid gives the following relationship

$$B = \frac{\mu_0}{2\pi} \cdot \frac{i_o N}{r} ,$$ (3.3.10)

where N is the total number of turns and r is the radius of the inner circular line where magnetic field is calculated. In contrast to a solenoid, B is not constant over the cross section of a toroid. Besides, for an ideal case, the magnetic field is equal to zero outside a toroid.

The density of a magnetic field, or the number of magnetic lines passing through a given surface, is defined as the magnetic flux Φ_B for that surface

$$\Phi_B = \int B dS \ . \tag{3.3.11}$$

The integral is taken over the surface and if the magnetic field is constant and is everywhere at a right angle to the surface, the solution of the integral is very simple: $\Phi_B = BA$, where A is the surface area. Flux, or flow of the magnetic field is analogous to flux of electric field. The SI unit for magnetic flux, as follows from the above, is tesla·meter^2, to which is name *weber*. It is abbreviated as Wb:

$$1 \ Wb = 1 \ T \cdot m^2 \tag{3.3.12}$$

3.3.6 Magnetic dipole

One of early meters of a magnetic field and electric current was a *galvanometer* — an instrument where a magnetic field produced by a magnet is interacting with that produced by an electric current. Fig. 3-3.7 shows a current loop in a magnetic field B. The loop axis is positioned with respect to the field at angle Θ. Interaction of the fields results in puling forces F_1 and F_2 and, subsequently, in rotating torque

$$\tau = BNia \sin\Theta \ , \tag{3.3.13}$$

where a is the area of the current loop (coil) and N is the number of turns. In a galvanometer, a spring provides a counter torque, that cancels out the magnetic torque. Without the spring, the loop would rotate until $\Theta = 0$. The spring balances the torque resulting in a steady deflection angle ϕ of the loop

$$Nia \sin\Theta = k\phi \ , \tag{3.3.14}$$

where k is the torsional constant of the spring. Then, a deflecting angle of the galvanometer needle can be found as

$$\phi = k \frac{1}{Nia \sin\Theta} \ . \tag{3.3.15}$$

In equation (3.3.13), Nia is called a magnetic dipole moment μ[1]

---

[1] Don't confuse it with magnetic permeability of free space μ_o.

$$\mu = Nia , \tag{3.3.16}$$

which is a unique description of a current loop producing a magnetic field. Then torque in a vector forms is

$$\tau = \mu B . \tag{3.3.17}$$

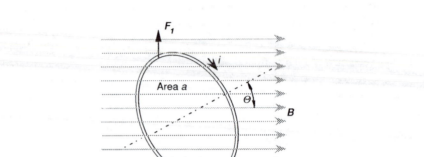

Fig. 3-3.7 Current loop in a magnetic field

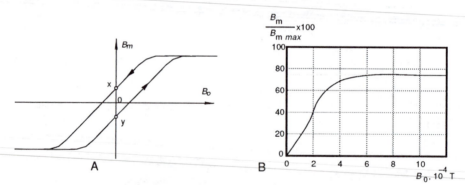

Fig. 3-3.8 Hysteresis in a magnetic material (A);
saturation of ferromagnetic material in an external magnetic field B)

A direction of a magnetic dipole moment must be taken to lie along an axis perpendicular to the plane of the loop and can be found from a mnemonic rule: *let the fingers of the right hand curl around the loop in the direction of the current, the extended right thumb will then point in the direction of* μ.

The current loop is called a magnetic dipole. Along its axis, it produces a magnetic field

$$B(x) = \frac{\mu_0 \mu}{2\pi x^3} \, ,$$

(3.3.18)

where x is distance from the loop plane.

The magnetic dipole has properties which are similar to the electric dipole (see section 3.1). This is illustrated by Table 3-2.

In all materials, atoms are in constant thermal agitation. However, for some elements and for many alloys this does not prevent a high degree of magnetic alignment. Examples of such materials are Co, Ni, Cd, and Dy (dysprosium). They are called *ferromagnetics* and characterized by a special form of interaction between adjacent atoms and molecules which is called exchange coupling. This is a purely quantum effect and can not be explained in terms of classical physics. If the temperature rises above a critical value, called the Curie temperature, the exchange coupling suddenly disappears. For iron, the Curie temperature is 770°C. If a ferromagnetic material is placed in a magnetic field, its elementary atomic dipoles become aligned with the external magnetic field B_o (B_o is the field which would be present without the ferromagnetic material). The combined magnetic field becomes

$$B = B_o + B_m \, ,$$

(3.3.19)

where B_m is the contribution of the ferromagnetic material. In most cases B_m is much stronger than B_o which determines the degree of magnetization. The increase in B_o results in an almost proportional increase in B_m up to the level where majority of atomic dipoles are aligned. After that level, which is called saturation, the external magnetic field can not further increase the setup alignment. Fig. 3-3.8B illustrates this for the saturation level at about 75% of the theoretically possible total exchange coupling. The use of ferromagnetic materials, especially iron, greatly increases the strength of magnetic fields. It should be noted that such a magnetization curve does not retrace itself when the external field B_o increases and then decreases. Fig. 3-3.8A shows this with two special points, x and y. At these points, the external magnetic field is absent, however the ferromagnetic material remains slightly magnetized. This phenomenon is called permanent magnetization. It is as though the magnetic material "remembers" the direction in which the magnetic field was changing. The magnetization curve is called a hysteresis loop.

Magnetic cores are used in many sensor applications. All equations which establish a relationship between electric current and a magnetic field incorporate quantity μ_0 which is magnetic permeability of free space. However, when the magnetic core is used, its magnetic permeability very much differs from that of free space. Like dielectric constant κ in equation (3.2.6), a dimensionless quantity called relative magnetic permitivity μ_r was introduced. It shows an increase in the magnetic field in a material with respect to free space (vacuum). From equation (3.3.19) it follows that

$$\mu_r = 1 + \frac{B_m}{B_o} \quad .$$

(3.3.20)

Table 3-2 Properties of electric and magnetic dipoles (adapted from [1])

Property	Dipole Type	Equation
Torque in an external field	electric	$\tau = pE$
	magnetic	$\tau = \mu B$
Field at a distant point along an axis	electric	$E(x) = \frac{1}{2\pi\varepsilon_o} \frac{p}{x^3}$
	magnetic	$B(x) = \frac{1}{2\pi} \frac{\mu}{x^3}$
Field at distant points in the median plane	electric	$E(x) = \frac{1}{4\pi\varepsilon_o} \frac{p}{x^3}$
	magnetic	$B(x) = \frac{\mu_o}{4\pi} \frac{\mu}{x^3}$

Permitivity of a material is defined as

$$\mu = \mu_o \mu_r \quad .$$

(3.3.21)

Contrary to κ, μ is not constant for a given material because its value depends on magnetic field.

3.3.7 Permanent magnets

Permanent magnets are useful components to fabricate magnetic sensors for the detection of motion, displacement, position, etc. To select the magnet for any particular application, the following characteristics should be considered:

✔ Residual inductance (B) in gauss - how strong the magnet is?

✔ Coercive force (H) in oersteds - how well will the magnet resist external demagnetization forces?

✔   Maximum energy product, MEP, $(B \times H)$ is gauss-oersteds times 10^6. A strong magnet that is also very resistant to demagnetization forces has a high MEP. Magnets with higher MEP are better, stronger, and more expensive.

✔   Temperature coefficient in %/°C shows how much B changes with temperature?

Magnets are produced from special alloys (Table 3-3). Examples are *rare earth* (e.g., samarium)-*cobalt* alloys. These are the best magnets, however, they are too hard for machining, and must be ground if shaping is required. Their maximum MEP is about $16 \cdot 10^6$. Another popular alloy is *Alnico*, which contains aluminum, nickel, cobalt, iron, and some additives. These magnets can be cast, or sintered by pressing metal powders in a die and heating them. Sintered Alnico is well suited to mass production. *Ceramic magnets* contain barium or strontium ferrite (or another element from that group) in a matrix of a ceramic material that is compacted and sintered. They are poor conductors of heat and electricity, are chemically inert, and have value of H. Another alloy for the magnet fabrication is *Cunife*, which contains copper, nickel, and iron. It can be stamped, swaged, drawn, or rolled into final shape. Its MEP is about $1.4 \cdot 10^6$. *Iron-chromium magnets* are soft enough to undergo machining before the final aging treatment hardens them. Their maximum MEP is $5.25 \cdot 10^6$. *Plastic and rubber* magnets consist of barium or strontium ferrite in a plastic matrix material. They are very inexpensive and can be fabricated in many shapes. Their maximum MEP is about $1.2 \cdot 10^6$.

Table 3-3 Properties of magnetic materials
(adapted from [2])

Material	MEP $(G \cdot Oe) \cdot 10^6$	Residual induction $(G) \cdot 10^3$	Coercive force $(Oe) \cdot 10^3$	Temperature coefficient %/°C	Cost
R.E.-Cobalt	16	8.1	7.9	-0.05	highest
Alnico 1,2,3,4	1.3-1.7	5.5-7.5	0.42-0.72	-0.02 to -0.03	medium
Alnico 5,6, 7	4.0-7.5	10.5-13.5	0.64-0.78	-0.02 to -0.03	medium/high
Alnico 8	5.0-6.0	7-9.2	1.5-1.9	-0.01 to 0.01	medium/high
Alnico 9	10	10.5	1.6	-0.02	high
Ceramic 1	1.0	2.2	1.8	-0.2	low
Ceramic 2,3,4,6	1.8-2.6	2.9-3.3	2.3-2.8	-0.2	low/medium
Ceramic 5,7,8	2.8-3.5	3.5-3.8	2.5-3.3	-0.2	medium
Cunife	1.4	5.5	0.53	-	medium
Fe-Cr	5.25	13.5	0.6	-	medium/high
Plastic	0.2-1.2	1.4	0.45-1.4	-0.2	lowest
Rubber	0.35-1.1	1.3-2.3	1-1.8	-0.2	lowest

3.4   INDUCTION

In 1831, Michael Faraday in England and Joseph Henry in the U.S.A. discovered one of the most fundamental effects of electromagnetism: an ability of a varying magnetic field to induce electric current in a wire. It is not important how the field is produced - either by a permanent magnet or by a solenoid - the effect is the same. Electric current is generated as long as the magnetic field *changes*. A stationary field produces no current. Faraday's law of induction says that the induced voltage, or electromotive force (*e.m.f.*), is equal to the rate at which the magnetic flux through the circuit changes. If the rate of change is in Wb/sec, the *e.m.f.* (*e*) will be in volts

$$e = -\frac{d\Phi_B}{dt} \ .$$

(3.4.1)

The minus sign is an indication of the direction of the induced *e.m.f.* If varying magnetic flux is applied to a solenoid, *e.m.f.* appears in every turn and all these *e.m.f.*'s must be added. If a solenoid, or other coil, is wound in such a manner as each turn has the same cross-sectional area, the flux through each turn will be the same, then the induced voltage is

$$V = -N\frac{d\Phi_B}{dt}$$

(3.4.2)

where N is the number of turns. This equation may be rewritten in a form which is of interest to a sensor designer or an application engineer

$$V = -N\frac{d(BA)}{dt}$$

(3.4.3)

The equation means that the voltage in a pick-up circuit can be produced by either changing amplitude of magnetic field (B) or area of the circuit (A). Thus, induced voltage depends on:

❖ Moving the source of the magnetic field (magnet, coil, wire, etc.);

❖ Varying the current in the coil or wire which produces the magnetic field;

❖ Changing the orientation of the magnetic source with respect to the pick-up circuit;

❖ Changing the geometry of a pick-up circuit, for instance, by stretching it or squeezing, or changing the number of turns in a coil.

Let us, for example, consider a motion sensor, where a pick-up coil having rectangular shape and N turns, moves into a gap of a permanent magnet (Fig. 3-4.1A). The flux enclosed by the loop is

$$\Phi_b = Blx \ , \tag{3.4.4}$$

where lx is the area that of that part of the loop which is entered into the gap of the magnet. The voltage induced in the loop can be found from Faraday's law

$$V = -\frac{d\Phi_B}{dt} = -N\frac{d}{dt}(BLx) = -nBl\frac{dx}{dt} = nBlv \ , \tag{3.4.5}$$

where $v = -dx/dt$ is the velocity of the coil movement. The output voltage is a linear function of the rate of coil movement. If the coil is not rectangular, its geometry coefficient as a function of x must be added as part of the equation. Fig. 3-4.1B shows voltages for the rectangular and circular coils. Note that as the output voltage becomes zero then the coil is either out of the magnet's gap or is fully inside the gap.

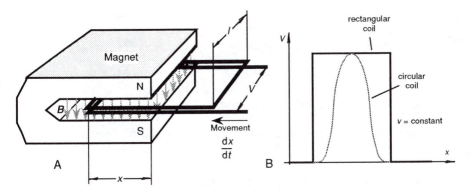

Fig. 3-4.1 Magnetic motion detector (A) and its transfer function (B).

If an electric current passes through a coil which is situated in close proximity with another coil, according to Faraday's law, *e.m.f.* in a second coil will appear. However, the magnetic field penetrates not only the second coil, but the first coil as well. Thus, the magnetic field sets *e.m.f.* in the same coil where it is originated. This is called *self-induction* and the resulting voltage is called a *self-induced e.m.f.* Faraday's law for a central portion of a solenoid is

$$v = -\frac{d(n\Phi_B)}{dt} \ . \tag{3.4.6}$$

The number in parenthesis is called the flux linkage and is an important characteristic of the device. For a simple coil with no magnetic material in the vicinity, this value is proportional to current through coil

$$n\Phi_B = Li \ , \tag{3.4.7}$$

where L is a proportionality constant, which is called the *inductance* of the coil. Then, equation (3.4.6) can be re-written as

$$v = \frac{d(n\Phi_B)}{dt} = -L\frac{di}{dt} \quad . \tag{3.4.8}$$

From this equation we can define inductance as

$$L = -\frac{v}{di/dt} \quad . \tag{3.4.9}$$

If no magnetic material is introduced in the vicinity of an *inductor* (a device possessing inductance), the value defined by equation (3.4.9) depends only on the geometry of the device. The SI unit for inductance is the volt·second/ampere, which was named after American physicist Joseph Henry (1797-1878): 1 henry = 1 volt·second/ampere. Abbreviation for henry is H.

Several conclusions can be drawn from equation (3.4.9):

❖ Induced voltage is proportional to the rate of change in current through the inductor;

❖ Voltage is essentially zero for dc;

❖ Voltage increases linearly with the current rate of change;

❖ Voltage polarity is different for increased and decreased currents flowing in the same direction;

❖ Induced voltage is always in the direction which opposes the change in current.

Like capacitance, inductance can be calculated from geometrical factors. For a closely packed coil it is

$$L = \frac{n\Phi_B}{i} \quad . \tag{3.4.10}$$

If n is the number of turns per unit length, the number of flux linkages in the length, l, is

$$N\Phi_B = (nl)\cdot(BA) \quad , \tag{3.4.11}$$

where A is the cross-sectional area of the coil. For the solenoid, $B = \mu_0 ni$, then the inductance is

$$L = \frac{N\Phi_B}{i} = \mu_0 n^2 lA \quad . \tag{3.4.12}$$

It should be noted that lA is the volume of a solenoid, Thus, having the same number of turns and changing the coil geometry, its inductance may be modulated (altered).

When connected into an electronic circuit, inductance may be represented as a "complex resistance"

$$\frac{V}{i} = j\omega L \; , \tag{3.4.13}$$

where $j = \sqrt{-1}$ and i is a sinusoidal current having a frequency of $\omega = 2\pi f$, meaning that the complex resistance of an inductor increases at higher frequencies. This is called Ohm's law for an inductor. Complex notation indicates that current lags behind voltage by 90°.

If two coils are brought in the vicinity of one another, one coil induces *e.m.f.* v_2, in the second coil

$$v_2 = -M_{21}\frac{di_1}{dt} \; , \tag{3.4.14}$$

where M_{21} is the coefficient of mutual inductance between two coils. The calculation of mutual inductance is not a simple exercise and in many practical cases can be easier performed experimentally. Nevertheless, for some relatively simple combinations mutual inductance have been calculated. For a coil (having N turns) which is placed around a long solenoid (Fig. 3-4.2A), having n turns per unit length, mutual inductance is

$$M = \mu_0 \pi R^2 nN \; . \tag{3.4.15}$$

For a coil placed around a toroid (Fig. 3-4.2B), mutual inductance is defined through numbers of turns, N_1 and N_2

$$M = \frac{\mu_0 N_1 N_2 h}{2\pi} \ln\frac{b}{a} \; . \tag{3.4.16}$$

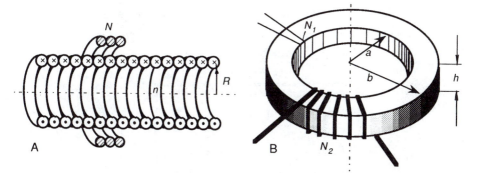

Fig. 3-4.2 Mutual inductances in solenoids (A) and in a toroid (B).

3.5 RESISTANCE

In any material, electrons move randomly like gas in a closed container. There is no preferred direction and an average concentration of electrons in any part of material is uniform (assuming that the material is homogeneous). Let us take a bar of an arbitrary material. The length of the bar is l. When the ends of the bar are connected to the battery having voltage V (Fig. 3-5.1), an electric field E will be setup within the material. It is easy to determine strength of the electric field

$$E = V/l \, . \tag{3.5.1}$$

For instance, if the bar has a length of 1 m and the battery delivers 1.5 V, the electric field has the strength of 1.5 V/m. The field acts on free electrons and sets them in motion against the direction of the field. Thus, the electric current starts flowing through the material. We can imagine a cross section of the material through which passes electric charge q. The rate of the electric charge flowing (unit of charge per unit of time) is called electric current

$$i = \frac{dq}{dt} \, . \tag{3.5.2}$$

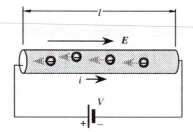

Fig. 3-5.1 Voltage across a material sets electric current

The SI unit of current is amperes (A): 1A=1 coulomb/1 sec. An ampere is quite strong electric current. In sensor technologies, generally much smaller currents are used, therefore, submultiples of A are often employed:

1 milliampere (mA)	10^{-3}	A
1 microampere (μA)	10^{-6}	A
1 nanoampere (nA)	10^{-9}	A
1 picoampere (pA)	10^{-12}	A
1 femtoampere (fA)	10^{-15}	A

No matter what is the cross section of the material, whether it is homogeneous or not, the electric current through any cross section is always the same for a given electric field. It is similar to water flow through a combination of serially connected pipes of different diameters - the rate of flow is the same throughout of the pipe combination. The water flows faster in the narrow sections and slower in the wide section, but amount of water passing through any cross section per unit of time is constant. The reason for that is very simple - water in the pipes is neither drained out, nor created. The same reason applies to electric current. One of the fundamental laws of physics is the law of conservation of charge. Under steady state conditions, charge in a material is neither created nor destroyed. Whatever comes in must go out. In this section, we do not consider any charge storages (capacitors), and all materials we discuss are said have pure *resistive* properties.

The mechanism of electrical conduction in a simplified form may be described as follows. A conducting material, say copper wire, can be modeled as a semi-rigid spring-like periodic lattice of positive copper ions. They are coupled together by strong electromagnetic forces. Each copper atom has one conduction electron which is free to move about the lattice. When electric field E is established within the conductor, force $-eE$ acts on each electron (e is the electron charge). The electron accelerates under the force and moves. However, the movement is very short as the electron collides with the neighboring copper atoms, which constantly vibrate with intensity which is determined by the material temperature. The electron transfers kinetic energy to the lattice and is often captured by the positive ion. It frees another electron which keeps moving in the electric field until, in turn, it collides with the next portion of the lattice. The average time between collisions is designated as τ. It depends on the material type, structure, and impurities. For instance, at room temperature, a conduction electron in pure copper moves between collisions for an average distance of 0.04 μm with $\tau = 2.5 \cdot 10^{-14}$s. In effect, electrons which flow into the material near the negative side of the battery are not the same which outflow to the positive terminal. However, the constant drift or flow of electrons is maintained throughout the material. Collisions of electrons with the material atoms further add to the atomic agitation and, subsequently, raise the material temperature. It was arbitrarily decided to define the direction of current flow along with the direction of the electric field, i.e., in the *opposite direction* of the electronic flow. Hence, the electric current flows from the positive to negative terminal of the battery while electrons actually move in the opposite direction.

3.5.1 Specific resistivity

If we fabricate two geometrically identical rods from different materials, say from copper and glass and apply to them the same voltage, the resulting currents will be quite different. A material may be characterized by its ability to pass electric current. It is called *resistivity* and material is said has electrical *resistance* which is defined by Ohm's law

$$R = \frac{V}{i} .$$

(3.5.3)

For the pure resistance (no inductance or capacitance) voltage and current are in-phase with each other, meaning that they are changing simultaneously.

Any material has electric resistivity[1] and therefore is called a *resistor*. The SI unit of resistance is 1 ohm (Ω) = 1 volt/1 ampere. Other multiples and submultiples of Ω are:

1 milliohm	(mΩ)	10^{-3} Ω
1 kilohm	(kΩ)	10^{3} Ω
1 megohm	(MΩ)	10^{6} Ω
1 gigohm	(GΩ)	10^{9} Ω
1 terohm	(TΩ)	10^{12} Ω

If we compare electric current with water flow, pressure across the pipe line (Pascal) is analogous of voltage (V) across the resistor, electric current (C/sec) is analogous of water flow (liters/sec) and electric resistance (Ω) corresponds to water flow resistance in the pipe. It is clear that resistance to water flow is smaller when the pipe is short, wide and empty. When the pipe has, for instance, a filter installed in it, resistance to water flow will be higher. Similarly, coronary blood flow may be restricted by cholesterol deposits on the inner lining of blood vessels. Flow resistance is increased and arterial blood pressure is not sufficient to provide the necessary blood supply rate for normal functioning of the heart. This may result in a heart attack.

Resistance is a characteristic of a device. It depends on both: the material and the geometry of the resistor. Material itself can be characterized by a *specific resistivity* ρ which is defined as

$$\rho = \frac{E}{j} ,$$

(3.5.4)

where j is current density: $j = i/a$ (a is the area of the material cross section). The SI unit of resistivity is $\Omega \cdot m$. Resistivities of some materials are given in Table 3-4. Quite often, a reciprocal quantity is used which is called *conductivity*: $\sigma = 1/\rho$.

Resistivity of a material can be expressed through mean time between collisions τ the electronic charge e the mass of electron m, and a number of conduction electrons per unit volume n

---

[1] Excluding superconductors which are beyond the scope of this book

$$\rho = \frac{m}{ne^2\tau} \; .$$

$$(3.5.5)$$

To find the resistance of a conductor the following formula may be used

$$R = \rho\frac{l}{a} \; ,$$

$$(3.5.6)$$

where a is the cross sectional area and l is the length of the conductor.

3.5.2 Temperature sensitivity

Conductivity of a material changes with temperature T and in a relatively narrow range it may be expressed through α, which is temperature coefficient of resistance (TCR)

$$\rho = \rho_0[1+\alpha(T-T_o)]$$

$$(3.5.7)$$

where ρ_0 is resistivity at reference temperature T_o (commonly either 0 or 25°C). In a broader range, resistivity is a nonlinear function of temperature.

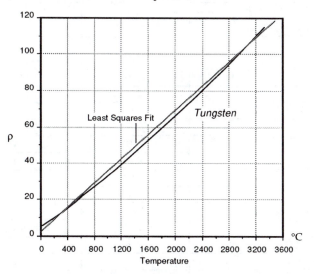

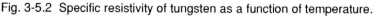

Fig. 3-5.2 Specific resistivity of tungsten as a function of temperature.

For nonprecision applications over a broad temperature range, resistivity of tungsten, as shown in Fig. 3-5.2 may be modeled by a best fit straight line with α=0.0058C^{-1}. However, this number will not be accurate at lower temperatures. For instance, near 25°C the slope of ρ is about 20% smaller: α=0.0045C^{-1}. When better accuracy is required, formula (3.5.7) should not be employed. Instead, higher order polynomials may be useful. For instance, over a broader temperature range, tungsten resistivity may be found from the second order equation

$$\rho = 4.45 + 0.0269 \cdot T + 1.914 \cdot 10^{-6} \cdot T^2 \ , \qquad (3.5.8)$$

where T is temperature in °C and ρ is in $\Omega \cdot m$.

Table 3-4 Resistivities and temperature coefficients of resistivity (TCR)
of some materials at room temperature

Material	Resistivity (ρ) $10^{-8}\Omega \cdot m$ (at room temperature)	TCR (α) 10^{-3}/°K
Aluminum (99.99%)	2.65	3.9
Beryllium	4.0	0.025
Bismuth	106	
Brass (70Cu, 30Zn)	7.2	2.0
Carbon	3500	−0.5
Constantan (60Cu, 40Ni)	52.5	0.01
Copper	1.678	3.9
Evanohm (75Ni, 20Cr, 2.5Al, 2.5Cu)	134	
Germanium (polycrystalline)	$46 \cdot 10^6$	
Gold	2.24	3.4
Iron (99.99%)	9.71	6.5
Lead	20.6	3.36
Manganin	44	0.01
Manganin (84Cu, 12Mn, 4Ni)	48	
Mercury	96	0.89
Nichrome	100	0.4
Nickel	6.8	6.9
Palladium	10.54	3.7
Platinum	10.42	3.7
Platinum + 10% Rhodium	18.2	
Silicon	$3.4 \cdot 10^6$	(very sensitive to purity)
Silicon bronze (96Cu, 3Si, 1Zn)	21.0	
Silver	1.6	6.1
Sodium	4.75	
Tantalum	12.45	3.8
Tin	11.0	4.7
Titanium	42	
Tungsten	5.6	4.5
Zinc	5.9	4.2

Metals have positive temperature coefficients (PTC) α, while many semiconductors and oxides have negative temperature coefficients of resistance (NTC). It is usually desirable to have very low TCR in resistors which are used in electronic circuits. On the other hand, a strong temperature coefficient of resistivity allows us to fabricate a temperature sensor, known as *thermistor* (a contraction of words *thermal resistor*) and the so-called resistive temperature detector (RTD)[1]. The most popular RTD is a platinum (Pt) sensor which operates over a broad temperature range from about $-200°C$ to over $600°C$. Resistance of Pt RTD is shown in Fig. 3-5.3. For a calibrating resistance R_o at $0°C$, the best fit straight line is given by equation

$$R = R_o(1.0036 + 36.79 \cdot 10^{-4}\, T) \quad , \qquad (3.5.9)$$

where T is temperature in $°C$ and R is in Ω.

The multiple at temperature (T) is the sensor's sensitivity (a slope) which may be expressed as $+0.3679\ \%/°C$.

There is a slight nonlinearity of the resistance curve which, if not corrected, may lead to an appreciable error. A better approximation of Pt resistance is a second order polynomial which gives accuracy better than $0.01°C$

$$R = R_o(1 + 39.08 \cdot 10^{-4}T - 5.8 \cdot 10^{-7}T^2)\, \Omega \qquad (3.5.10)$$

It should be noted, however, that the coefficients in the polynomial somewhat depend on the material purity and manufacturing technologies.

If a Pt RTD sensor at $0°C$ has resistivity $R_o=100\Omega$, at $+150°C$ the linear approximation gives

$$R = 100 \cdot (1.0036 + 36.79 \cdot 10^{-4} \cdot 150) = 155.55\ \Omega \quad ,$$

while for the second-order approximation (formula 3.5.10)

$$R = 100 \cdot (1 + 39.08 \cdot 10^{-4}150 - 5.8 \cdot 10^{-7}150^2) = 157.32\ \Omega.$$

The difference between the two is 1.76Ω. This is equivalent to an error of $-4.8°C$ at $+150°$

Thermistors are resistors with large either negative (NTC) or positive (PTC) temperature coefficients. The thermistors are ceramic semiconductors commonly made of oxides of one or more of the following metals: nickel, manganese, cobalt, titanium, iron. Oxides of other metals are occasionally used. Resistances vary from a fraction of an ohm to many megohms. Thermistors can be produced in the form of disks, droplets, tubes, flakes, or thin films deposited on ceramic substrates. Recent progress in thick film technology allows us to print thermistor on ceramic substrates.

---

[1] See Section 16.1.

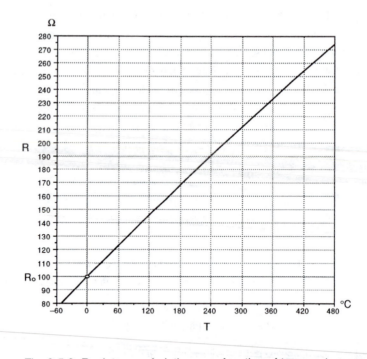

Fig. 3-5.3 Resistance of platinum as function of temperature

The NTC thermistors often are fabricated in a form of beads. Usually bead thermistors have platinum alloy lead wires which are sintered into the ceramic body. During the fabrication process, a small portion of mixed metal oxides is placed with a suitable binder onto a pair of platinum alloy wires, which are under slight tension. After the mixture has been allowed to set, the beads are sintered in a tubular furnace. The metal oxides shrink around the platinum lead wires and form intimate electrical bonds. The beads may be left bare or they may be given organic or glass coatings.

Thermistors possess nonlinear temperature-resistance characteristics (Fig. 3-5.4), which are generally approximated by one of several different equations. The most popular of them is the exponential form

$$R_t = R_{to} \cdot e^{\beta \cdot (1/T - 1/T_o)}, \qquad (3.5.11)$$

where T_o is the calibrating temperature which is commonly 25°C. R_{to} is the resistance at calibrating temperature, and β is a material's characteristic temperature. All temperatures and β are in kelvin. Commonly, β ranges between 3,000 and 5,000 K and for a relatively narrow temperature range it can be considered temperature independent, which makes equation (3.5.11) a reasonably good approximation. When higher accuracy is required, a polynomial approximation is generally employed. Fig. 3-5.4 shows resistance/temperature dependence of thermistors having β=3000 and 4000°K and that for the platinum RTD. The platinum characteristic is substantially less sensitive and more liner with a positive slope while thermistors are nonlinear with high sensitivity and a negative slope.

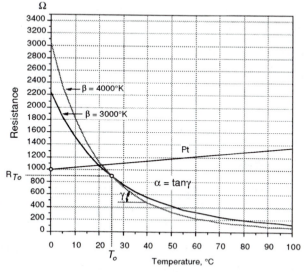

Fig. 3-5.4 Resistance/temperature characteristics for two thermistors and Pt RTD (R_o=1k); thermistors are calibrated at T_o=25°C and RTD at 0°C

3.5.3   Strain sensitivity

Usually, electrical resistance changes when the material is mechanically deformed. This is called the *piezoresistive effect*. In some cases, the effect is a source of error. On the other hand, it is successfully employed in sensors which are responsive to stress σ:

$$\sigma = \frac{F}{a} = E\frac{dl}{l} ,$$ (3.5.12)

where E is Young's modulus of the material and F is the applied force. In this equation, the ratio $dl/l = e$ is called *strain* which is a normalized deformation of the material.

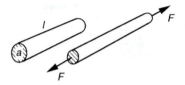

Fig. 3-5.5 Strain changes geometry of a conductor and its resistance

Fig. 3-5.5 shows a cylindrical conductor (wire) which is stretched by applied force, F. Volume v of the material stays constant, while the length increases and the cross sectional area becomes smaller. As a result, equation (3.5.6) can be rewritten as

$$R = \frac{\rho}{v} l^2 \ . \qquad\qquad (3.5.13)$$

After differentiating, we can define sensitivity of resistance with respect to wire elongation

$$\frac{dR}{dl} = 2\frac{\rho}{v} l \ . \qquad\qquad (3.5.14)$$

It follows from this equation that the sensitivity becomes higher for the longer and thinner wires with high specific resistance. Normalized incremental resistance of the strained wire is a linear function of strain e and it can be expressed as

$$\frac{dR}{R} = S_e e \ , \qquad\qquad (3.5.15)$$

where S_e is known as the *gauge factor* or *sensitivity* of the strain gauge element. For metallic wires it ranges from 2 to 6. It is much higher for the semiconductor gauges where it is between 40 and 200.

Early strain gauges where metal filaments. The gauge elements were formed on a backing film of electrically isolating material. Today, they are manufactured from constantan (copper/nickel alloy) foil or single crystal semiconductor materials (silicon with boron impurities). The gauge pattern is formed either by mechanical cutting or photochemical etching. When a semiconductor material is stressed, its resistivity changes depending on the type of the material and the doping dose (see Section 3-17). However, the strain sensitivity in semiconductors is temperature dependent which requires a proper compensation when used over a broad temperature range.

3.6 PIEZOELECTRIC EFFECT

The piezoelectric effect is generation of electric charge by a crystalline material upon subjecting it to stress. The effect exists in natural crystals, such as quartz (chemical formula SiO_2), and poled (artificially polarized) man-made ceramics and some polymers, such as PVDF. It is said that piezoelectric material possess ferroelectric properties. The name was given by an analogy with ferromagnetic properties. Word *piezo* comes from the Greek *piezen* meaning *to press*. The Curie brothers discovered the piezoelectric effect in quartz in 1880, but very little practical use was made until 1917 when another Frenchman, professor P. Langevin used x-cut plates of quartz to generate and detect sound waves in water. His work led to the development of sonar.

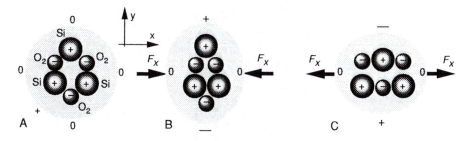

Fig. 3-6.1 Piezoelectric effect in a quartz crystal

A simplified, yet quite explanatory model of the piezoelectric effect was proposed in 1927 by A. Meissner [3]. A quartz crystal is modeled as a helix (Fig. 3-6.1A) with one silicon Si and two oxygen O_2 atoms alternating around the helix. A quartz crystal is cut along its axes x, y, and z, thus Fig. 3-6.1A is a view along the z-axis. In a single crystal-cell there are three atoms of silicon and six oxygen atoms. Oxygen is being lumped in pairs. Each silicon atom carries four positive charges and a pair of oxygen atoms carries four negative charges (two per atom). Therefore a quartz cell is electrically neutral under the no-stress conditions. When external force F_x is applied along the x-axis, the hexagonal lattice becomes deformed. Fig. 3-6.1B shows a compressing force which shifts atoms in a crystal in such a manner as a positive charge is built up at the silicon atom side and the negative at the oxygen pair side. Thus, the crystal develops an electric charge along the y-axis. If the crystal is stretched along the x-axis (Fig. 3-6.1C), a charge of opposite polarity is built along the y-axis which is a result of a different deformation. This simple model illustrates that crystalline material can develop electric charge on its surface in response to a mechanical deformation. A similar explanation may be applied to the pyroelectric effect which is covered in the next section of this chapter.

To pickup an electric charge, conductive electrodes must be applied to the crystal at the opposite sides of the cut (Fig. 3-6.2). As a result, a piezoelectric sensor becomes a capacitor with a dielectric material which is a piezoelectric. The dielectric acts as a generator of electric charge, resulting in voltage V across the capacitor. Although charge in a crystalline dielectric is formed at the location of an acting force, metal electrodes equalize charges along the surface making the capacitor not selectively sensitive. However, if electrodes are formed with a complex pattern, it is possible to determine the exact location of the applied force by measuring the response from a selected electrode.

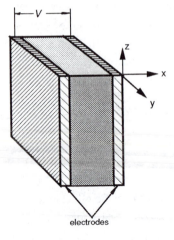

Fig. 3-6.2 Piezoelectric sensor is formed by applying electrodes
to a poled crystalline material

The piezoelectric effect is a reversible physical phenomenon. That means that applying voltage across the crystal produces mechanical strain. It is possible by placing several electrodes on the crystal to use one pair of electrodes to deliver voltage to the crystal and the other pair of electrodes to pick up charge resulting from developed strain. This method is used quite extensively in various piezoelectric transducers.

The magnitude of the piezoelectric effect in a simplified form can be represented by the vector of polarization [4]

$$\mathbf{P} = \mathbf{P}_{xx} + \mathbf{P}_{yy} + \mathbf{P}_{zz} \;, \tag{3.6.1}$$

where x, y, z refer to a conventional ortogonal system related to the crystal axes. In terms of axial stress σ we can write[1]

$$\mathbf{P}_{xx} = d_{11}\sigma_{xx} + d_{12}\sigma_{yy} + d_{13}\sigma_{zz} \;,$$

$$\mathbf{P}_{yy} = d_{21}\sigma_{xx} + d_{22}\sigma_{yy} + d_{23}\sigma_{zz} \;, \tag{3.6.2}$$

$$\mathbf{P}_{zz} = d_{31}\sigma_{xx} + d_{32}\sigma_{yy} + d_{33}\sigma_{zz} \;,$$

where constants d_{mn} are the piezoelectric coefficients along the ortogonal axes of crystal cut. Dimensions of these coefficients are C/N (coulomb/newton), i.e., charge unit per unit force.

---

[1] The complete set of coefficients also includes shear stress and the corresponding d-coefficients.

For the convenience of computation, two additional units have been introduced. The first is a g-coefficient which is defined by a division of corresponding d_{mn}-coefficients by the absolute dielectric constant

$$g_{mn} = \frac{d_{mn}}{\varepsilon_0 \varepsilon_{mn}} \quad . \tag{3.6.3}$$

This coefficient represents a voltage gradient (electric field) generated by the crystal per unit applied pressure, i.e., its dimension is

$$\frac{V}{m} \bigg/ \frac{N}{m^2} \quad .$$

Another coefficient which is designated h is obtained by multiplying the g-coefficients by the corresponding Young's moduli for the corresponding crystal axes. Dimension of the h-coefficient is

$$\frac{V}{m} \bigg/ \frac{m}{m} \quad .$$

Piezoelectric crystals are direct converters of mechanical energy into electrical. Efficiency of the conversion can be determined from the so-called *coupling coefficients* k_{mn}:

$$k_{mn} = \sqrt{d_{mn} \cdot h_{mn}} \quad . \tag{3.6.4}$$

The k-coefficient is an important characteristic for applications where energy efficiency is of a prime importance, like in acoustics and ultrasonics.

Charge generated by the piezoelectric crystal is proportional to applied force, for instance, in the x-direction the charge is

$$Q_x = d_{11} F_x \quad . \tag{3.6.5}$$

Since a crystal with deposited electrodes forms a capacitor having capacitance C, voltage V, which develops across between the electrodes is

$$V = \frac{Q_x}{C} = \frac{d_{11}}{C} F_x \quad . \tag{3.6.6}$$

In turn, the capacitance can be represented (see formula 3.2.4) through the electrode surface area[1] a and the crystal thickness l

$$C = \varepsilon \varepsilon_0 \frac{a}{l} \quad . \tag{3.6.7}$$

---

[1] Not the crystal area. Piezo induced charge can be collected only over the area covered by the electrode.

Then, the output voltage is

$$V = \frac{d_{11}}{C} F_x = \frac{d_{11}l}{\varepsilon\varepsilon_0 a} F_x \quad .$$

(3.6.8)

Manufacturing of ceramic PZT sensors[1] begins with high purity metal oxides (lead oxide, zirconium oxide, titanium oxide, etc.) in the form of fine powders having various colors. The powders are milled to a specific fineness, and mixed thoroughly in chemically correct proportions. In a process called "calcining", the mixtures are then exposed to an elevated temperature, allowing the ingredients to react to form a powder, each grain of which has a chemical composition close to the desired final composition. At this stage, however, the grain doesn't have yet the desired crystalline structure.

The next step is to mix the calcined powder with solid and/or liquid organic binders (intended to burn out during firing) and mechanically form the mixture into a "cake" which closely approximates a shape of the final sensing element. To form the "cakes" of desired shapes, several methods can be used. Among them are pressing (under force of a hydraulic powered piston), casting (pouring viscous liquid into molds and allowing to dry), extrusion (pressing the mixture through a die, or a pair of rolls to form thin sheets), tape casting (pulling viscous liquid onto a smooth moving belt).

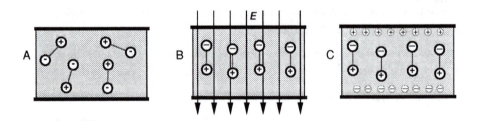

Fig. 3-6.3 Thermal poling of a piezo- and pyroelectric material

After the "cakes" have been formed, they are placed into a kiln and exposed to a very carefully controlled temperature profile. After burning out of organic binders, the material shrinks by about 15%. The "cakes" are heated to a red glow and maintained at that state for some time, which is called the "soak time", during which the final chemical reaction occurs. The crystalline structure is formed when the material is cooled down. Depending on the material, the entire firing may take 24 hours.

---

[1] Information on fabrication methods of piezoelectric materials is courtesy of Piezo Electric products, Inc., Metuchen, NJ.

When the material is cold, contact electrodes are applied to its surface. This can be done by several methods. The most common of them are: a fired-on silver (a silk-screening of silver-glass mixture and re-firing), an electroless plating (a chemical deposition in a special bath), and a sputtering (an exposure to metal vapor in a partial vacuum).

Crystallities (crystal cells) in the material can be considered electric dipoles. In some materials, like quartz, these cells are naturally oriented along the crystal axes, thus giving the material sensitivity to stress. In other materials, the dipoles are randomly oriented and the materials need to be "poled" to possess piezoelectric properties. To give a crystalline material piezoelectric properties, several poling techniques can be used. The most popular poling process is a thermal poling, which includes the following steps:

1. A crystalline material (ceramic or polymer film) which has randomly oriented dipoles (Fig. 3-6.3A) is warmed up slightly below its Curie temperature. In some cases (for a PVDF film) the material is stressed. High temperature results in stronger agitation of dipoles and permits us to more easily orient them in a desirable direction.

2. Material is placed in strong electric field E (Fig. 3-6.3B) where dipoles align along the field lines. The alignment is not total. Many dipoles deviate from the filed direction quite strongly, however, statistically predominant orientation of the dipoles is maintained.

3. The material is cooled down while the electric field across its thickness is maintained.

4. The electric field is removed and the poling process is complete. As long as the poled material is maintained below the Curie temperature, its polarization remains permanent. The dipoles stay "frozen" in the direction which was given to them by the electric field at high temperature. The above method is used to manufacture ceramic and plastic pyroelectric materials.

Another method called a corona discharge poling, is also used to produce polymer piezo/pyroelectric films. The film is subjected to a corona discharge from an electrode at several million volts per cm of film thickness for 40-50 seconds [5, 6]. Corona polarization is uncomplicated to perform and can be easily applied before electric breakdown occurs, making this process useful at room temperature.

The final operation in preparation the sensing element is shaping and finishing. This includes cutting, machining and grinding. After the piezo (pyro) element is prepared, it is installed into a sensor's housing, where its electrodes are bonded to electrical terminals and other electronic components.

After poling, the crystal remains permanently polarized, however, it is electrically charged for a relatively short time. There is a sufficient amount of free carriers which move in the electric field setup inside the bulk material and there are plenty charged ions in the surrounding air. The charge carriers move toward the poled dipoles and neutralize their charges (Fig. 3-6.3C). Hence, after a while, the poled piezoelectric material becomes electrically discharged as long as it remains under steady-state conditions. When stress is applied, or air blows near its surface (Section 10.7) the balanced state is degraded and the piezoelectric material develops an electric charge. If the stress is maintained for a while, the charges again will be neutralized by the internal leakage. Thus, a piezoelectric sensor is responsive only to a changing stress rather than to a steady level of it. In other words, a piezoelectric sensor is an ac device, rather than a dc device.

Piezoelectric sensitivities (d-coefficients) are temperature dependent. For some materials (quartz), sensitivity drops with a slope of –0.016%/°C. For the others (PVDF films and ceramics) at temperatures below 40°C it may drop and at higher temperatures it increases with a raise in temperature. Nowadays, the most popular materials for fabrication of piezoelectric sensors are ceramics [7-9]. The earliest of the ferroelectric ceramics was barium titanate, a polycrystalline substance having the chemical formula $BaTiO_3$. The stability of permanent polarization relies on the coercive force of the dipoles. In some materials, polarization may decrease with time. To improve stability of poled material, impurities have been introduced in the basic material with the idea that the polarization may be "locked" into position [4]. While the piezoelectric constant changes with operating temperature, a dielectric constant ε exhibits a similar dependence. Thus, according to formula (3.6.8), variations in these values tend to cancel each other as they are entered into nominator and denominator. This results in a better stability of the output voltage V over a broad temperature range.

In 1969, H. Kawai discovered a strong piezoelectricity in PVDF (polyvinylidene fluoride) and in 1975 the Japanese company Pioneer, Ltd. developed the first commercial product with the PVDF as a piezoelectric loudspeakers and earphones [10]. PVDF is a semicrystalline polymer with an approximate degree of crystallinity of 50% [11]. Like other semicrystalline polymers, PVDF consists of a lamellar structure mixed with amorphous regions. The chemical structure of it contains the repeat unit of doubly fluorinated ethene CF_2-CH_2:

$$\left[\begin{array}{cc} H & F \\ | & | \\ -C & -C- \\ | & | \\ H & F \end{array} \right]_n$$

PVDF molecular weight is about 10^5, which corresponds to about 2000 repeat units. The film is quite transparent in the visible and near-IR region, and is absorptive in the far infrared portion of electromagnetic spectrum. The polymer melts at about 170°C. Its density is about 1780 kg/m^3. PVDF is a mechanically durable and flexible material. In piezoelectric applications, it is usually drawn, uniaxially or biaxially, to several times its length. Elastic constants, for example, Young modulus, depend on this draw ratio. Thus, if the PVDF film was drawn at 140°C to the ratio of 4:1, the modulus value is 2.1 GPa, while for the draw ratio of 6.8:1 it was 4.1 GPa. Resistivity of the film also depends on the stretch ratio. For instance, at low stretch it is about 6.3·10^{15} Ω·cm, while for the stretch ratio 7:1 it is 2·10^{16} Ω·cm.

PVDF does not have a higher, or even as high piezoelectric coefficient as other commonly used materials, like BaTiO$_3$ or PZT. However, it has a unique quality not to depolarize while being subjected to very high alternating electric fields. This means that even though the value of d$_{31}$ of PVDF is about 10% of PZT, the maximum strain observable in PVDF will be 10 times larger than in PZT since the maximum permissible field is a hundred times greater for PVDF. The film exhibits good stability: when stored at 60°C it loses its sensitivity by about 1-2% over 6 months. Comparative characteristics for various piezoelectric materials are given in Table 3-5.

Table 3-5 Properties of piezoelectric materials at 20°C

	PVDF	BaTiO$_3$	PZT	Quartz	TGS
Density (x10^3 kg/m^3)	1.78	5.7	7.5	2.65	1.69
Dielectric constant, ε_r	12	1700	1200	4.5	45
Elastic modulus (10^{10} N/m)	0.3	11	8.3	7.7	3
Piezoelectric constant (pC/N)	d_{31}=20 d_{32}=2 d_{33}=-30	78	110	2.3	25
Pyroelectric constant (10^{-4} C/m^{2}K)	4	20	27	-	30
Electromechanical coupling constant (%)	11	21	30	10	-
Acoustic impedance (10^6 kg/m^{2}s)	2.3	25	25	14.3	-

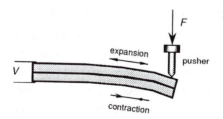

Fig. 3-6.4 Laminated two-layer piezoelectric sensor

The piezoelectric elements may be used as a single crystal, or in a multilayer form where several plates (films) of the material are laminated together. This must be done with electrodes placed in-between. Fig. 3-6.4 shows a two-layer force sensor. When an external force is applied, the upper part of the sensor expands while the bottom compresses. If the layers are laminated correctly, this produces a double output signal. Double sensors can have either a parallel connection as shown in Fig. 3-6.5A, or a serial connection as in Fig. 3-6.5C. The electrical equivalent circuit of the piezoelectric sensor is a parallel connection of a stress-induced current source (i), leakage resistance (r), and capacitance (C). Depending on the layer connection equivalent circuits for the laminated sensors are as shown in Figs. 3-6.5B and D. The leakage resistors r are very large - on the orders of 10^{12}–$10^{14}\Omega$ which means that the sensor has an extremely high output impedance. This requires special interface circuits, such as charge and current-to-voltage converters, or voltage amplifiers with high input resistances.

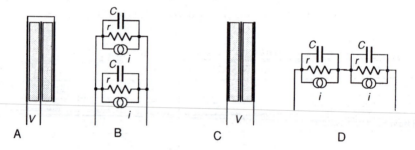

Fig. 3-6.5 Parallel (A) and serial (C) laminated piezoelectric sensors
and their corresponding equivalent circuits (B and D).

3.7 PYROELECTRIC EFFECT

The pyroelectric materials are crystalline substances capable of generating an electrical charge in response to heat flow. The pyroelectric effect is very closely related to the piezoelectric effect. Before going further, we recommend that the reader familiarizes oneself with Section 3.6.

Like piezoelectrics, the pyroelectric materials are used in the form of thin slices or films with electrodes deposited on the opposite sides to collect the thermally induced charges (Fig. 3-7.1). The pyroelectric detector is essentially a capacitor which can be charged by an influx of heat. The detector does not require any external electrical bias (excitation signal). It needs only an appropriate electronic interface circuit to measure the charge. Contrary to thermoelectrics (thermocouples) which produce a steady voltage when two dissimilar metal junctions are held at steady but different temperatures (see Section 3.9), pyroelectrics generate charge in response to a *change* in temperature. Since a change in temperature essentially requires propagation of heat, a pyroelectric device is a heat *flow* detector rather than heat detector. Sometimes it is called a dynamic sensor which reflects the nature of its response. When the pyroelectric crystal is exposed to a heat flow (for instance, from an infrared radiation source), its temperature elevates and it becomes a source of heat, in turn. Hence, there is an outflow of heat from the opposite side of the crystal as it is shown in Fig. 3-7.1.

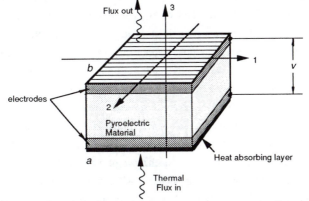

Fig. 3-7.1 Pyroelectric sensor has two electrodes at the opposite sides of the crystal
Thermal radiation is applied along axis 3.

A crystal is considered to be pyroelectric if it exhibits a spontaneous temperature dependent polarization. Of the 32 crystal classes, twenty-one are noncentrosymmetric and ten of these exhibit pyroelectric properties. Beside pyroelectric properties, all these materials exhibit some degree of piezoelectric properties as well - they generate an electrical charge in response to mechanical stress.

Pyroelectricity was observed for the first time in tourmaline crystals in the eighteenth century (some claim that the Greeks noticed it 23 centuries ago). Later, in the nineteenth century, Rochelle salt was used to make pyroelectric sensors. A large variety of materials became available after 1915: KDP (KH_2PO_4), ADP ($NH_4H_2PO_4$), $BaTiO_3$, and a composite of $PbTiO_3$ and $PbZrO_3$ known as PZT. Presently, more than 1,000 materials with reversible polarization are known. They are called ferroelectric crystals. The most important among them are triglycine sulfate (TGS) and lithium tantalate ($LiTaO_3$). In 1969 H. Kawai discovered strong piezoelectricity in the plastic materials, polyvinyl fluoride (PVF) and polyvinylidene fluoride (PVDF) [12]. These materials also possess substantial pyroelectric properties.

A pyroelectric material can be considered as a composition of a large number of minute crystallities, where each behaves as a small electric dipole. All these dipoles are randomly oriented (Figs. 3-2.5 and 3-6.3A). Above a certain temperature, known as the Curie point, the crystallities have no dipole moment. Manufacturing (poling) of pyroelectric materials is analogous to that of piezoelectrics (see Section 3.6).

There are several mechanisms by which changes in temperature will result in pyroelectricity. Temperature changes may cause shortening or elongation of individual dipoles. It may also affect the randomness of the dipole orientations due to thermal agitation. These phenomena are called *primary* pyroelectricity. There is also *secondary* pyroelectricity which, in a simplified way, may be described as a result of the piezoelectric effect, that is, a development of strain in the material due to thermal expansion. Fig. 3-7.1 shows a pyroelectric sensor whose temperature, T_o, is homogeneous over its volume. Being electrically polarized the dipoles are oriented (poled) in such a manner as to make one side of the material positive and the opposite side - negative. However, under steady state conditions, free charge carriers (electrons and holes), neutralize the polarized charge and the capacitance between the electrodes appears to be not charged. That is, the sensor generates zero charge. Now, let us assume that heat is applied to the bottom side of the sensor. Heat may enter the sensor in a form of thermal radiation which is absorbed by the bottom electrode and propagates toward the pyroelectric material via the mechanism of thermal conduction. The bottom electrode may be given a heat absorbing coating, such as goldblack or organic paint. As a result of heat absorption, the bottom side becomes warmer (the new temperature is T_1) which causes the bottom side of the material to expand. The expansion leads to flexing of the sensor which, in turn, produces stress and a change in dipole orientation. Being piezoelectric, stressed material generates electric charges of opposite polarities across the electrodes. Hence, we may regard a secondary pyroelectricity as a sequence of events: a thermal radiation – a heat absorption – a thermally induced stress – an electric charge.

The dipole moment M of the bulk pyroelectric sensor is

$$M = \mu A h \ ,$$

(3.7.1)

where μ - the dipole moment per unit volume, A is the sensor's area and h is the thickness. The charge Q_a which can be picked up by the electrodes, develops the dipole moment across the material

$$M_o = Q_a \cdot h \ .$$

(3.7.2)

M must be equal to M_o, so that

$$Q_a = \mu \cdot A$$

(3.7.3)

As the temperature varies, the dipole moment also changes, resulting in an induced charge.

Thermal absorption may be related to a dipole change, so that μ must be considered as a function of both temperature T_a and an incremental thermal energy ΔW absorbed by the material

$$\Delta Q_a = A \cdot \mu(T_a, \Delta W) \ . \tag{3.7.4}$$

Fig. 3-7.2 shows a pyroelectric detector (pyro-sensor) connected to a resistor R_b which represents either internal leakage resistance or a combined input resistance of the interface circuit which is connected to the sensor. The equivalent electrical circuit of the sensor is shown at right. It consists of three components: 1) the current source generating a heat induced current i (remember that a current is a movement of electric charges), 2) the sensor's capacitance C, and 3) the leakage resistance R_b.

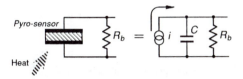

Fig. 3-7.2 Pyroelectric sensor and its equivalent circuit

The output signal from the pyroelectric sensor can be taken in the form of either charge (current) or voltage, depending on the application. Being a capacitor, the pyroelectric device is discharged when connected to a resistor R_b. Electric current through the resistor and voltage across the resistor represent the heat flow induced charge. It can be characterized by two pyroelectric coefficients [13]

$$P_Q = \frac{dP_s}{dT} \qquad \text{Pyroelectric charge coefficient,}$$

$$\tag{3.7.5}$$

$$P_V = \frac{dE}{dT} \qquad \text{Pyroelectric voltage coefficient,}$$

where P_s is the spontaneous polarization (which is the other way to say: "*electric charge*"), E is the electric field strength, and T is the temperature in K. Both coefficients are related by way of the electric permitivity ε_r and dielectric constant ε_0

$$\frac{P_Q}{P_V} = \frac{dP_s}{dE} = \varepsilon_r \cdot \varepsilon_0 \ , \tag{3.7.6}$$

The polarization is temperature dependent and, as a result, both pyroelectric coefficients (3.7.5) are also functions of temperature.

If a pyroelectric material is exposed to a heat source, its temperature rises by ΔT and the corresponding charge and voltage changes can be described by the following equations

$$\Delta Q = P_Q A \Delta T .$$ (3.7.7)

$$\Delta V = P_V h \Delta T .$$ (3.7.8)

Remembering that the sensor's capacitance can be defined as

$$C_e = \frac{\Delta Q}{\Delta V} = \varepsilon_r \varepsilon_0 \frac{A}{h} ,$$ (3.7.9)

then, from (3.7.6), (3.7.8), and (3.7.9) it follows that

$$\Delta V = P_Q \frac{A}{C_e} \Delta T = P_Q \frac{\varepsilon_r \varepsilon_0}{h} \Delta T .$$ (3.7.10)

It is seen, that the peak output voltage is proportional to the sensor's temperature rise and pyroelectric charge coefficient and inversely proportional to its thickness.

When the pyroelectric sensor is subjected to a thermal gradient its polarization (electric charge developed across the crystal) varies with the temperature of the crystal. A typical polarization-temperature curve is shown in Fig. 3-7.3. The voltage pyroelectric coefficient P_V is a slope of the polarization curve. It increases dramatically near the Curie temperature where the polarization disappears and the material permanently loses its pyroelectric properties. The curves imply that the sensor's sensitivity increases with temperature at the expense of nonlinearity.

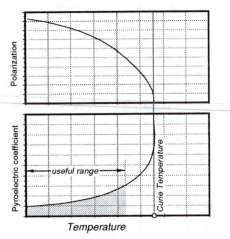

Fig. 3-7.3 Polarization of a pyroelectric crystal
The sensor must be stored and operated below the Curie temperature

To select the most appropriate pyroelectric material energy conversion efficiency should be considered. It is, indeed, the function of the pyroelectric sensor to convert thermal energy into electrical. "How effective is the sensor" - is a key question in the design practice. A measure of efficiency is k_p^2 which is called the pyroelectric coupling coefficient[1] [13,14]. It shows the factor by which the pyroelectric efficiency is lower than the Carnot limiting value $\Delta T/T_a$. Numerical values for k_p^2 are shown in Table 3-6.

Table 3-6 represents that triglycine sulfate (TGS) crystals are the most efficient pyroelectric converters. However, for a long time they were quite impractical for use in the sensors because of a low Curie temperature. If the sensors' temperature is elevated above that level, it permanently loses its polarization. In fact, TGS sensors proved to be unstable even below the Curie temperature with a signal being lost quite spontaneously [15]. It was discovered that doping of TGS crystals with L-alanine (LATGS process patented by Philips) during its growth stabilizes the material below the Curie temperature. The Curie temperature was raised to 60°C which allows us to use it with the upper operating temperature of 55°C which is sufficient for many applications.

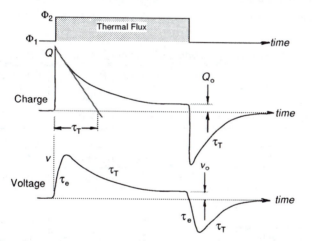

Fig. 3-7.4 Response of a pyroelectric sensor to a thermal step function
The magnitudes of charge Q_o and voltage v_o are exaggerated for clarity

Other materials, such as lithium tantalate and pyroelectric ceramics are also used to produce the pyroelectric sensors. Polymer films, SOLEF from Solvay (Belgium) and KYNAR from AUTOCHEM (France/USA) become increasingly popular for variety of applications. During recent years, a deposition of pyroelectric thin films have been intensively researched. Especially

---

[1] Coefficient k_p is analogous to the piezoelectric coupling coefficient k.

promising is use of lead titanate ($PbTiO_3$) which is a ferroelectric ceramic having both a high py-roelectric coefficient and a high Curie temperature of about 490°C. This material can be easily deposited on silicon substrates by the so called sol-gel spin casting deposition method [16].

Fig. 3-7.4 shows timing diagrams for a pyroelectric sensor when it is exposed to a step function of heat. It is seen that the electric charge reaches its peak value almost instantaneously, and then decays with a *thermal time constant* τ_T. This time constant is a product of the sensors' thermal capacitance C and thermal resistance R, which defines a thermal loss from the sensing element to its surroundings

$$\tau_T = CR = cAhR,\qquad\qquad (3.7.11)$$

where c is the specific heat of the sensing element. The thermal resistance R is a function of all thermal losses to the surroundings through convection, conduction, and thermal radiation. For the low frequency applications, it is desirable to use sensors with τ_T as large as practical, while for the high speed applications (for instance, to measure laser pulses), a thermal time constant should be dramatically reduced. For that purpose, the pyroelectric material may be laminated with a heat sink: a piece of aluminum or copper.

Table 3-6 Physical properties of pyroelectric materials (from [13])

Material	Curie Temperature °C	Thermal Conductivity W/(mK)	Relative Permitivity ε_r	Pyroelectric Charge Coef. $C/(m^2K)$	Pyroelectric Voltage Coef. $V/(mK)$	Coupling k_p^2 (%)
Single Crystals						
TGS	49	0.4	30	$3.5 \cdot 10^{-4}$	$1.3 \cdot 10^6$	7.5
LiTaO$_3$	618	4.2	45	$2.0 \cdot 10^{-4}$	$0.5 \cdot 10^6$	1.0
Ceramics						
BaTiO$_3$	120	3.0	1000	$4.0 \cdot 10^{-4}$	$0.05 \cdot 10^6$	0.2
PZT	340	1.2	1600	$4.2 \cdot 10^{-4}$	$0.03 \cdot 10^6$	0.14
Polymers						
PVDF	205	0.13	12	$0.4 \cdot 10^{-4}$	$0.40 \cdot 10^6$	0.2
Polycrystalline Layers						
PbTiO$_3$	470	2 (monocrystal)	200	$2.3 \cdot 10^{-4}$	$0.13 \cdot 10^6$	0.39

Notes: The above figures may vary depending on manufacturing technologies

When a pyroelectric sensor is exposed to a target, we consider a thermal capacity of a target very large (an infinite heat source), and the thermal capacity of the sensor small. Therefore, the surface temperature T_b of a target can be considered constant during the measurement, while temperature of the sensor T_s is a function of time. That function is dependent on the sensing element: its density, specific heat and thickness. If the input thermal flux has the shape of a step function, for the sensor freely mounted in air, the output current can be approximated by an exponential function, so that

$$i = i_o e^{-t/\tau_T} \ , \tag{3.7.12}$$

where i_o is peak current.

In Fig. 3-7.4, charge Q and voltage V do not completely return to zero, no matter how much time has elapsed. Thermal energy enters the pyroelectric material from side a (Fig. 3-7.1), resulting in a material temperature increase. This causes the sensor's response which decays with a thermal time constant τ_T. However, since the other side b of the sensor faces a cooler environment, part of the thermal energy leaves the sensor and is lost to its surroundings. Because the sides a and b face objects of different temperatures (one is a temperature of a target and the other is a temperature of the environment), a continuous heat flow exists through the pyroelectric material. Electric current generated by the pyroelectric sensor has the same shape as the thermal current through its material. An accurate measurement can demonstrate that as long as the heat continues to flow, the pyroelectric sensor will generate a constant voltage v_o whose magnitude is proportional to the heat flow. This may lead us to a conclusion that the electric current generated by the pyroelectric material can be represented by equation

$$i = p_1 \frac{dQ}{dt} + p_2 \frac{d^2Q}{dt^2} \ , \tag{3.7.13}$$

where p_1 and p_2 are constants.

Several conclusions can be drawn from the above equation:

1. The sensor's output is zero when there is no heat flow through the sensor. The first conclusion states that the sensor can potentially generate a dc response.

2. The sensor's output is constant when the heat flow is constant.

3. The sensor's output is proportional to a first derivative of heat flow.

3.8 HALL EFFECT

This physical effect was discovered in 1879 in Johns Hopkins University by E. H. Hall. Initially, the effect had a limited, however, a very valuable application as a tool for studying electrical conduction in metals, semiconductors, and other conductive materials. Nowadays, the Hall sensors are used to detect magnetic fields, position and displacement of objects [17,18].

The effect is based on the interaction between moving electric carriers and an external magnetic field. In metals, these carriers are electrons. When an electron moves through a magnetic field, upon it acts a sideways force

$$F = qvB \ , \tag{3.8.1}$$

where $q=1.6 \times 10^{-19}$ C is an electronic charge, v is the speed of an electron, and B is the magnetic field. Vector notations (bold face) are an indication that the force direction and its magnitude depend on the spatial relationship between the magnetic field and the direction of the electron movement. The unit of B is 1 tesla=1newton/(ampere·meter)=10^4 gauss.

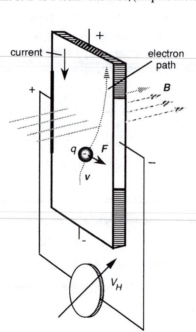

Fig. 3-8.1 Hall effect sensor
A magnetic field deflects movement of electric charges

Let us assume that the electrons move inside a flat conductive strip which is placed in magnetic field B (Fig. 3-8.1). The strip has two additional contacts at its left and right sides which are connected to a voltmeter. Two other contacts are placed at the upper and lower ends of the strip. These are connected to a source of electric current. Due to the magnetic field, the deflecting force shifts moving electrons toward the right side of the strip which becomes more negative than the left side. That is, the magnetic field and the electric current produce the so called *transverse Hall potential difference* V_H. The sign and amplitude of this potential depends on both magnitude and directions of magnetic field and electric current. At a fixed temperature it is given by

$$V_H = hiB\sin\alpha \ , \tag{3.8.2}$$

where α is the angle between the magnetic field vector and the Hall plate (Fig. 3-8.2), and h is the coefficient of overall sensitivity whose value depends on the plate material, its geometry (active area), and its temperature.

The overall sensitivity depends on the *Hall coefficient* which can be defined as the transverse electric potential gradient per unit magnetic field intensity per unit current density. According to the free electron theory of metals, the Hall coefficient should be given by

$$H = \frac{1}{Ncq} \ , \tag{3.8.3}$$

where N is the number of free electrons per unit volume and c is the speed of light. Depending on the material crystalline structure, charges may be either electrons (negative) or holes (positive). As a result, the Hall effect may be either negative or positive.

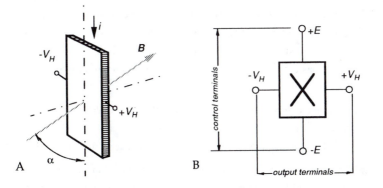

Fig. 3-8.2 Output signal of a Hall sensor depends on the angle between the magnetic field vector and the plate (A); four terminals of a Hall sensor (B)

A linear Hall effect sensor is usually packaged in a four-terminal housing. Terminals for applying the control current are called the *control* terminals and a resistance between them is called the *control resistance* R_i (Fig. 3-8.3). Terminals where the output voltage is observed are called the *differential output terminals* and a resistance between them is called the *differential output resistance* R_o. The sensor's equivalent circuit (Fig. 3-8.4) may be represented by cross-connected resistors and two voltage sources connected in series with the output terminals. The cross $\otimes$ in Figs. 3-8.2B and 3-8.3 indicates the direction of the magnetic field from the viewer to the symbol plane.

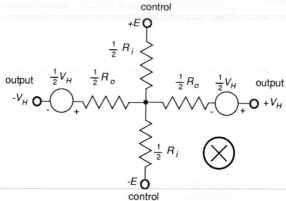

Fig. 3-8.3 Equivalent circuit of a Hall sensor

The sensor is specified by its resistances R_i and R_o across both pairs of terminals, the offset voltage at no magnetic field applied, the sensitivity and the temperature coefficient of sensitivity. Many Hall effect sensors are fabricated from silicon and fall into two general categories - the basic sensors and the integrated sensors. Other materials used for the element fabrication include InSb, InAs, Ge and GaAs. In the silicon element, an interface electronic circuit can be incorporated into the same wafer. This integration is especially important since the Hall effect voltage is quite small. For instance, a linear basic silicon sensor UGN-3605K manufactured by Sprague® has typical characteristics presented in Table 3-7.

A built-in electronic interface circuit may contain a threshold device thus making an integrated sensor a two-state device. That is, its output is "zero" when the magnetic field is below the threshold, and it is "one" when the magnetic field is strong enough to cross the threshold.

Because of a piezoresistivity of silicon, all Hall effect sensors are susceptible to mechanical stress effects. Caution should be exercised to minimize the application of stress to the leads or the housing. The sensor is also sensitive to temperature variations because temperature effects a resistance of the element. If the element is fed by a voltage source, temperature will change the control resistance, and subsequently the control current. Hence, it is preferable to connect the control terminals to a current source rather than to a voltage source.

Table 3-7 Typical characteristics of a linear Hall Effect sensor (source [19])

Control current	3 mA
Control Resistance, R_i	2.2 kΩ
Control Resistance vs. Temperature	+0.8 %/°C
Differential Output Resistance, R_o	4.4 kΩ
Output Offset Voltage	5.0 mV (at $B=0$ Gauss)
Sensitivity	60 µV/Gauss
Sensitivity vs. Temperature	+0.1%/°C
Overall Sensitivity	20 V/ΩkG
Maximum magnetic flux density, B	unlimited

One way to fabricate the Hall sensor is to use a silicon p-substrate with ion-implanted n-wells (Fig. 3-8.4A). Electrical contacts provide connections to the power supply terminals and form the sensor outputs. A Hall element is a simple square where well with four electrodes attached to the diagonals (Fig. 3-8.4B). A helpful way of looking at the Hall sensor is to picture it as a resistive bridge depicted in Fig. 3-8.4C. This representation makes its practical applications more conventional because the bridge circuits are the most popular networks with well established methods of design (Section 4.7).

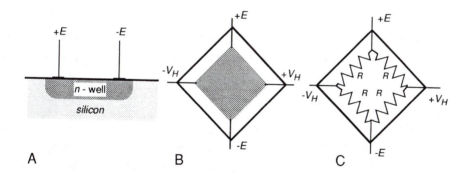

Fig. 3-8.4 Silicon Hall effect sensor with n-well (A and B) and its equivalent resistive bridge circuit (C)

3.9 SEEBECK AND PELTIER EFFECTS

In 1821, Thomas Johann Seebeck (1770-1831), an Estonian born and Berlin and Göttingen educated physician, accidentally joined semicircular pieces of bismuth and copper while studying thermal effects on galvanic arrangements [20]. A nearby compass indicated a magnetic disturbance (Fig. 3-9.1). Seebeck experimented repeatedly with different metal combinations at various temperatures, noting related magnetic field strengths. Curiously, he did not believe that an electric current was flowing, and preferred to describe that effect as "thermomagnetism" [21].

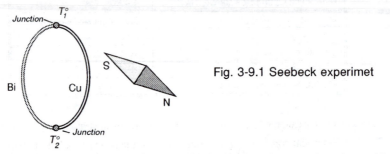

Fig. 3-9.1 Seebeck experimet

If we take a conductor a and place one end of it into a cold place and the other end into a warm place, energy will flow from the warm to cold part. The energy takes the form of heat. The intensity of the heat flow is proportional to the thermal conductivity of the conductor. Besides, the thermal gradient sets an electric field inside the conductor (this directly relates to Thompson effect[1]). The field results in incremental voltage

$$dV_a = \alpha_a \frac{dT}{dx} dx \ ,$$
(3.9.1)

where dT is the temperature gradient across small length, dx and α_a is the *absolute* Seebeck coefficient of the material [22]. If the material is homogeneous, α_a is not a function of length and (3.9.1) reduces to

$$dV_a = \alpha_a dT \ .$$
(3.9.2)

To observe the electric current we must form a closed loop with a meter connected in series with the wire (Fig. 3-9.2A). If the loop is made of a uniform material, say cooper, then no current will be observed. Electric fields in the left and right arms of the loop produce equal currents $i_a = i_b$ which cancel one another resulting in zero net current [23]. In order to observe *thermoelectricity*, it

---

[1] A Thompson effect was discovered by William Thompson around 1850. It consists of absorption or liberation of heat by passing current through a homogeneous conductor which has a temperature gradient across its length. The heat is linearly proportional to current. Heat is absorbed when current and heat flow in opposite directions, and heat is produced when they flow in the same direction.

is in fact necessary to have a circuit composed of two different materials[1], and we can then measure the net difference between their thermoelectric properties. Fig. 3-9.2B shows a loop of two dissimilar metals which produces net current $\Delta i = i_a - i_b$. The actual current depends on many factors, including the shape and size of the conductors. If, on the other hand, instead of current we measure the net voltage across the broken conductor, the potential will depend only on the materials and the temperature difference. It does not depend on any other factors. A thermally induced potential difference is called the *Seebeck potential*.

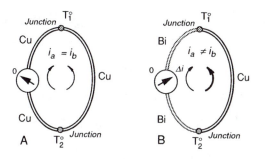

Fig. 3-9.2 Thermoelectric loop
A: joints of identical metals produce zero net current at any temperature difference;
B: joints of dissimilar metals produce a net current Δi

What happens when two conductors are joined together? Free electrons in metal may behave as an ideal gas. Kinetic energy of electrons is a function of the material temperature. However, in different materials, energies and densities of free electrons are not the same. When two dissimilar materials at the same temperature are brought into a contact, free electrons diffuse through the junction [22]. The electric potential of the material accepting electrons becomes more negative at the interface, while the material emitting electrons becomes more positive. Different electronic concentrations across the junction sets up an electric field which balances the diffusion process and the equilibrium is established. If the loop is formed and both junctions are at the same temperature, the electric fields at both junctions cancel each other, which is not the case when the junctions are at different temperatures.

A subsequent investigation has shown the Seebeck effect to be fundamentally electrical in nature. It can be stated that the thermoelectric properties of a conductor are in general just as much bulk properties as are the electrical and thermal conductivities. Coefficient α_a is a unique property of a material. When a combination of two dissimilar materials (A and B) is used, the Seebeck potential is determined from a *differential* Seebeck coefficient

---

[2] Or perhaps the same material in two different states, for example, one under strain, the other is not.

$$\alpha_{AB} = \alpha_A - \alpha_B \ , \tag{3.9.3}$$

and the net voltage of the junction is

$$dV_{AB} = \alpha_{AB}dT \ . \tag{3.9.4}$$

The above equation can be used to determine a differential coefficient

$$\alpha_{AB} = \frac{dV_{AB}}{dT} \tag{3.9.5}$$

For example, voltage as function of a temperature gradient for a T-type thermocouple with a high degree of accuracy can be approximated by a second order equation

$$V_{AB} = a_o + a_1T + a_2T^2 = -0.0543 + 4.094 \cdot 10^{-2}T + 2.874 \cdot 10^{-5}T^2, \tag{3.9.6}$$

then a differential Seebeck coefficient for the T-type thermocouple is

$$\alpha_T = \frac{dV_{AB}}{dT} = a_1 + 2a_2T = 4.094 \cdot 10^{-2} + 5.748 10^{-5}T \tag{3.9.7}$$

Table 3-8 Characteristics of some thermocouple types

Junction Materials	Sensitivity $\mu V/°C$ (@ 25°C)	Temperature Range (°C)	Applications	Designation
Copper/Constantan	40.9	-270 to +600	Oxidation, reducing, inert, vacuum. Preferred below 0°C. Moisture resistant	T
Iron/Constantan	51.7	-270 to +1000	Reducing and inert atmosphere. Avoid oxidation and moisture	J
Chromel/Alumel	40.6	-270 to 1300	Oxidation and inert atmospheres	K
Chromel/Constantan	60.9	-200 to 1000		E
Pt (10%)/Rh-Pt	6.0	0 to 1550	Oxidation and inert atmospheres, avoid reducing atmosphere and metallic vapors	S
Pt (13%)/Rh-Pt	6.0	0 to 1600	Oxidation and inert atmospheres, avoid reducing atmosphere and metallic vapors	R

It is seen that the coefficient is a linear function of temperature. Sometimes, it is called the *sensitivity* of a thermocouple junction. A junction which is kept at a cooler temperature is called a *cold junction* and the warmer is a *hot junction*. The Seebeck coefficient does not depend on the nature of the junction: metals may be pressed together, welded, fused, etc. What counts, is the temperature of the junction and the actual metals. In effect, the Seebeck effect is a direct conversion of thermal energy into electric energy.

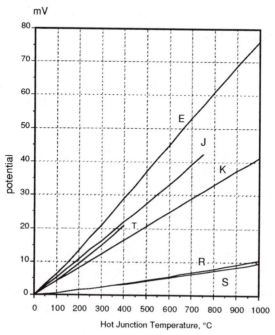

Fig. 3-9.3 Output voltage from standard thermocouples
as functions of a cold-hot temperature gradient

In 1826, A. C. Becquerel suggested to use the Seebeck's discovery for temperature measurements. Nevertheless, the first practical thermocouple was constructed by Henry LeChatelier almost 60 years later [24]. He had found that the junction of platinum and platinum-rhodium alloy wires produce "the most useful voltage". Thermoelectric properties of many combinations have been well documented and for many years used for measuring temperature. Table 3-8 gives sensitivities of some thermocouples (@25°C) and Fig. 3-9.3 shows Seebeck voltages for the standard types of thermocouples over a broad temperature range. It should be emphasized that a thermoelectric sensitivity is not constant over the temperature range and it is customary to reference thermocouples at 0°C. Besides the thermocouples, the Seebeck effect also is employed in *thermopiles* which are, in essence, multiple serially connected thermocouples. Nowadays, thermopiles are most extensively used for the detection of thermal radiation (Section 13.7.1). The original thermopile was made of wires and intended for increasing the output voltage. It was invented by James Joule (1818-89) [25].

In the early 19th century, a French watchmaker turned physicist, Jean Charles Athanase Peltier (1785-1845) discovered that if electric current passes from one substance to another (Fig. 3-9.4), then heat may be given or absorbed at the junction [26]. Heat absorption or production is a function of the current direction

$$dQ_\mathrm{P} = \pm pidt \ , \qquad\qquad (3.9.8)$$

where i is the current and t is time. The coefficient p has a dimension of voltage and represents thermoelectric properties of the material. It should be noted that heat does not depend on temperature at the other junction.

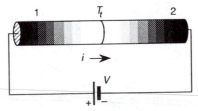

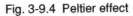

Fig. 3-9.4 Peltier effect

The Peltier effect concerns the reversible absorption of heat which usually takes place when an electric current crosses a junction between two dissimilar metals. The effect takes place whether the current is introduced externally, or is induced by the thermocouple junction itself (due to Seebeck effect).

The Peltier effect is used for two purposes: it can produce heat or "produce" cold, depending on the direction of electric current through the junction. This makes it quite useful for the devices where precision thermal control is required. Apparently, the Peltier effect is of the same nature as the Seebeck effect. It should be well understood that the Peltier heat is different from that of the Joule. The Peltier heat depends *linearly* on the magnitude of the current flow as contrasted to Joule heat[1]. The magnitude and direction of Peltier heat do not depend in any way on the actual nature of the contact. It is purely a function of two different bulk materials which have been brought together to form the junction and each material makes its own contribution depending on its thermoelectric

---

[1] Joule heat is produced when electric current passes in any direction through a conductor having finite resistance. Released thermal power of Joule heat is proportional to squared current: $P = i^2/R$, where R is resistance of a conductor.

properties. The Peltier effect is a basis for operation of thermoelectric coolers which are used for the cooling of photon detectors operating in the far infrared spectral range (Section 13.6) and chilled mirror hygrometers (Fig. 12-9).

In summary, thermoelectric currents may exist whenever the junctions of a circuit formed of at least two dissimilar metals are exposed to different temperatures. This temperature difference is always accompanied by irreversible Fourier heat conduction, while the passage of electric currents is always accompanied by irreversible Joule heating effect. At the same time, the passage of electric current always is accompanied by reversible Peltier heating or cooling effects at the junctions of the dissimilar metals, while the combined temperature difference and passage of electric current always is accompanied by reversible Thomson heating or cooling effects along the conductors. The two reversible heating-cooling effects are manifestations of four distinct e.m.f.s which make up the net Seebeck e.m.f.

$$E_s = p_{AB}|_{T_2} - p_{AB}|_{T_1} + \int_{T_1}^{T_2} \sigma_A dT - \int_{T_1}^{T_2} \sigma_A dT = \int_{T_1}^{T_2} \alpha_{AB} dT \qquad (3.9.9)$$

where σ is a quantity called the Thomson coefficient, which Thomson referred to as the specific heat of electricity, because of an apparent analogy between σ and the usual specific heat c of thermodynamics. The quantity of σ represents the rate at which heat is absorbed, or liberated, per unit temperature difference per unit mass [27, 28].

3.10 MECHANICAL MEASUREMENTS

Mechanical properties of objects were the first ever measured for the engineering purposes, and of all mechanical characteristics, the events of motion were the oldest ever studied. Motion analysis constitutes a part of physics which is known as the *kinematics*. When motion is related to forces, it is studied by *dynamics*. Here we briefly review some fundamental properties both of which can be directly measured by sensors.

Using a vector notation, a *position* of an object with respect to a given system of coordinates can be described by a vector r (Fig. 3-10.1A) which mathematically is defined as

$$r = i x + j y , \qquad (3.10.1)$$

where i and j are the unit vectors in x and y directions. This means that a position of an object can be defined by a distance and a direction from a reference point c. When an object moves along a dotted line (trajectory) in Fig. 3-10.1A, a position of the vector **r** changes with rate **v** which is called an instantaneous *velocity*

$$\mathbf{v} = \frac{d\mathbf{r}}{dt} = \mathbf{i}\frac{dx}{dt} + \mathbf{j}\frac{dy}{dt} \qquad (3.10.2)$$

Scalar values $v_x = \dfrac{dx}{dt}$ and $v_y = \dfrac{dy}{dt}$ are ortogonal components of a velocity vector **v**. Its value may change as well with the velocity of a velocity (Fig. 3-10.1B). It was named the instantaneous *acceleration*

$$\mathbf{a} = \frac{d\mathbf{v}}{dt} = \mathbf{i}\frac{dv_x}{dt} + \mathbf{j}\frac{dv_y}{dt} \qquad (3.10.3)$$

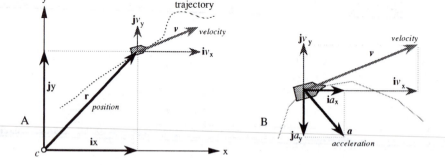

Fig. 3-10.1 Vector representation for position (A), velocity (B), and acceleration (B) of an object

The kinematics deals not only with the instantaneous expressions for a position, velocity, and acceleration. Some other expressions, like an average velocity and acceleration may be required for the system monitoring and analysis. However, of importance to sensing devices is just a monitoring of instantaneous (or near-instantaneous) kinematic values. The other values may be easily derived from the instantaneous by appropriate data processing. The above equations represent a two-dimensional mathematical model of a moving body. Naturally, similar equations can be written for a movement in a three-dimensional space.

The fact that position, velocity, and acceleration are vector values indicates that a corresponding sensor must be properly oriented in space. A unidirectional sensor responds only to the vector component which is parallel to it sensory axis. Fig. 3-10.2A illustrates an accelerometer which is sensitive only in the direction *x-x'*. It is positioned on an object whose acceleration, **a**, is in the direction *y-y'*. The sensor's signal will be a measure of a component $\boldsymbol{a}_x$, rather than of an entire acceleration, **a**. When a direction of a moving object is unpredictable, it is necessary to use omnidirectional sensors. Usually, such a sensor is a combination of three unidirectional sensors whose axes are positioned ortogonally, that is, at $90°$ with respect to each other.

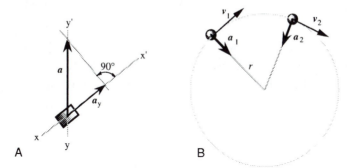

Fig. 3-10.2 Unidirectional accelerometer (A); velocity and acceleration of a rotating object (B)

Of practical importance is a uniform circular motion where the direction of velocity vector v changes continuously (Fig. 3-10.2B). If a rotation is stationary, a magnitude of the velocity vector stays constant. The same can be said about the acceleration. The acceleration vector a is always directed toward the center of a rotation. If a unidirectional accelerometer is positioned on a rotating object and is directed along the velocity vector, such an accelerometer will indicate a zero acceleration. If it is aimed toward the center of rotation, its signal will represent the true acceleration magnitude, which can be calculated from the following formula

$$a = \frac{v^2}{r} \, ,$$

(3.10.4)

where r is the radius of rotation.

In SI units, velocity is measured in meters per second and the corresponding multiple or fractions. The acceleration, linear or centripetal (center-seeking), is measured in meters per sec^2.

While the kinematics study positions of objects and their motions, the dynamics answers the question - what causes the motion? Classical mechanics deal with moving objects whose velocities are substantially smaller than the speed of light. Moving particles, such as atoms and electrons, are the subjects of quantum mechanics and the theory of relativity. A typical problem of classical mechanics is the question: "What is motion of an object, which initially had a given mass, charge, dipole moment, position, etc. and was subjected to external objects having known mass, charge, velocity, etc.?" That is, classical mechanics deals with interactions of macro-objects. In a general form, this problem was solved by Sir Isaac Newton (1642-1727) who was born in the year when

Galileo died. He brilliantly developed ideas of Galileo and other great mechanics. Newton stated his first law as: *"Every body persists in its state of rest or of uniform motion in a straight line unless it is compelled to change that state by forces impressed on it."* Sometimes, this is called a law of inertia. Another way to state the first law is to say that: "If no net force acts on a body, its acceleration a is zero."

When force is applied to a body, it gives the body an acceleration in a direction of force. Thus, we can define force as a vector value. Newton had found that acceleration is proportional to the acting force F and inversely proportional to the property of a body called the mass m which is a scalar value:

$$a = \frac{F}{m} \; .$$

(3.10.5)

This equation is known as *Newton's second law* - the name given by the great Swiss mathematician and physicist Leonhard Euler in 1752, 65 years after the publication of Newton's *Principia* [29]. The first law is contained in the second law as a special case: when net acting force $F=0$, acceleration $a=0$.

Newton's second law allows us to establish the mechanical units. In SI terms, mass (kg), length (m), and time (s) are the base units (see Table 1-7). Force and acceleration are derivative units. The force unit is the force which will accelerate 1 kg mass to acceleration 1 m/s^2. This unit is called a *newton*.

In the British and US Customary systems of units, however, force (lb), length (ft), and time (s) are selected as the base units. The mass unit is defined as the mass which is accelerated at 1ft/s^2 when it is subjected to force of 1 pound. The British unit of mass is *slug*. Hence, the mechanical units are as shown in Table 3-9.

Table 3-9 Mechanical units (bold face indicates base units)

System of Units	Force	Mass	Acceleration
SI	newton (N)	**kilogram** (kg)	m/s^2
British	**pound** (lb)	slug	ft/s^2

Newton's third law establishes a principle of a mutual interaction between two bodies: *"To every action there is always opposed an equal reaction; or, the mutual actions of two bodies upon each other are always equal, and directed to contrary parts."*

In engineering measurements, its is often necessary to know the density of a medium which is amount of matter per unit volume. Density is defined through mass m and volume V as

$$\rho = \frac{m}{V} \; . \tag{3.10.6}$$

The unit of density is kg/m^3 or lb/ft^3 (British system). Densities of some materials are given in Table 3-10.

Table 3-10 Densities (kg/m^3) of some materials
at 1 atm pressure and 0°C

Best laboratory vacuum	10^{-17}
Hydrogen	0.0899
Helium	0.1785
Methane	0.7168
Carbon monoxide	1.250
Air	1.2928
Oxygen	1.4290
Carbon dioxide	1.9768
Styrofoam	100
Benzene	680-740
Alcohol	789.5
Turpentine	860
Mineral oil	900-930
Ice	920
Water	1000
Hydrochloric acid (20%)	1100
Coal tar	1200
Glycerin	1260
Sulfuric acid (20%)	1700
Aluminum	2700
Mercury (0°C)	13596
Platinum	21400

3.11 MECHANICAL COMPONENTS

Virtually all sensors use mechanical components. Some of them are essential for a specific detection, like a seismic mass, damping media, etc. In other sensors, mechanical parts play auxiliary roles - they serve as supporting, structural, and housing components. A detailed analysis of these parts can be found in specialized books, for instance in [4, 30]. Here, we briefly summarize essential properties of mechanical components.

3.11.1 Stress

Stress is usually defined as force per unit area which does or may deform an object. There are three normal stresses: σ_a, σ_y, and σ_z, all positive; and six shear stresses τ_{xy}, τ_{yx}, τ_{yz}, τ_{zy}, τ_{zx} and τ_{xz}, also all positive. If the element is in a static equilibrium, then $\tau_{xy}=\tau_{yx}$, $\tau_{yz}=\tau_{zy}$ and $\tau_{zx}=\tau_{xz}$. Fig. 3-11.1 shows a biaxial, or plane, stress diagram. The two normal stresses are shown in the positive direction. Shear stresses will be taken as positive when they are in the clockwise direction. Thus, τ_{yx} is positive and τ_{xy} is negative.

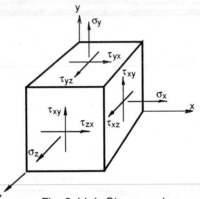

Fig. 3-11.1 Stress vectors

In mechanical components, the assumption that stress is distributed uniformly is made quite often. The normal stress is called *pure tension* if it is directed outwards and it is called *pure compression* if it is directed inwards. The word *simple* sometimes is used instead of *pure* to indicate that there is no other complicating effects. A general expression for the normal stress is

$$\sigma = \frac{F}{a} ,$$

(3.11.1)

where F is the force and a is the area.

3.11.2 Strain

When a straight body is subjected to a tensile load, the body becomes longer. The amount of stretch, or elongation is called *strain*. The elongation per unit length of the body is called *unit strain*. In spite of this definition, however, it is customary to use the word *strain* to mean unit strain and the expression *total strain* to mean total elongation, or deformation of the body. The expression for strain is

$$\varepsilon = \frac{\delta}{l} \, , \qquad\qquad\qquad (3.11.2)$$

where δ is the total elongation (total strain) of the object having initial length l.

3.11.3 Hysteresis

Hysteresis (see Section 2.6 for a general definition) in a mechanical device is associated with friction, backlash, fractured or corroded components, and pollutants. Also, hysteresis is related to the crystalline structure of the material. When the material is stressed, the crystals are partly dislocated and do not return to their original state after the load has been removed. The magnitude of hysteresis in mechanical components are load dependent but they are not functions of time.

3.11.4 Viscous flow

Viscous flow (creep) is related to the flow in the material (Fig. 3-11.2A). As compared with hysteresis, creep is time dependent. When load remains constant (point a) the material keeps flowing, although with a decreasing rate. The creep magnitude increases with an increasing load and temperature. Creep is especially pronounced in materials with low melting temperatures, such as plastics. It results in permanent deformation of the component.

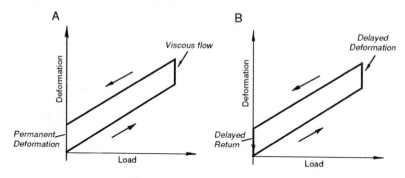

Fig. 3-11.2 Graphs for viscous (A) and elastic (B) flows

A creep-related phenomenon is called *elastic after-effect*. When the load is applied to the device and kept constant there is still movement resulting in deformation (Fig. 3-11.2B). In contrast to creep, the load-induced flow does not result in a permanent distortion. After some delay, the material returns to its original shape, leaving no residual deformation.

3.11.5 Elasticity

Elasticity is that property of a material which enables it to regain its original shape and dimensions when the load is removed. Hooke's law states that, within certain limits, the stress in a material is proportional to the strain which produced it. For the materials which obey Hooke's law, we can write

$$\sigma = E\varepsilon \ , \tag{3.11.3}$$

where E is constant. Since ε is dimensionless, E has the dimension of stress and is called the *modulus of elasticity*, or sometimes, the *modulus of rigidity*. It is indicative of the stiffness and represents one of the fundamental properties of a material. By substituting (3.11.1) and (3.11.2) into (3.11.3) and after rearranging, we obtain the equation for the total deformation of the bar loaded in axial tension or compression [30]

$$\delta = \frac{Fl}{aE} \ . \tag{3.11.4}$$

Experiments demonstrate that when a material is placed in tension, there exists not only an axial strain, but also a lateral strain. Poisson demonstrated that these two strains were proportional to each other within the range of Hooke's law. This constant is expressed as

$$v = -\frac{\text{lateral strain}}{\text{axial strain}} \ , \tag{3.11.5}$$

and is known as Poisson's ratio. The same relations apply for compression, except that a lateral expansion takes place instead. Poisson ratios for different materials are given in Table 3-11.

Table 3-11 Mechanical properties of some materials

Material	Modulus of elasticity (GPa)	Poisson's ratio (v)	density (KN/m³)
Aluminum	71	0.334	26.6
Beryllium copper	124	0.285	80.6
Brass	106	0.324	83.6
Copper	119	0.326	87.3
Glass	46.2	0.245	25.4
Lead	36.5	0.425	111.5
Molybdenum	331	0.307	100.0
Phosphor bronze	11	0.349	80.1
Steel (Carbon)	207	0.292	76.5
Steel (Stainless)	190	0.305	76.0

3.11.6 Hardness

The resistance of a material to penetration by a pointed tool is called *hardness*. Though there is a lot of hardness-measuring systems, we will briefly mention only two of them. The *Rockwell hardness* is easy to measure. It is measured with scales which are designated A, B, C,..., etc. The indenting tools are numbered 1, 2 or 3. Applied load is 60, 100 or 150 kg. Thus, the Rockwell B-scale (designated R_B), uses a 100-kg load and #2 indenter, which is a 1/16-inch diameter ball. The R_C scale uses a diamond cone (#2 indenter), and a load 150 kg. Hardness numbers so obtained are relative. Thus, a hardness $R_C = 50$ has meaning only in relation to another hardness number using the same scale. The *Brinell hardness* is another method. It defines hardness as a ratio of applied force (to an indenting ball) to spherical surface area of the indentation.

3.12 SOUND WAVES

Alternate physical compression and expansion of medium (solids, liquids, and gases) with certain frequencies are called sound waves. The medium contents oscillate in the direction of wave propagation, hence these waves are called longitudinal mechanical waves. The name *sound* is associated with the hearing range of a human ear which is approximately from 20 to 20,000 Hz. Longitudinal mechanical waves below 20 Hz are called *infrasound* and above 20,000 Hz (20 kHz) *ultrasound*. If the classification is made by other animals, like dogs, the range of sound waves surely would be wider.

Detection of infrasound is of interest with respect to analysis of building structures, earthquake prediction and other geometrically large sources. When infrasound is of a relatively strong magnitude it can be if not heard at least felt by humans, producing quite irritating psychological effects (panic, fear, etc.). Audible waves are produced by vibrating strings (string music instruments), vibrating air columns (wind music instruments), and vibrating plates (some percussion instruments, vocal cords, loudspeaker). Whenever sound is produced, air is alternatively compressed and rarefied. These disturbances propagate outwardly. A spectrum of waves may be quite different - from a simple monochromatic sounds from a metronome or an organ pipe, to a reach violin music. Noise may have a very broad spectrum. It may be of a uniform distribution of density or it may be "colored" with predominant harmonics at some of its portions.

When a medium is compressed, its volume changes from V to $V-\Delta V$. The ratio of change in pressure Δp to relative change in volume is called the bulk modulus of elasticity of medium:

$$B = -\frac{\Delta p}{\Delta V/V} = \rho_0 v^2 ,$$

(3.12.1)

where ρ_0 is the density outside the compression zone and v is the speed of sound in the medium. Then speed of sound can be defined as

$$v = \sqrt{\frac{B}{\rho_0}} \ . \tag{3.12.2}$$

Hence, the speed of sound depends on the elastic (B) and inertia (ρ_0) properties of the medium. Since both variables are functions of temperature, the speed of sound also depends on temperature. This feature forms a basis for operation of the acoustic thermometers (Section 16.5). For solids, longitudinal velocity can be defined through its Young modulus E and Poisson ratio v

$$v = \sqrt{\frac{E(1-v)}{\rho_0(1+v)(1-2v)}} \ . \tag{3.12.3}$$

Table. 3-12 provides speeds of longitudinal waves in some media. It should be noted that the speed depends on temperature which always must be considered for the practical purposes.

Table 3-12 Speed of sound waves
Gases at 1 atm pressure, solids in long thin rods

Medium	speed (m/s)
Air (dry at 20°C)	331
Steam (134°C)	494
Hydrogen (20°C)	1,330
Water (fresh)	1,486
Water (sea)	1,519
Lead	1,190
Copper	3,810
Aluminum	6,320
Pyrex® glass	5,170
Steel	5,200
Beryllium	12,900

If we consider the propagation of a sound wave in an organ tube, each small volume element of air oscillates about its equilibrium position. For a pure harmonic tone, the displacement of a particle from the equilibrium position may be represented by

$$y = y_m \cos \frac{2\pi}{\lambda} (x - vt) \ , \tag{3.12.4}$$

where x is the equilibrium position of a particle and y is a displacement from the equilibrium position, y_m is the amplitude, and λ is the wavelength. In practice, it is more convenient to deal with pressure variations in sound waves rather than with displacements of the particles. It can be shown that the pressure exerted by the sound wave is

$$p = (k\rho_0 v^2 y_m) \sin (kx-\omega t) \ , \tag{3.12.5}$$

where $k = \dfrac{2\pi}{\lambda}$ is a wave number, ω is angular frequency, and the first parentheses represent an amplitude p_m of the sound pressure. Therefore, a sound wave may be considered as a pressure wave. It should be noted that sin and cos in (3.12.4) and (3.12.5) indicate that the displacement wave is 90° out of phase with the pressure wave.

Pressure at any given point in media is not constant and changes continuously, and the difference between the instantaneous and the average pressure is called an *acoustic pressure P*. During the wave propagation, vibrating particles oscillate near a stationary position with the instantaneous velocity ξ. The ratio of the acoustic pressure and the instantaneous velocity (do not confuse it with a wave velocity) is called an acoustic impedance

$$Z = \frac{P}{\xi} \ , \tag{3.12.6}$$

which is a complex quantity, that is characterized by an amplitude and a phase. For an idealized media (no loss) the Z is real and is related to the wave velocity as

$$Z = \rho_0 v \ . \tag{3.12.7}$$

We can define intensity I of a sound wave as the power transferred per unit area. Also, it can be expressed through the acoustic impedance

$$I = P\xi = \frac{p^2}{Z} \ . \tag{3.12.8}$$

It is common, however, to specify sound not by intensity but rather by a related parameter β, called the sound level and defined with respect to a reference intensity $I_0 = 10^{-12}$ W/m^2

$$\beta = 10 \log_{10} \frac{I}{I_0} \tag{3.12.9}$$

The magnitude of I_0 was chosen because it is the lowest ability of a human ear. The unit of β is a decibel (dB), named after Alexander Graham Bell. If $I = I_0$, $\beta = 0$.

Pressure levels also may be expressed in decibels as

$$\Pi = 20 \log_{10}\frac{p}{p_o} \ ,$$

(3.12.10)

where $p_o = 2 \cdot 10^{-5}$ N/m^2 (0.0002 µbar) = 2.9·10^{-9} psi.

Examples of some sound levels are given in Table 3-13. Since the response of a human ear is not the same at all frequencies, sound levels are usually referenced to I_o at 1 kHz where the ear is most sensitive.

Table 3-13 Sound levels (β) referenced to I_o at 1,000 Hz

Sound source	dB
Rocket engine at 50 m	200
Supersonic boom	160
Hydraulic press at 1 m	130
Threshold of pain	120
10W Hi-Fi speaker at 3 m	110
Unmuffled motorcycle	110
Rock-n-roll band	100
Subway train at 5 m	100
Pneumatic drill at 3 m	90
Niagara Falls	85
Heavy traffic	80
Automobiles at 5 m	75
Dishwashers	70
Conversation at 1 m	60
Accounting office	50
City street (no traffic)	30
Whisper at 1 m	20
Rustle of leaves	10
Threshold of hearing	0

3.13   TEMPERATURE
AND THERMAL PROPERTIES OF MATERIALS

Our bodies have a sense of temperature which by no means is an accurate method to measure outside heat. Human senses are not only nonlinear, but relative with respect to our previous experience. Nevertheless, we can easily tell the difference between warmer and cooler objects. Then, what is going on with these objects that they produce different perceptions?

Every single particle in this Universe exists in perpetual motion. Temperature, in the simplest way, can be described as a measure of kinetic energy of vibrating particles. The stronger the movement, the higher the temperature of that particle. Of course, molecules and atoms in a given volume of material do not move with equal intensities. That is, microscopically, they all are at different temperatures. The average kinetic energy of a large number of moving particles determines *macroscopic* temperature of an object. These processes are studied by statistical mechanics. Here, however, we are concerned with methods and devices which are capable of measuring the macroscopic average kinetic energy of material particles, which is the other way to say the temperature of the material. Since temperature is related to the movement of molecules, it is closely associated with pressure, which is defined as the force applied by moving molecules per unit area.

When atoms and molecules in a material move, they interact with other materials which happen to be brought in contact with them. Furthermore, every vibrating atom acts as a microscopic radio-transmitter which emanates electromagnetic radiation to the surrounding space. These two types of activities form a basis for heat transfer from warmer to cooler objects. The stronger the atomic movement the hotter the temperature and the stronger the electromagnetic radiation. A special device (we call it a *thermometer*), which either contacts the object or receives its electromagnetic radiation, produces a physical reaction, or signal. That signal becomes a measure of the object's temperature.

The word *thermometer* first appeared in literature in 1624 in a book by J. Leurechon, entitled *La Récréation Mathématique* [22]. The author described a glass water-filled thermometer whose scale was divided by 8 degrees. The first pressure-independent thermometer was built in 1654 by Ferdinand II, Grand Duke of Tuscany in a form of an alcohol-filled hermetically sealed tube.

3.13.1 Temperature scales

There are several scales to measure temperature. A first zero for a scale was established in 1664 by Robert Hooke at a point of freezing distilled water. In 1694 Carlo Renaldi of Padua suggested to take a melting point of ice and a boiling point of water to establish two fixed points on a linear thermometer scale. He divided the span by 12 equal parts. Unfortunately, his suggestion had been forgotten for almost 50 years. In 1701, Newton also suggested to use two fixed points to define a temperature scale. For one point he selected the temperature of melting ice (the zero point) and for the second point he chose the armpit temperature of a healthy Englishman (he labeled that point 12). At Newton's scale, water was boiling at point No. 34. Daniel Gabriel Fahrenheit, a Dutch instrument maker, in 1706 selected *zero* for his thermometer at the coldest temperature produced by a mixture of water, ice, and sal-ammoniac or household salt. For the sake of a finer division, he established the other point at 96 degrees which is "...found in the blood of a healthy man...".[1] On his scale, the melting point of water was at 32° and boiling at 212°. In 1742, Andreas Celsius, professor of astronomy at the University of Uppsala, proposed a scale with zero as the melting point of ice and 100 at boiling point of water.

Nowadays, in science and engineering, Celsius and Kelvin scales are generally employed. The Kelvin scale is arbitrarily based on the so-called *triple point of water*. There is a fixed temperature at a unique pressure of 4.58 mm Hg where water vapor, liquid and ice can coexist. This unique temperature is 273.16°K (degrees kelvin) which approximately coincides with 0°C. The Kelvin scale is linear with zero intercept (0°K) at a lowest temperature where kinetic energy of all moving particles is equal to zero. This point can not be achieved in practice and is a strictly theoretical value. It is called the *absolute zero*. Kelvin and Celsius scales have the same slopes[2], i.e., 1°C=1°K and °0K = -273.15°C:

$$°C = °K - 273.15°$$
(3.13.1)

The boiling point of water is at 100°C = 373.15°K. A slope of the Fahrenheit scale is steeper, because 1°C = 1.8°F. The Celsius and Fahrenheit scales cross at – 40°C and F. The conversion between the two scales is

$$°F = 32 + 1.8°C$$
(3.13.2)

which means that at 0°C, temperature in the Fahrenheit scale is +32°F.

---

[1] Not necessarily Englishman. Besides, now it is known that blood temperature of a healthy person varies between approximately 97 and 100°F.

[2] There is a difference of 0.01° between Kelvin and Celsius scales, as Celsius' zero point is defined not at a tripple point of water as for the Kelvin, but at temperature where ice and air-saturated water are at equilibrium at atmospheric pressure.

3.13.2 Thermal expansion

Essentially, all solids expand in volume with an increase in temperature. This is a result of vibrating atoms and molecules. When the temperature goes up, an average distance between the atoms increases, which leads to an expansion of a whole solid body. The change in any linear dimension: length, width, or height is called a *linear expansion*. A length l_2 at temperature T_2 depends on length l_1 at initial temperature T_1

$$l_2 = l_1[1 + \alpha (T_2 - T_1)] \ , \tag{3.13.3}$$

where α called the coefficient of linear expansion has different values for different materials. It is defined as

$$\alpha = \frac{\Delta l}{l} \frac{1}{\Delta T} \ , \tag{3.13.4}$$

where $\Delta T = T_2 - T_1$. Table 3-14 gives values of α for different materials[1]. Strictly speaking, α depends on the actual temperature. However, for most engineering purposes, small variations in α may be neglected. For the so-called *isotropic* materials, α is the same for any direction. The fractional change in area of an object and its volume with a high degree of accuracy can be represented, respectively, by

$$\Delta A = 2\alpha A \Delta T \ , \tag{3.13.5}$$

$$\Delta V = 3\alpha V \Delta T \ . \tag{3.13.6}$$

Heat is measured in *calories*[2]. One calorie (cal) is equal to amount of heat which is required to warm up by 1°C one gram of water at normal atmospheric pressure. In the U.S.A., a British unit of heat is generally used, which is 1 Btu (British thermal unit): 1Btu=252.02 cal.

3.13.3 Heat capacity

When an object is warmed up, its temperature increases. By warming we mean transfer of a certain amount of heat (thermal energy) into the object. Heat is stored in the object in a form of kinetic energy of vibration atoms. The amount of heat which an object can store is analogous to the amount of water that a water tank can store. Naturally, it can not store more than its volume, which

---

[1] More precisely, thermal expansion can be modeled by higher order polynomials:
$l_2 = l_1[1 + \alpha_1(T_2 - T_1) + \alpha_2((T_2 - T_1)^2 + \alpha_3(T_2 - T_1)^3 + \cdots]$, however, for the majority of practical purposes, a linear approximation is usually sufficient.
[2] A calorie which measures energy in food is actually equal to 1,000 physical calories which is called a kilocalorie.

is a measure of a tank's capacity. Similarly, every object may be characterized by a heat capacity which depends on both: the material of the object and its mass m

$$C = cm \;, \tag{3.13.7}$$

where c is a constant which characterizes thermal properties of material. It is called the *specific heat* and is defined as

$$c = \frac{Q}{m\Delta T} \cdot \tag{3.13.8}$$

Table 3-14 Coefficient (α) of linear thermal expansion of some materials (per $°C \times 10^{-6}$)

Material	α	Material	α
Alnico I (Permanent magnet)	12.6	Nylon	90
Alumina (polycrystalline)	8.0	Phosphor-bronze	9.3
Aluminum	25.0	Platinum	9.0
Brass	20.0	Plexiglas (Lucite)	72
Cadmium	30.0	Polycarbonate (ABS)	70
Chromium	6.0	Polyethylene (high density)	216
Comol (Permanent magnet)	9.3	Silicon	2.6
Copper	16.6	Silver	19.0
Fused quartz	0.27	Solder 50-50	23.6
Glass (Pyrex®)	3.2	Steel (SAE 1020)	12.0
Glass (regular)	9.0	Steel (stainless: type 304)	17.2
Gold	14.2	Teflon	99
Indium	18.0	Tin	13.0
Invar	0.7	Titanium	6.5
Iron	12.0	Tungsten	4.5
Lead	29.0	Zinc	35.0
Nickel	11.8		

The specific heat describes the material while a thermal capacity describes an object which is made of that material. Strictly speaking, specific heat is not constant over an entire temperature range of a specific phase of the material. It may change dramatically, when a phase of the material changes, say from solid to liquid. Microscopically, specific heat reflects structural changes in the material. For instance, the specific heat of water is almost constant between 0 and 100°C (liquid phase). Almost, but not exactly: it is higher near freezing, and decreases slightly when the temperature goes to about 35°C and then slowly rises again from 38° to 100°. Remarkably, the specific heat of water is the lowest near 37°C: a biologically optimal temperature of the warm-blooded animals.

Table 3-15 gives the specific heat for various materials in cal/(g°C). Some other tables provide specific heat in SI units of energy which is joule/g °C. The relationship between cal/(g°C) and j/(g°C) is as follows

$$1 \frac{j}{g°C} = 0.2388 \frac{cal}{g°C} \ . \tag{3.13.9}$$

It may be noted, that generally the heavier the material the lower is its specific heat.

3.14 HEAT TRANSFER

There are two fundamental properties of heat which should be well recognized: (1) the heat is totally *not specific*, that is, once it is produced, it is impossible to say what origin it has, and (2) the heat *can not be contained*, which means that it flows spontaneously from warmer to the cooler part of the system.

Thermal energy may be transferred from one object to another by three ways: conduction, convection, and radiation. Naturally, one of the objects which gives or receives heat may be a thermal detector. Its purpose would be to measure the amount of heat which represents some information about the object producing that heat. Such information may be the temperature of an object, chemical reaction, location or movement of the object, etc.

Let us consider a sandwich-like multilayer entity, where each layer is made of a different material. When heat moves through the layers, a temperature profile within each material depends on its thickness and thermal conductivity. Fig. 3-14.1 shows three laminated layers where the first layer is attached to a heat source (a device having an "infinite" heat capacity and a high thermal conductivity). One of the best solid materials to act as an infinite heat source is a thermostatically controlled bulk copper. Temperature within the source is higher and constant, except of a very thin region near the laminated materials. Heat propagates from one material to another by conduction. The temperature within each material drops with different rates depending on the thermal properties of the material. The last material loses heat to air through the natural convection and to the surrounding objects through the infrared radiation. Thus, Fig. 3-14.1 illustrates all three possible ways to transfer heat from one object to another.

Table 3-15 Specific heat and thermal conductivity of some materials (at 25°C)

Material	Specific heat $\frac{cal}{g \cdot °C}$	Thermal conductivity $\frac{W}{m \cdot °C}$	Density $\frac{g}{cm^3}$
Air (1 atm)	0.238	0.024	0.0012
Alumina	0.238	6	4.0
Aluminum	0.115	237	2.7
Bakelite	0.382	0.23	1.3
Brass	0.091	120	8.5
Chromium	0.110	91	
Constantan	0.095	22	8.8
Copper	0.092	401	8.9
Diamond		99-232	
Fiberglass	0.190	0.002-0.4	0.06
Germanium		60	
Glass (Pyrex)	0.191	0.1	2.2
Glass (regular)		1.0	
Gold	0.031	296	19.3
Graphite		79-92	
Iron	0.108	79	7.8
Lead	0.031	35	11.4
Manganin	0.098	21	8.5
Mercury	0.033	8.4	13.5
Nickel	0.106	90	8.9
Nylon	0.406	0.24	1.1
Platinum	0.032	70	21.4
Polyurethane foam		0.024	0.04
Silicon	0.17	149	
Silicone oil	0.4	0.1	0.9
Silver	0.057	429	10.5
Stainless steel		14	8.02
Styrofoam	0.310	0.003-0.03	0.05
Teflon TFE	0.238	0.4	2.1
Tin	0.054	64	7.3
Tungsten	0.032	199	19.0
Water	1.0	0.6	1.0
Zinc	0.093	115	7.1

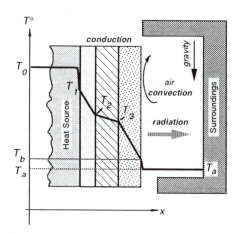

3-14.1 Temperature profile in laminated materials

3.14.1 Thermal conduction

Heat conduction requires a physical contact between two bodies. Thermally agitated particles in a warmer body jiggle and transfer kinetic energy to a cooler body by agitating its particles. As a result, the warmer body loses heat while the cooler body gains heat. Heat transfer by conduction is analogous to water flow or to electric current. For instance, heat passage through a rod is governed by a law which is similar to Ohm's law. A heat flow rate (thermal "current") is proportional to a thermal gradient (thermal "voltage") across the material (dT/dx) and a cross-sectional area A

$$H = \frac{dQ}{dt} = - kA\frac{dT}{dx} \; , \tag{3.14.1}$$

where k is called *thermal conductivity*. The minus sign indicates that heat flows in the direction of temperature decrease (a negative derivative is required to cancel the minus sign). A good thermal conductor has a high k (most of metals) while thermal insulators (most of dielectrics) have a low k. Thermal conductivity is considered constant, however it somewhat increases with temperature. To calculate a heat conduction through, say, an electric wire, temperatures at both ends (T_1 and T_2) must be used in equation

$$H = kA\frac{T_1 - T_2}{L} \; , \tag{3.14.2}$$

where L is the length of the wire. Quite often, a thermal resistance is used instead of a thermal conductivity

$$R = \frac{L}{k} \ , \tag{3.14.3}$$

then, (3.14.2) can be rewritten as

$$H = A\frac{T_1 - T_2}{R} \ . \tag{3.14.4}$$

Values of thermal conductivities for some materials are shown in Table 3-15.

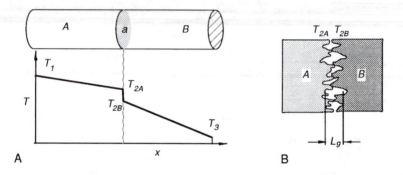

Fig. 3-14.2 Temperature profile in a joint (A)
and a microscopic view of a surface contact (B)

Fig. 3-14.1 shows an idealized temperature profile within the layers of laminated materials having different thermal conductivities. In the real world, heat transfer through an interface of two adjacent materials may be different from that idealized case. If we join together two materials and observe the heat propagation through the assembly, a temperature profile may look like the one shown in Fig. 3-14.2A. If the sides of the materials are well insulated, under steady-state conditions, the heat flux must be the same through both materials. The sudden temperature drop at the interface, having surface area a is the result of a thermal *contact resistance*. Heat transfer through the assembly can be described as

$$H = \frac{T_1 - T_3}{R_A + R_C + R_B} \ , \tag{3.14.5}$$

where R_A and R_B are thermal resistances of two materials and R_C is the contact resistance

$$R_C = \frac{1}{h_c a} \ . \tag{3.14.6}$$

The quantity h_c is called the contact coefficient. This factor can be very important in a number of sensor applications because of many heat-transfer situations which involve the mechanical joining of two materials. Microscopically, the joint may look like the one shown in Fig. 3-14.2B. No real surface is perfectly smooth, and the actual surface roughness is believed to play a central role in determining the contact resistance. There are two principal contributions to the heat transfer at the joint:

1. The material-to-material conduction through the actual physical contact

2. The conduction through trapped gases (air) in the void spaces created by the rough surfaces.

Since thermal conductivity of gases is very small as compared with many solids, the trapped gas creates the most resistance to heat transfer. Then, the contact coefficient can be defined as [31]

$$h_C = \frac{1}{L_g} \left(\frac{a_c}{a} \frac{2k_A k_A}{k_A + k_B} + \frac{a_v}{a} k_f \right), \quad (3.14.7)$$

where L_g is the thickness of the void space, k_f is the thermal conductivity of the fluid (for instance, air) filling the void space, a_c and a_v are areas of the contact and void, respectively, and k_A and k_B are the respective thermal conductivities of the materials. The main problem with this theory is that it is very difficult to determine experimentally areas a_c and a_v, and distance L_g. This analysis, however, allows us to conclude that the contact resistance should increase with a decrease in the ambient gas pressure. On the other hand, contact resistance decreases with an increase in the joint pressure. This is a result of a deformation of the high spots of the contact surface which leads to enlarging a_c and creating a greater contact area between the materials. To decrease the thermal resistance, a dry contact between materials should be avoided. Before joining, surfaces may be coated with fluid having low thermal resistance. For instance, silicone thermal grease is often used for the purpose.

3.14.2 Thermal convection

Another way to transfer heat is convection. It requires an intermediate agent (gas or liquid) which takes heat from a warmer body, carries it to a cooler body, releases heat and then may or may not return back to a warmer body to pick up another portion of heat. Heat transfer from a solid body to a moving agent or within the moving agent is also called convection. Convection may be natural (gravitational) or forced (produced by a mechanism). With the natural convection of air, buoyant forces produced by gravitation act upon air molecules. Warmed up air rises carrying heat away from a warm surface. Cooler air descends toward the warmer object. Forced convection of air is produced by a fan or blower. Forced convection is used in liquid thermostats to maintain the temperature of a device at a predetermined level. Efficiency of a convective heat transfer depends on the rate of media movement, temperature gradient, surface area of an object and thermal properties of moving medium. An object whose temperature is different from the surroundings will lose (or receive) heat which can be determined from equation similar to that of a thermal conduction

$$H = \alpha A \, (T_1 - T_2) \; , \qquad\qquad (3.14.8)$$

where convective coefficient α depends on the fluid's specific heat, viscosity, and a rate of movement. For an object freely mounted in air, α is approximately equal to 11 W/m^2K.

3.14.3 Thermal radiation

It was mentioned above that in any object every atom and every molecule vibrate. The average kinetic energy of vibrating particles is represented by the absolute temperature. According to laws of electrodynamics, a moving electric charge is associated with a variable electric field which produces an alternating magnetic field. In turn, when the magnetic field changes, it results in a coupled with it changing electric field, and so on. Thus, a vibrating particle is a source of electromagnetic field which propagates outwardly with the speed of light and is governed by the laws of optics. Electromagnetic waves can be reflected, filtered, focused, etc. Fig. 3-14.3 shows the total electromagnetic radiation spectrum which spreads from γ rays to radio waves.

The wave length directly relates to frequency ν by means of speed of light c in a particular media

$$\lambda = \frac{c}{\nu} \; . \qquad\qquad (3.14.9)$$

A relationship between λ and temperature is more complex and is governed by Plank's law which was discovered in 1901[1]. It establishes radiant flux density W_λ as a function of wavelength λ and absolute temperature T. Radiant flux density is power of electromagnetic radiation per unit of wavelength:

$$W_\lambda = \frac{\varepsilon(\lambda) \cdot C_1}{\pi \lambda^5 (e^{C_2/\lambda T} - 1)} \; , \qquad\qquad (3.14.10)$$

where $\varepsilon(\lambda)$ is the emissivity of an object, $C_1 = 3.74 \times 10^{-12}$ $W \cdot cm^2$ and $C_2 = 1.44$ $cm \cdot K$ are constants, and e is the base of natural logarithms.

---

[1] In 1918, Max K. E. L. Plank (Germany, Berlin University) was awarded Nobel Prize "in recognition of his services he rendered to the advancement of Physics by his discovery of energy quanta".

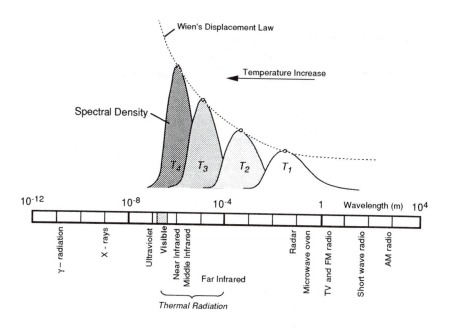

Fig. 3-14.3 Spectrum of electromagnetic radiation

Temperature is a result of averaged kinetic energies of an extremely large number of vibrating particles. However, all particles do not vibrate with the same frequency or magnitude. Different permissive frequencies (also wavelengths and energies) are spaced very close to one another which makes the material capable of radiating in a virtually infinite number of frequencies spreading from very long to very short wavelengths. Since temperature is a statistical representation of an average kinetic energy, it determines the highest probability for the particles to vibrate with a specific frequency and to have a specific wavelength. This most probable wavelength is established by the Wien's law[1], which can be found by equating to zero a first derivative of (3.14.10). The result of the calculation is a wavelength near which most of the radiant power is concentrated

$$\lambda_m = \frac{2898}{T} \ , \tag{3.14.11}$$

where λ_m is in μm and T in K. Wien's law states that the higher the temperature the shorter the wavelength (Fig. 3-14.3). In view of (3.14.9), the law also states that the most probable frequency in the entire spectrum is proportional to the absolute temperature

---

[1] In 1911, Wilhelm Wien (Germany, Würtzburg University) was awarded Nobel Prize "for his discoveries regarding the laws governing the radiation of heat".

$$\nu_m = 10^{11}T \quad [\text{Hz}] \; .$$
(3.14.12)

For instance, at normal room temperature most of the far infrared energy is radiated from objects near 30 THz ($30 \cdot 10^{12}$Hz). Radiated frequencies and wavelengths depend only on temperature, while the magnitude of radiation also depends on the emissivity $\varepsilon(\lambda)$ of the surface.

Theoretically, a thermal radiation bandwidth is infinitely wide. However, when detecting that radiation, properties of the real world sensors must be accounted for. The sensors are capable of measuring only a limited range of radiation. In order to determine the total radiated power within a particular bandwidth, Eq. (3.14.10) is integrated within the limits from λ_1 to λ_2:

$$\Phi_{bo} = \frac{1}{\pi} \int\limits_{\lambda_1}^{\lambda_2} \frac{\varepsilon(\lambda)C_1\lambda^{-5}}{e^{C_2/\lambda T}-1} \; .$$
(3.14.13)

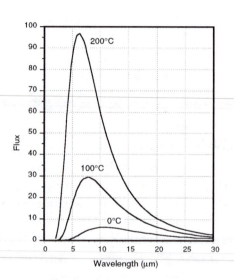

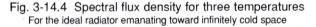

Fig. 3-14.4 Spectral flux density for three temperatures
For the ideal radiator emanating toward infinitely cold space

Fig. 3-14.4 shows the radiant flux density for three different temperatures for the infinitely wide bandwidth ($\lambda_1=0$ and $\lambda_2=\infty$). It is seen, that the radiant energy is distributed over the spectral range highly nonuniformly, with clearly pronounced maximum defined by Wien's law. A hot object radiates a significant portion of its energy in the visible range, while the power radiated by the cooler objects is concentrated in the infrared and far infrared portion of the spectrum.

Equation (3.14.13) is quite complex and can not be solved analytically for any particular bandwidth. A solution can be found either numerically or by an approximation. An approximation for a broad bandwidth (when λ_1 and λ_2 embrace well over 50% of the total radiated power) is a 4th order parabola which is known as the *Stefan-Boltzmann law*

$$\Phi_{bo} = A\varepsilon \cdot \sigma \cdot T^4 \quad . \tag{3.14.14}$$

Here $\sigma = 5.67 \times 10^{-8}$ W/m²K⁴ (Stefan-Boltzmann constant) and ε is assumed to be wavelength independent [32]. While wavelengths of the radiated flux are temperature dependent, the magnitude of radiation is also a function of the surface property which is called *emissivity* ε. Emissivity is measured on a scale from 0 to 1. It is a ratio of flux which is emanated from a surface to that emanated from the ideal emitter having the same temperature. There is a fundamental equation which connects emissivity ε, transparency γ, and reflectivity ρ:

$$\varepsilon + \gamma + \rho = 1 \quad . \tag{3.14.15}$$

In 1860, Kirchhoff had found that emissivity and absorptivity, α, is the same thing. As a result, for an opaque object ($\gamma=0$), reflectivity ρ and emissivity ε are connected by a simple relationship: $\rho=1-\varepsilon$.

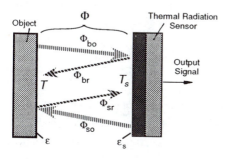
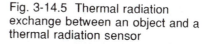

Fig. 3-14.5 Thermal radiation exchange between an object and a thermal radiation sensor

The Stefan-Boltzmann law specifies flux which would be emanated from a surface of temperature T toward an infinitely cold space (at absolute zero). When thermal radiation is detected by a thermal sensor[1], the opposite radiation from the sensor toward the object must also be accounted for. A thermal sensor is capable of responding only to a net thermal flux, i.e., flux from the object minus flux from itself. The surface of the sensor which faces the object has emissivity ε_s (and, subsequently reflectivity $\rho=1-\varepsilon_s$). Since the sensor is only partly absorptive, not the entire flux Φ_{bo} is absorbed and utilized. A part of it Φ_{ba} is absorbed by the sensor while another part

---

[1] Here we discuss the so-called *thermal* sensors as opposed to quantum sensors which are described in Chapter 13.

Φ_{br} is reflected (Fig. 3-14.5) back toward to object[1]. The reflected flux is proportional to the sensor's coefficient of reflectivity

$$\Phi_{br} = -\rho_s\Phi_{bo} = -A\varepsilon(1-\varepsilon_s)\sigma T^4. \qquad (3.14.16)$$

A negative sign indicates an opposite direction with respect to flux Φ_{bo}. As a result, the net flux originated from the object is

$$\Phi_b = \Phi_{bo} + \Phi_{br} = A\varepsilon\varepsilon_s\sigma T^4 . \qquad (3.14.17)$$

Depending on its temperature T_s, the sensor's surface radiates its own net thermal flux toward the object in a similar way

$$\Phi_s = -A\varepsilon\varepsilon_s\sigma T_s^{4} . \qquad (3.14.18)$$

Two fluxes propagate in the opposite directions and are combined into a final net flux existing between two surfaces

$$\Phi = \Phi_b + \Phi_s = A\varepsilon\varepsilon_s\sigma(T^4 - T_s^{4}) . \qquad (3.14.19)$$

This is a mathematical model of a net thermal flux which is converted by a thermal sensor into the output signal. It establishes a connection between thermal power Φ absorbed by the sensor and the absolute temperatures of the object and the sensor.

Emissivity

The emissivity of a media is a function of its dielectric constant and, subsequently, refractive index n. The highest possible emissivity is 1. It is attributed to the so-called blackbody – an ideal emitter of electromagnetic radiation. The name implies its appearance at normal room temperatures. If the object is opaque ($\gamma=0$) and nonreflective ($\rho=0$) according to (3.14.15), it becomes an ideal emitter and absorber of electromagnetic radiation (since $\varepsilon=\alpha$). It should be noted, however, that emissivity is generally wavelength dependent (Fig. 3-14.6). For example, a white sheet of paper is very much reflective in the visible spectral range and emits virtually no visible light. However, in the far infrared spectral range its reflectivity is very low and emissivity is high (about 0.92), thus making paper a good emitter of far infrared radiation. Polyethylene, which is widely used for fabrication of far infrared lenses, heavily absorbs (emits) in narrow bands around 3.5, 6.8, and 13.5μm, while being quite transparent (nonemissive) in other bands.

---

[1] This analysis assumes that there are no other objects in the sensor's field of view.

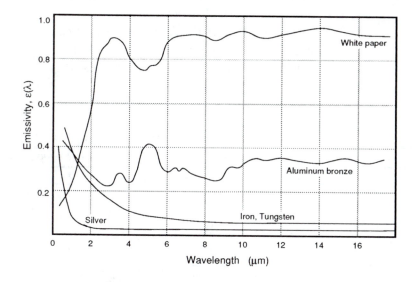

Fig. 3-14.6 Wavelength dependence of emissivities

For many practical purposes, emissivity in a relatively narrow spectral range of thermal radiation (for instance, from 8 to 16 μm) may be considered constant. However, for precision, measurements, when thermal radiation must be detected with accuracy better than 1%, surface emissivity either must be known, or the so called dual band IR detectors should be employed[1].

For a nonpolarized far infrared light in normal direction emissivity may be expressed by the equation

$$\varepsilon = \frac{4n}{(n+1)^2} \cdot \qquad (3.14.20)$$

All nonmetals are very good diffusive emitters of thermal radiation with a remarkably constant emissivity defined by (3.14.20) within a solid angle of about ±70°. Beyond that angle, emissivity begins to decrease rapidly to zero with the angle approaching 90°. Near 90° emissivity is very low. A typical calculated graph of the directional emissivity of nonmetals into air is shown in Fig. 3-14.7A. It should be emphasized that the above considerations are applicable only to wavelengths in the far infrared spectral range and are not true for the visible light, since emissivity of thermal radiation is a result of electromagnetic effects which occur at an appreciable depth.

---

[1] A dual band detectors use two narrow spectral ranges to detect IR flux. Then, by using a ratiometric technique of signal processing, temperature of an object is calculated. During the calculation emissivity and other multiplicative constants are cancelled out.

Metals behave quite differently. Their emissivities greatly depend on surface finish. Generally, polished metals are poor emitters within the solid angle of ±70° while their emissivity increases at larger angles (Fig. 3-14.7B). Table 3-16 gives typical emissivities of some materials in a temperature range between 0 and 100°C.

Table 3-16 Typical emissivities of different materials (from 0 to 100°C)

Material	Emissivity	Material	Emissivity
Blackbody (ideal)	1.00	Green leaves	0.88
Cavity Radiator	0.99-1.00	Ice	0.96
Aluminum (anodized)	0.70	Iron or Steel (rusted)	0.70
Aluminum (oxidized)	0.11	Nickel (oxidized)	0.40
Aluminum (polished)	0.05	Nickel (unoxidized)	0.04
Aluminum (rough surface)	0.06-0.07	Nichrome (80Ni-20Cr) (oxidized)	0.97
Asbestos	0.96	Nichrome (80Ni-20Cr) (polished)	0.87
Brass (dull tarnished)	0.61	Oil	0.80
Brass (polished)	0.05	Silicon	0.64
Brick	0.90	Silicone Rubber	0.94
Bronze (polished)	0.10	Silver (polished)	0.02
Carbon filled latex paint	0.96	Skin (human)	0.93-0.96
Carbon Lamp Black	0.96	Snow	0.85
Chromium (polished)	0.10	Soil	0.90
Copper (oxidized)	0.6-0.7	Stainless Steel (buffed)	0.20
Copper (polished)	0.02	Steel (flat rough surface)	0.95-0.98
Cotton cloth	0.80	Steel (ground)	0.56
Epoxy Resin	0.95	Tin plate	0.10
Glass	0.95	Water	0.96
Gold	0.02	White Paper	0.92
Gold-black	0.98-0.99	Wood	0.93
Graphite	0.7-0.8	Zinc (polished)	0.04

Unlike most solid bodies, gases in many cases are transparent to thermal radiation. When they absorb and emit radiation, they usually do so only in certain narrow spectral bands. Some gases, such as N_2, O_2 and others of nonpolar symmetrical molecular structure, are essentially transparent at low temperatures, while CO_2, H_2O, and various hydrocarbon gases radiate and absorb to an appreciable extent. When infrared flux enters a layer of gas, its absorption has an exponential decay profile, governed by *Beer's law*

$$\frac{\Phi_x}{\Phi_0} = e^{-\alpha_\lambda x} , \tag{3.14.21}$$

where Φ_0 is the incident thermal flux, Φ_x is the flux at thickness x, α_λ is the spectral coefficient of absorption. The above ratio is called a monochromatic transmissivity γ_λ at a specific wavelength λ. If gas is nonreflecting, then its emissivity is defined as

$$\varepsilon_\lambda = 1 - \gamma_\lambda = 1 - e^{-\alpha_\lambda x} . \tag{3.14.22}$$

It should be emphasized that since gasses absorb only in narrow bands, emissivity and transmissivity must be specified separately for any particular wavelength. For instance, water vapor is highly absorptive at wavelengths of 1.4, 1.8, and 2.7 μm and is very transparent at 1.6, 2.2 and 4 μm.

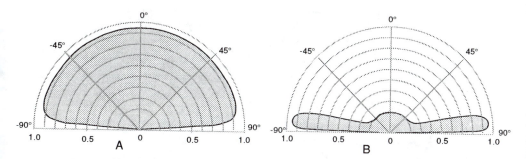

Fig. 3-14.7 Spatial emissivities for nonmetal (A) and a polished metal (B)

Cavity effect

An interesting effect develops when electromagnetic radiation is measured from a cavity. For this purpose, a cavity means an opening in a concave body of a generally irregular shape whose inner wall temperature is uniform over an entire surface (Fig. 3-14.8A). Emissivity of a cavity opening dramatically increases approaching unity at any wavelength, as compared with a flat surface. The cavity effect is especially pronounced when its inner walls have relatively high emissivity. Let us consider a nonmetal cavity. All nonmetals are diffuse emitters. Also, they are diffuse reflectors. We assume that temperature and surface emissivity of the cavity are homogeneous over an entire area. The ideal emitter (blackbody) would emanate from area a the infrared photon flux $\Phi_0 = a\sigma T_b^4$. However, the object has the actual emissivity ε_b and, as a result, the flux radiated from that area is smaller: $\Phi_r = \varepsilon_b \Phi_0$ (Fig. 3-14.8A). Flux which is emitted by other parts of the object toward area a is also equal to Φ_r (since the object is thermally homogeneous we may disregard spatial distribution of flux). A substantial portion of that incident flux Φ_r is absorbed by the surface of area a, while a smaller part is diffusely reflected:

$$\Phi_\rho = \rho\Phi_r = (1 - \varepsilon_b)\varepsilon_b\Phi_o, \qquad (3.14.23)$$

and the combined radiated and reflected flux from area a is

$$\Phi = \Phi_r + \Phi_\rho = \varepsilon_b\Phi_o + (1 - \varepsilon_b)\varepsilon_b\Phi_o = (2 - \varepsilon_b)\varepsilon_b\Phi_o . \qquad (3.14.24)$$

As a result, the effective emissivity may be expressed as

$$\varepsilon_e = \frac{\Phi}{\Phi_o} = (2 - \varepsilon_b)\varepsilon_b \qquad (3.14.25)$$

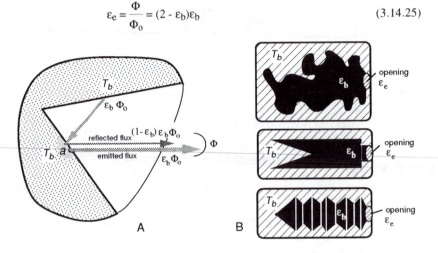

Fig. 3-14.8 A: Cavity effect enhances emissivity; B: shapes of blackbody cavities
Note that $\varepsilon_e > \varepsilon_b$

It follows from the above that due to a single reflection, a perceived (effective) emissivity of a cavity is equal to the surface emissivity magnified by a factor of $(2-\varepsilon_b)$. Of course, there may be more than one reflection of radiation before it exits the cavity. In other words, the incident on area a flux could already be a result of a combined effect from the reflectance and emittance at other parts of the cavity's surface. The flux intensity will be higher than the originally emanated flux Φ_r. For any number of reflections, the effective emissivity may be calculated from formula

$$\varepsilon_e = 1 - (1-\varepsilon_b)^{\xi+1} , \qquad (3.14.26)$$

where ξ is the number of reflections. For a relatively high ε_b (0.9 and higher), the second summand very quickly converges to zero with an increase in ξ, thus resulting in the effective emissivity equal to unity, which is almost independent of the actual emissivity of the inner walls. For a cavity effect to work, the effective emissivity must be attributed to the cavity opening from which radiation escapes. If a sensor is inserted into the cavity facing its wall directly, the cavity effect may disappear and the emissivity will be equal to that of a wall surface. The cavity effect is used for the fabrication of blackbody thermal radiation sources. Also, it can be employed when measuring temperature of an object whose emissivity is too low to assure good accuracy. If an object has a shape where a cavity is an integral part, thermal radiation should be measured from that cavity.

For research purposes and calibration of thermal radiation detectors, a blackbody cavity is usually required. A practical blackbody can be fabricated in several ways. Copper is the best choice for the cavity material, thanks to its high thermal conductivity. Fig. 3-14.8B shows several practical shapes of blackbody cavities. It is extremely important to reduce thermal gradients within the metal mass around the cavity. This assures uniform temperatures of the surface. The walls of a cavity portion should be surrounded with a good thermal insulator which in combination with a large thermal capacity of the metal, creates a high thermal constant of the device. Heating of a blackbody may be provided from electric heating elements, while cooling can be done either by a forced convection of the outside air, or, for a broader control range, by thermoelectric cooling elements. The inner portion of a cavity should be painted with organic paint. The color of paint is not important.

A cavity radiator with a liquid heating/cooling can be fabricated in a form of a tube whose inner radius is at least ten times smaller than its length, one end is closed and the inner surface is painted with high emissivity paint. Such a tubular cavity is immersed into a liquid bath to provide a uniform temperature along its walls. The temperature of the liquid (usually water) is precisely controlled, thus assuring a constant and uniform temperature around the tube.

3.15 LIGHT AND OPTICAL PROPERTIES OF MATERIALS

Light has an electromagnetic nature. It may be considered a propagation of either quanta of energy or electromagnetic waves. Different portions of the wave frequency spectrum are given special names, for example: ultraviolet (UV), visible, near, mid and far infrared (IR), microwaves, radio-waves, etc. The name "light" was arbitrarily given to electromagnetic radiation which occupies wavelengths from approximately 0.1 to 100 μm. Light below the shortest wavelength that we can see (violet) is called ultraviolet and farther than the longest that we can see (red) is called infrared. The infrared range is subdivided into three regions: near-infrared (from about 0.9 to 1.5 mμ), mid-infrared (1.5 to 4 μm), and far-infrared (4 to 100 μm).

Different portions of the radiation spectrum are studied be separate branches of physics. An entire electromagnetic spectrum is represented in Fig. 3-14.3. It spreads from γ rays (the shortest) to radio-waves (the longest). In this Section, we will review those properties of light which are mostly concerned with the visible and near infrared portions of the electromagnetic spectrum. Thermal radiation (mid and far infrared regions) are covered in Section 3.14.

The velocity of light a in vacuum c_o is independent of wavelengths and can be expressed through $\mu_0 = 4\pi \cdot 10^{-7}\ \frac{henry}{m}$ and $\varepsilon_0 = 8.854 \cdot 10^{-12}\ \frac{farad}{m}$, which are the magnetic and electric permitivities of free space:

$$c_o = \frac{1}{\sqrt{\mu_0 \varepsilon_0}} = 299,792,458.7 \pm 1.1 \text{ m/s} \quad . \tag{3.15.1}$$

The frequency of light waves in vacuum or any particular medium relates to its wavelength λ by the equation (3.14.9) which we rewrite here as

$$v = \frac{c}{\lambda} \quad , \tag{3.15.2}$$

where c is the speed of light in a medium.

The energy of a photon relates to its frequency as

$$E = hv \quad , \tag{3.15.3}$$

where $h = 6.63 \cdot 10^{-34} J \cdot s$ ($4.13 \cdot 10^{-15} eV \cdot s$) is Plank's constant. The energy E is measured in $1.602 \cdot 10^{-19} J = 1 eV$ (electron-volt). A near-infrared photon having a wavelength of 1μm has the energy of 1.24 eV. Hence, an optical quantum detector operating in the range of 1μm must be capable of responding to that level of energy. Human skin (at 37°C) radiates far infrared photons with energies near 0.13eV which is an order of magnitude lower, making them much more difficult to detect. This is the reason, why low energy radiation is often detected by thermal detectors rather than quantum detectors (Section 13.7).

3.15.1 Radiometry

Let us consider light traveling through a three layer material. All layers are made of different substances called media. Fig 3-15.1 shows what happens to a ray of light which travels from the first medium into a flat plate of a second medium, and then to a third medium. Part of the incident light is reflected from a planar boundary between the first and second media according to the *law of reflection* which historically is attributed to Euclid

$$\Theta_1 = \Theta'_1 \tag{3.15.4}$$

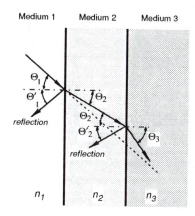

Fig. 3-15.1 Light passing through materials with different refractive indices

A part of light enters the plate (medium 2) at a different angle. The new angle Θ_2 is governed by the *refraction law* which was discovered by Willebrod Snell and is known as *Snell's law*

$$n_1 \sin\Theta_1 = n_2 \sin\Theta_2 \; , \tag{3.15.5}$$

where n_1 and n_2 are the indices of refraction of two media.

An *index of refraction* is a ratio of velocity of light in vacuum c_o to that in a medium c

$$n = \frac{c_o}{c} \; , \tag{3.15.6}$$

Since $c < c_o$, the refractive index of a medium is always more than unity. The velocity of light in a medium directly relates to a dielectric constant ε_r of a medium, which subsequently determines the refractive index

$$n = \sqrt{\varepsilon_r} \; , \tag{3.15.7}$$

Generally, n is a function of wavelength. A wavelength dependence of index of refraction is manifested in a prism which was used by Sir Isaac Newton in his experiments with the light spectrum. In the visible range, the index of refraction n is often specified at a wavelength of 0.58756 µm, the yellow-orange helium line. Indices of refraction for some materials are presented in Table 3-17.

A refractive index dependence of wavelengths is called a dispersion. The change in n with the wavelength is usually very gradual, and often negligible, unless the wavelength approaches a region where the material is not transparent. Fig. 3-15.2 shows transparency curves of some optical materials.

A portion of light reflected from a boundary at angle Θ'_1, depends on light velocities in two adjacent media. Amount of reflected flux Φ_ρ relates to incident flux Φ_0 through the *coefficient of reflection* ρ which can be expressed by means of refractive indices

$$\rho = \frac{\Phi_\rho}{\Phi_0} = \left(\frac{n_1 - n_2}{n_1 + n_2}\right)^2 .$$

(3.15.8)

Equations (3.14.20) and (3.15.8) indicate that both the reflection and the absorption (emissivity) depend solely on the refractive index of the material at a particular wavelength.

Table 3-17 Refractive indices (*n*) of some materials

Material	n	wavelength (μm)	note
Vacuum	1		
Air	1.00029		
Acrylic	1.5	0.41	
As$_2$S$_3$	2.4	8.0	
CdTe	2.67	10.6	
Crown glass	1.52		
Diamond	2.42	0.54	Excellent thermal conductivity
Fused silica (SiO$_2$)	1.46	3.5	
GaAs	3.13	10.6	Laser windows
Germanium	4.00	12.0	
Heaviest flint glass	1.89		
Heavy flint glass	1.65		
Irtran 2 (ZnS)	2.25	4.3	Windows in IR sensors
KBr	1.46	25.1	Hygroscopic
KCl	1.36	23.0	Hygroscopic
KRS-5	2.21	40.0	Toxic
KRS-6	2.1	24	Toxic
NaCl	1.89	0.185	Hygroscopic, corrosive
Polyethylene	1.54	8.0	Low cost IR transparent
Polystyrene	1.55		
Pyrex 7740	1.47	0.589	Good thermal and optical properties
Quartz	1.54		
Sapphire (Al$_2$O$_3$)	1.59	5.58	Chemically resistant
Silicon	3.42	5.0	Windows in IR sensors
Silver Bromide (AgBr)	2.0	10.6	Corrosive
Silver Chloride (AgCl)	1.9	20.5	Corrosive
Water [20°C]	1.33		
ZnSe	2.4	10.6	IR windows, brittle

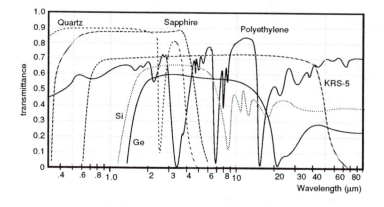

Fig. 3-15.2 Transparency characteristics for various optical materials

If the light flux enters from air into an object having refractive index n, Eq. (3.15.8) is simplified

$$\rho = \left(\frac{n-1}{n+1}\right)^2 , \qquad (3.15.9)$$

Before light exits the second medium (Fig. 3-15.1) and enters the third medium having refractive index n_3 another part of it is reflected internally from the second boundary between the n_2 and n_3 media at angle Θ'_2. The remaining portion of light exits at angle Θ_3 which is also governed by the Snell's law. If media 1 and 3 are the same (for instance, air) at both sides of the plate, then $n_1=n_3$ and $\Theta_1=\Theta_3$. This case is illustrated in Fig. 3-15.3. It follows from formula (3.15.8) that coefficients of reflection are the same for light striking a boundary from either direction - approaching from the higher or lower index of refraction.

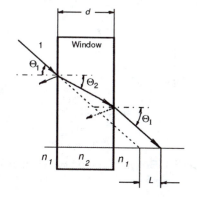

Fig. 3-15.3 Light passing through an optical plate

A combined coefficient of two reflections from both surfaces of a plate can be found from a simplified formula

$$\rho_2 \approx \rho_1(2 - \rho_1),$$
(3.15.10)

where ρ_1 is the reflective coefficient from one surface. In reality, the light reflected from the second boundary is reflected again from the first boundary back to the second boundary, and so on. Thus, assuming that there is no absorption in the material, the total reflective loss within the plate can be calculated through the refractive index of the material

$$\rho_2 = 1 - \frac{2n}{n^2 + 1}$$
(3.15.11)

Reflection increases for higher differences in refractive indices. For instance, if visible light travels without absorption from air through a heavy flint glass plate, two reflectances result in loss of about 11%, while for the air-germanium-air interfaces (in the far infrared spectral range) the reflective loss is about 59%. To reduce losses, optical materials are often given antireflective coatings which have refractive indices and thickness geared to specific wavelengths.

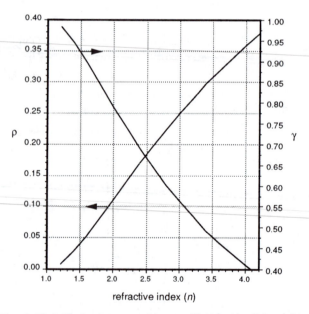

refractive index (*n*)

Fig. 3-15.4   Reflectance and transmittance of a thin plate
as functions of a refractive index

The radiant energy balance equation (3.14.15) should be modified to account for two reflections in an optical material:

$$\rho_2 + \alpha + \gamma = 1 \; , \tag{3.15.12}$$

where α is a coefficient of absorption and γ is a coefficient of transmittance. In a transparency region, $\alpha \approx 0$, therefore, transmittance is:

$$\gamma = 1 - \rho_2 \approx \frac{2n}{n^2 + 1} \; . \tag{3.15.13}$$

The above specifies the maximum theoretically possible transmittance of the optical plate.

In the above example, transmittance of a glass plate is 88.6% (visible) while transmittance of a germanium plate is 41% (far IR). In the visible range, germanium transmittance is zero, which means that 100% of flux is reflected and absorbed. Fig. 3-15.4 shows reflectance and transmittance of a thin plate as functions of refractive indices. Here a plate means any optical device (like a window or a lens) operating within its useful spectral range, that is, where its absorptive loss is small ($\alpha \approx 0$).

Fig. 3-15.5 shows an energy distribution within an optical plate when incident light flux Φ_0 strikes its surface. A part of incident flux Φ_ρ is reflected, another part Φ_α is absorbed by the material and the third part Φ_γ is transmitted through. The absorbed portion of light is converted into heat, a portion of which ΔP is lost to a supporting structure and surroundings through thermal conduction and convection. The rest of the absorbed flux, raises the temperature of the material. The temperature increase may be of concern when the material is used as a window in a powerful laser. Another application where temperature increase may cause problems is in far infrared detectors. The problem is associated with the flux $\Phi_\varepsilon = \Phi_\alpha - \Delta P$ which is radiated by the material due to its temperature change. This is called a secondary radiation. Naturally, a radiated spectrum relates to a temperature of the material and is situated in the far infrared region of the spectrum. The spectral distribution of the secondary radiation corresponds to the absorption distribution of the material because absorptivity and emissivity are the same thing.

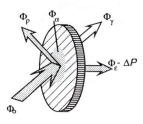

Fig. 3-15.5 Radiant energy distribution at optical plate

For materials with low absorption, the absorption coefficient can be determined through a temperature rise in the material

$$\alpha = \frac{mc}{\Phi\gamma}\,\frac{2n}{n^2+1}\left(\frac{dT_g}{dt} + \frac{dT_L}{dt}\right)T_o \quad , \tag{3.15.14}$$

where m and c are the mass and the specific heat of the optical material, T_g and T_L are the slopes of the rising and lowering parts of the temperature curve of the material, respectively, at test temperature T_o. Strictly speaking, light in the material is lost not only due to absorption but to scattering as well. A combined loss within material depends on its thickness and can be expressed through the so-called *attenuation coefficient g* and the thickness of the sample h. The transmission coefficient can be determined from equation (3.15.13) which is modified to account for the attenuation:

$$\gamma \approx (1 - \rho_2)e^{-gh} \quad . \tag{3.15.15}$$

The attenuation (or extinction) coefficient g is usually specified by manufacturers of optical materials.

3.15.2 Photometry

When using light sensitive devices (photodetectors), it is critical to take into a consideration both the sensor and the light source. In some applications, light is received from independent sources, while in others the light source is a part of the measurements system. In any event, the so-called photometric characteristics of the optical system should be accounted for. Such characteristics include flux, emittance, luminance, brightness, etc.

To measure radiant intensity and brightness, special units have been devised. Radiant flux (energy emitted per unit time) which is situated in a visible portion of the spectrum is referred to as luminous flux. This distinction is due to the inability of the human eye to respond equally to like power levels of different visible wavelengths. For instance, one red and one blue light of the same intensity will produce very different sensations - the red will be perceived as much brighter. Hence, comparing lights of different colors, the watt becomes a poor measure of brightness and a special unit called a *lumen* was introduced. It is based on a standard radiation source with molten platinum formed in a shape of a blackbody and visible through a specified aperture within a solid angle of one steradian. A solid angle is defined in a spherical geometry as

$$\omega = \frac{A}{r^2} \quad , \tag{3.15.16}$$

where r is the spherical radius and A is the spherical surface of interest. When $A=r$, then the unit is called a spherical radian or *steradian* (see Table 1-8).

Illuminance is given as

$$E = \frac{dF}{dA} , \qquad\qquad (3.15.17)$$

that is, a differential amount of luminous flux (F) over a differential area. It is most often expressed in lumens per square meter (square foot), or foot-meter (foot-candle). The luminous intensity specifies flux over solid angle:

$$I_L = \frac{dF}{d\omega} , \qquad\qquad (3.15.18)$$

most often it is expressed in lumens per steradian or candela. If the luminous intensity is constant with respect to the angle of emission, Eq. (3.15.18) becomes

$$I_L = \frac{F}{\omega} . \qquad\qquad (3.15.19)$$

If the wavelength of the radiation varies, but the illumination is held constant, the radiative power in watts is found to vary. A relationship between illumination and radiative power must be specified at a particular frequency. The point of specification has been taken to be at a wavelength of 0.555μm, which is the peak of the spectral response of a human eye. At this wavelength, 1 watt of radiative power is equivalent to 680 lumens. For the convenience of the reader, some useful terminology is given in Table 3-18.

In the selection of electro-optical sensors, design considerations of light sources are of prime concern. A light source will effectively appear as either a *point source*, or as an *area source*, depending upon the relationship between the size of the source and the distance between the source and the detector. Point sources are arbitrarily defined as those whose diameter is less than 10% of the distance between the source and the detector. While it is usually desirable that a photodetector is aligned such that its surface area is tangent to the sphere with the point source at its center, it is possible that the plane of the detector can be inclined from the tangent plane. Under this condition, the incident flux density (irradiance) is proportional to the cosine of the inclination angle φ:

$$H = \frac{I_r}{\cos\varphi} , \qquad\qquad (3.15.20)$$

and the illuminance

$$E = \frac{I_L}{r^2} \cos\varphi . \qquad\qquad (3.15.21)$$

Table 3-18 Radiometric and Photometric Terminology
(adapted from [33])

Description	Radiometric	Photometric
Total flux	radiant flux (F) in watts	luminous flux (F) in lumens
Emitted flux density at a source surface	radiant emittance (W) in watts/cm^2	luminous emittance (L) in lumens/cm^2 (lamberts) or lumens/ft^2 (foot-lamberts)
Source intensity (point source)	radiant intensity (I_r) in watts/steradian	luminous intensity (I_L) in lumens/steradian (candela)
Source intensity (area source)	radiance (B_r) in watts/steradian/cm^2	luminance (B_L) in lumens/steradian/cm^2 (lambert)
Flux density incident on a receiver surface	Irradiance (H) in watts/cm^2	illuminance (E) in lumens/cm^2 (candle) or lumens/ft^2 (footcandle)

The area sources are arbitrarily defined as those whose diameter is greater than 10% of the separation distance. A special case that deserves some consideration occurs when radius R of the light source is much larger than the distance r to the sensor. Under this condition

$$H = \frac{B_r A_s}{r^2 + R^2} \approx \frac{B_r A_s}{R^2} \, , \qquad (3.15.22)$$

where A_s is the area of the light source and B_r is the radiance. Since the area of the source $A_s = \pi R^2$, irradiance is:

$$H \approx B_r \pi = W \, , \qquad (3.15.23)$$

that is, the emitted and incident flux densities are equal. If the area of the detector is the same as area of the source, and R>>r, the total incident energy is approximately the same as the total radiated energy, that is, unity coupling exists between the source and the detector. When the optical system is comprised of channeling, collimating, or focusing components, its efficiency and, subsequently, coupling coefficient must be considered. Important relationships for point and area light sources are given in Tables 3-19 and 3-20.

Table 3-19 Point source relationships
(adapted from [33])

Description	Radiometric	Photometric
Point source intensity	I_r, watts/steradian	I_L lumens/steradian
Incident flux density	irradiance, $H = \dfrac{I_r}{r^2}$ watts/m^2	illuminance, $E = \dfrac{I_L}{r^2}$ lumens/m^2
Total flux output of a point source	$P = 4\pi I_r$, watts	$F = 4\pi I_L$, lumens

Table 3-20 Area source relationships
(adapted from [33])

Description	Radiometric	Photometric
Point source intensity	B_r, watts/(cm^2·steradian)	B_L lumens/(cm^2·steradian)
Emitted flux density	$W = \pi B_r$, watts/cm^2	$L = \pi B_L$, lumens/cm^2
Incident flux density	$H = \dfrac{B_r A_s}{r^2 + R^2}$, watts/cm^2	$E = \dfrac{B_L A_s}{r^2 + R^2}$, lumens/cm^2

3.16 OPTICAL COMPONENTS

Light phenomena, such as reflection, refraction, absorption, interference, polarization and speed are the powerful utensils in a designer's toolbox. Optical components help to manipulate light in many ways. In this section, we discuss these components from a standpoint of a geometrical optics. When using geometrical optics we omit properties of light which are better described by quantum mechanics and quantum electrodynamics. We will ignore not only quantum properties of light but the wave properties as well. We consider light as a moving front or a ray which is perpendicular (normal) to that front. To do so, we should not discuss any optical elements whose dimensions are too small as compared with the wavelength. For example, if a glass window is impregnated with small particles of sub-micron sizes, we should completely ignore them for any geometrical calculations from the near-infrared to longer wavelengths[1]. Geometrical optics techniques are useful only for the components whose sizes are much larger than a wavelength of interest. Here, we briefly summarize those optical elements which are applicable for the sensor design. For more detailed discussions of geometrical optics we refer the reader to special texts, for example [34].

3.16.1 Windows

The main purpose of windows is to protect interiors of sensors and detectors from the environment. A good window should transmit light rays in a specific wavelength range with minimal distortions. Therefore, windows should possess appropriate characteristics depending on a particular application. For instance, if an optical detector operates under water, possibly its window should have the following properties: a mechanical strength to withstand water pressure, a low

---

[1]They must be taken into account when energy properties of light passing through such a window are considered. The particles absorb radiant energy resulting in window temperature increase.

water absorption, a transmission band corresponding to the wavelength of interest and an appropriate refractive index. A useful window which can withstand high pressures is spherical as shown in Fig. 3-16.1. To minimize optical distortions, three limitations should be applied to a spherical window: an aperture D (its largest dimension) must be smaller than the window's spherical radius R_1, a thickness d of the window must be uniform and much smaller than radius R_1. If these conditions are not met, the window becomes a concentric spherical lens.

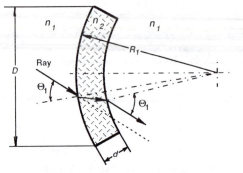

Fig. 3-16.1 Spherical window

A surface reflectivity of a window must be considered for its overall performance. To minimize a reflective loss, windows may be given special antireflective coatings which may be applied on either one or both sides of the window. These are the coatings which give bluish and amber appearances to popular photographic lenses and filters. Due to a refraction in the window (see Fig. 3-15.3), a passing ray is shifted by a distance L which for small angles Θ_1 may be found from formula:

$$L = d\frac{n-1}{n} \,,$$

(3.16.1)

where n is the refractive index of the material.

Sensors operating in the far infrared range require special windows which are opaque in the visible and ultraviolet spectral regions and quite transparent in the wavelength of interest. Several materials are available for fabrication of such windows. Spectral transmittances of some materials are shown in Fig. 3-15.2. When selecting the material for a far infrared window, refractive index must be seriously considered because it determines the coefficient of reflectivity (3.15.8), absorptivity (3.14.20), and eventually transmittance (3.15.12). Fig. 3-16.2 shows spectral transmittances of silicon windows having two different thicknesses. Total radiation (100%) at the window is divided into three portions: reflected (about 50% over the entire spectral range), absorptive (varies at different wavelengths), and transmitted, which is whatever is left after the reflection and absorption.

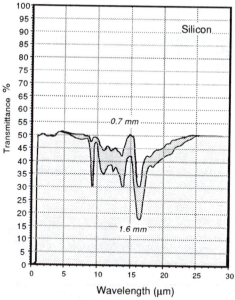

Fig. 3-16.2 Spectral transmittance of a silicon window
Note that majority of loss is due to a reflection from two surfaces

3.16.2 Mirrors

A mirror is the oldest optical instrument ever used or designed. Whenever light passes from one medium to another, there is some reflection. To enhance a reflectivity, a single or multilayer reflecting coating is applied either on the front (first surface) or the rear (second surface) of a plane-parallel plate or other substrate of any desirable shape. The first surface mirrors are the most accurate. In the second surface mirror, light must enter a plate having generally a different index of refraction than the outside medium. Reflecting coatings applied to a surface for operation in the visible and near infrared range can be silver, aluminum, chromium, and rhodium. Gold is preferable for the far infrared spectral range devices. By selecting an appropriate coating, the reflectance may be achieved of any desired value from 0 to 1 (Fig. 3-16.4).

Several effects in the second surface mirror must be taken into consideration. First, due to the refractive index n of a plate, a reflective surface appears closer (Fig. 3-16.3). A virtual thickness L of the carrier for smaller angles Θ_1 may be found from a simple formula:

$$L \approx \frac{d}{n} \, .$$

$$(3.16.2)$$

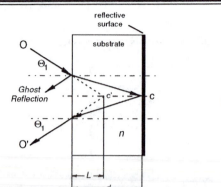

Fig. 3-16.3 Second surface mirror

A front surface of the second surface mirror may also reflect a substantial amount of light creating the so-called ghost reflection. For instance, a glass plate reflects about 4% of visible light. Further, a carrier material may have a substantial absorption in the wavelength of interest. For instance, if a mirror operates in a far infrared spectral range, it should use either first surface metallization or a second surface where the substrate is fabricated of ZnSe or other long wavelength transparent materials. Materials such as Si or Ge have too strong a surface reflectivity to be useful for the fabrication of the second surface mirrors.

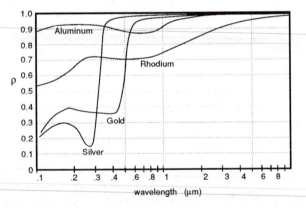

Fig. 3-16.4 Spectral reflectances of some mirror coatings

The best mirrors for broadband use have pure metallic layers, vacuum or electrolytically deposited on glass, fused silica, or metal substrates. Before the reflective layer deposition, to achieve a leveling effect a mirror may be given an under-coat of copper, zirconium-copper or molybdenum.

Another useful reflector which may serve as a second surface mirror without the need for reflective coatings is a prism where the effect of total internal reflection is used. The angle of a total internal reflection is a function of a refractive index:

$$\Theta_o = \arcsin \frac{1}{n} \; . \tag{3.16.3}$$

The total internal reflectors are the most efficient in the visible and near-infrared spectral ranges as the reflectivity coefficient is close to unity.

A reflective surface may be formed practically in any shape to divert the direction of light travel. In the optical systems, curved mirrors produce effects equivalent to that of lenses. The advantages they offer include (1) higher transmission, especially in the longer wavelength spectral range, (2) absence of distortions incurred by refracting surfaces due to dispersion (chromatic aberrations), (3) lower size and weight as compared with many types of lenses. Spherical mirrors are used whenever light must be collected and focused (*focus* is from the Latin meaning *fireplace* - a gathering place in a house). However, spherical mirrors are good only for the parallel or near parallel beams of light that strike a mirror close to normal. These mirrors suffer from imaging defects called aberrations. Fig. 3-16.5A shows a spherical mirror with the center of curvature in point C. A focal point is located at a distance of 1/2 of the radius from the mirror surface. A spherical mirrors is astigmatic, that means that the off-axis rays are focused away from its focal point. Nevertheless, such mirrors prove very useful in detectors where no quality imaging is required, for instance in infrared motion detectors which are covered in detail in Section 6.5.

A parabolic mirror is quite useful for focusing light off-axis. When it is used in this way, there is complete access to the focal region without shadowing, as shown in Fig. 3-16.5B.

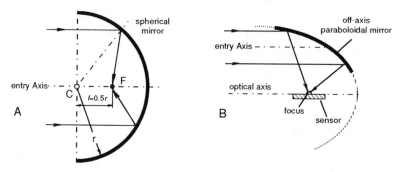

Fig. 3-16.5 Spherical (A) and parabolic (B) first surface mirrors

3.16.3 Lenses

Lenses are useful in sensors and detectors to divert the direction of light rays and arrange them in a desirable fashion. Fig. 3-16.6 shows a plano-convex lens which has one surface spherical and the other is flat. The lens has two focuses at both sides: F and F' which are positioned at equal distances $-f$ and f from the lens. When light rays from object G enters the lens, their directions change according to Snell's law.

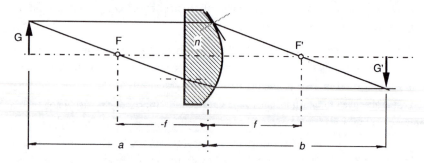

Fig. 3-16.6 Geometry of a plano-convex lens

To determine the size and the position of an image created by the lens, it is convenient to draw two rays that have special properties. One is parallel to the optical axis which is a line passing through the sphere's center of curvature. After exiting the lens that ray goes through focus F'. The other ray first goes though focus F and upon exiting the lens, propagates in parallel with the optical axis. A thin lens whose radius of curvature is much larger than thickness of the lens has a focal distance f which may be found from the equation

$$\frac{1}{f} = (n-1)\left(\frac{1}{r_1} + \frac{1}{r_2}\right) , \tag{3.16.4}$$

where r_1 and r_2 are radii of lens curvature. Image G' is inverted and positioned at a distance b from the lens. That distance may be found from a thin lens equation

$$\frac{1}{f} = \frac{1}{a} + \frac{1}{b} . \tag{3.16.5}$$

For the thick lenses where thickness t is comparable with the radii of curvature a focal distance may be found from formula

$$f = \frac{nr_1r_2}{(n-1)\ln(r_1+r_2) - t(n-1)l} . \tag{3.16.6}$$

Several lenses may be combined into a more complex system. For two lenses separated by a distance d, a combination focal length may be found from equation

$$f = \frac{f_1f_2}{f_1 + f_2 - d} . \tag{3.16.7}$$

3.16.4 Fresnel Lenses

Fresnel lenses are optical elements with step-profiled surfaces. They prove to be very useful in sensors and detectors where a high quality of focusing is not required. Major applications include light condensers, magnifiers, and focusing element in occupancy detectors. Fresnel lenses may be fabricated of acrylic (visible and near infrared range) or polyethylene (far infrared range). History of Fresnel lenses began in 1748, when Count Buffon proposed to grind out of a solid piece of glass a lens in steps of the concentric zones in order to reduce the thickness of the lens to a minimum. In 1822, Augustin Fresnel constructed a lens in which the centers of curvature of the different rings receded from the axis according to their distances from the center, so as to practically eliminate spherical aberration.

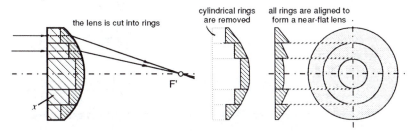

Fig. 3-16.7 Concept of a Fresnel lens

A concept of that lens is illustrated in Fig. 3-16.7 where a regular plano convex lens is depicted. The lens is sliced into several concentric rings. After slicing, all rings still remain lenses which refract incident rays into a common focus defined by (3.16.4). A change in an angle occurs when a ray exits a curved surface. A section of a ring which is marked by letter x doesn't contribute into focusing properties. If all such sections are removed, the lens will look like it is shown in Fig. 3-16.7. Now, all the rings may be shifted with respect to one another to align their flat surfaces. A resulting near-flat lens is called Fresnel which has the same focusing properties as the original plano-convex lens. A Fresnel lens basically consists of a series of concentric prismatic grooves, designed to cooperatively direct incident light rays into a common focus.

The Fresnel lens has several advantages over a conventional lens, such as low weight, thin size, ability to be curved (for a plastic lens) to any desirable shape, and the most important - lower loss of light flux. The last feature is very important for fabrication of far infrared lenses where absorption in the material may be very high. This is the reason, why polymer Fresnel lenses are used almost exclusively in the far infrared motion detectors.

Two common types of Fresnel lenses are presently manufactured. One is a constant step lens (Fig. 3-16.8A) and the other is a constant depth lens (Fig. 3-16.8B). In practice, it is difficult to maintain a curved surface of each small groove, hence the profile of a groove is approximated by a flat surface. This demands that the steps are positioned close to each other. In fact, the closer the steps the more accurate is the lens.

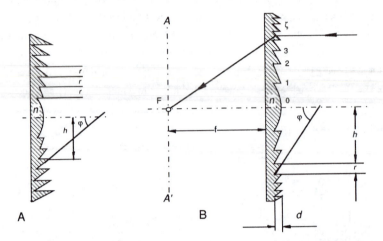

Fig. 3-16.8 Constant step (A) and constant depth (B) Fresnel lenses

In a constant step lens, a slope angle φ of each groove is a function of its distance h from the optical axis. As a result, the depths of the grooves increase with the distance from the center. A central portion of the lens may be flat if its diameter is at least 20 times smaller than the focal length. For the shorter focal lengths, it is a good practice to maintain a spherical profile of a central portion. A slope angle of each step may be determined from the following formula which is valid for small values of h

$$\varphi = \arctan \frac{hn}{f(n-1)} \ , \tag{3.16.8}$$

where f is the focal length.

For a constant depth lens, both the slope angle φ and the step distance r vary with the distance from the center. The following equations may be useful for the lens calculation. A distance of a groove from the center may be found through the groove number ξ (assuming the center portion has number 0)

$$h = \sqrt{2f(n-1)\xi d - \xi^2 d^2} \ , \tag{3.16.9}$$

and the slope angle is

$$\phi = \arcsin \frac{h}{(n-1)f} \qquad (3.16.10)$$

A total number of grooves in the lens may be found through a Fresnel lens aperture (maximum dimension) D:

$$\Gamma = \frac{(n-1)f - \sqrt{f^2(n-1)^2 - D^2}}{d} \quad . \qquad (3.16.11)$$

A Fresnel lens may be slightly bent if it is required for a sensor design. However, a bend changes positions of the focal points. If a lens is bent with its groves inside the curvature, a focal distance decreases.

It is known that a spherical surface of a lens will produce a spherical aberration. Therefore, for the applications, where high quality focusing is required, the continuous surface from which the contours of the groves are determined should not be spherical, but aspherical. The profile of a continuous aspherical surface can often be closely described by a standard equation of a conic, axially symmetrical about the z axis (Fig. 3-16.9):

$$Z = \frac{CY^2}{1 + \sqrt{1 - (K + 1)C^2 Y^2}} \, , \qquad (3.16.12)$$

where Z and Y are the coordinates of the surface C is the vertex curvature, and K is the conic constant. The vertex curvature and the conic constant can be chosen depending upon the desired characteristics of the lens, and the contours of each grove can be figured using this equation. C and K will depend upon several factors, such as the desired focal length, index of refraction, and the particular application.

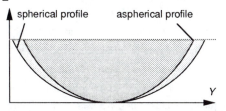

Fig. 3-16.9 Comparison of spherical and aspheric lens profiles

3.16.5 Fiber optics and waveguides

While light doesn't go around the corner, it can be channeled along complex paths by the use of waveguides. To operate in the visible and near-infrared spectral range, the guides may be fabricated of glass or polymer fibers. For the middle and far infrared ranges the waveguides are made as hollow tubes with highly reflective inner surfaces. A waveguide operates on a principle of the internal reflections where light beams travel in a zigzag pattern. A fiber can be used to transmit light energy in the otherwise inaccessible areas without any transport of heat from the light source. The surface and ends of a round or other cross section fiber are polished. An outside cladding may be added. When glass is hot, the fibers can be bent to curvature radii of 20 to 50 their section diameter and after cooling, to 200-300 diameters. Plastic fibers fabricated of polymethyl methacrylate may be bent at much smaller radii than glass fibers. A typical attenuation for a 0.25 mm polymer fiber is in the range of 0.5dB per each meter of length. Light propagates through a fiber by means of a total internal reflection as shown in Fig. 3-16.10B. It follows from formula (3.16.3) that light passing to air from medium having refractive index n is subject to the limitation of an angle of total internal reflection. In a more general form, light may pass to another medium (cladding) having refractive index n_1, then Eq. (3.16.3) becomes

$$\Theta_0 = \arcsin \frac{n_1}{n} . \qquad (3.16.13)$$

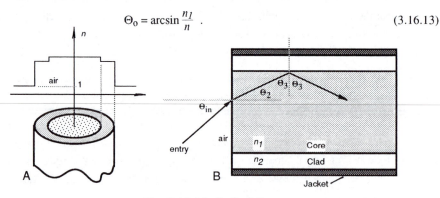

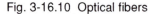

Fig. 3-16.10 Optical fibers
A: a step-index multiple fiber; B: to determination of maximum angle of entry

Fig. 3-16.10A shows a profile of index of refraction for a single fiber with the cladding where the cladding must have a lower index of refraction to assure a total internal reflection at the boundary. For example, a silica-clad fiber may have compositions set so that the core (fiber) material has an index of refraction of 1.5, and the clad has an index of refraction of 1.485. To protect the clad fiber, it is typically enclosed in some kind of a protective rubber or plastic jacket. This type of the fiber is called a "step index multimode" fiber which refers to the profile of the index of refraction.

When light enters the fiber, it is important to determine the maximum angle of entry which will result in total internal reflections (Fig. 3-16.10B). If we take that minimum angle of an internal reflection $\Theta_0 = \Theta_3$, then maximum angle Θ_2 can be found from Snell's law:

$$\Theta_{2(max)} = \arcsin \frac{\sqrt{n^2 - n_1^2}}{n} .$$

(3.16.14)

Applying Snell's law again and remembering that for air $n \approx 1$ we arrive at

$$\sin\Theta_{in(max)} = n_1 \sin\Theta_{2(max)} .$$

(3.16.15)

Combining Eqs. (3.16.14) and (3.16.15) we obtain the largest angle with the normal to the fiber end for which the total internal reflection will occur in the core

$$\Theta_{in(max)} = \arcsin \sqrt{n^2 - n_1^2}$$

(3.16.16)

Light rays entering the fiber at angles greater than $\Theta_{in(max)}$ will pass through to the jacket and will be lost. If for data transmission, this would be an undesirable event, in a specially designed fiber optic sensor, the maximum entry angle can be a useful phenomenon for modulating light intensity.

Sometimes, value $\Theta_{in(max)}$ is called a numerical aperture of the fiber. Due to variations in the fiber properties, bends and skewed paths, the light intensity doesn't drop to zero abruptly but rather gradually diminishes to zero while approaching $\Theta_{in(max)}$. In practice, the numerical aperture is defined as the angle where light intensity drops by some arbitrary number, for instance -10dB of the maximum value.

One of the useful properties of the fiber optic sensors is that they can be formed into a variety of geometrical shapes depending on the desired application. They are very useful for the design of miniature optical sensors which are responsive to such stimuli, as pressure, temperature, chemical concentration, etc. The basic idea for use of fiber optics in sensing is to modulate one or several characteristics of light in a fiber, and subsequently, to optically demodulate the information by conventional methods. A stimulus may act on a fiber either directly or it can be applied to a component attached to the fiber's outer surface or the polished end to produce an optically detectable signal.

To make a fiber chemical sensor, a special solid phase of a reagent may be formed in the optical path coupled to the fiber. The reagent interacts with the analyte to produce an optically detectable effect, for instance, to modulate the index of refraction or coefficient of absorption. A cladding on a fiber may be created of a chemical substance whose refractive index may be changed in the presence of some fluids [35]. When angle of total internal reflection changes, the light intensity varies as well.

Optical fibers may be used in two modes. In the first mode (Fig. 3-16.11A), the same fiber is used to transmit the excitation signal and to collect and conduct an optical response back to the processing device. In the second mode, two or more fibers are employed where excitation (illumination) function and collection function are carried out by separate fibers (Fig. 3-16.11B).

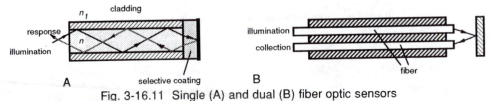

Fig. 3-16.11 Single (A) and dual (B) fiber optic sensors

The most commonly used type of a fiber optic sensor is an intensity sensor, where light intensity is modulated by an external stimulus [36]. Fig. 3-16.12 shows a displacement sensor where a single-fiber waveguide emits light toward the reflective surface. Light travels along the fiber and exits in a conical profile toward the reflector. If the reflector is close to the fiber end (distance d), most of light is reflected into the fiber and propagates back to the light detector at the other end of the fiber. If the reflector moves away, some of the rays are reflected outside of the fiber end, and less and less photons are returned back. Due to a conical profile of the emitted light, a quasi-linear relationship between the distance d and the intensity of the returned light can be achieved over a limited range.

Fig. 3-16.12 Fiber-optic displacement sensor

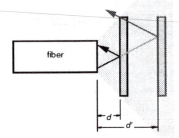

The so-called microbend strain gauge can be designed with an optical fiber which is squeezed between two deformers, as shown in Fig. 3-16.13A. External force applied to the upper deformer bends the fiber affecting a position of an internal reflective surface. Thus, a light beam which normally would be reflected in direction x approaches the lower part of the fiber at an angle which is less than Θ_0 - the angle of total internal reflection [Eq. (3.16.13)]. Thus, instead of being reflected, light is refracted and moves in the direction y through the fiber wall. The closer the deformers come to each other, the more light goes astray and less light is transmitted along the fiber.

An interesting new class of optical fibers fabricated from silicone rubber (SR) may find use for design of various sensors. The experimental work at Oak Ridge National Laboratory (Oak

Ridge, Tenn.) has demonstrated that fibers made of a silicone rubber can be used for sensing weight, pressure, humidity, force, strain, displacement and other variables. The SR fiber can be stretched up to 100% of its original length modulating the light intensity almost proportionally. That is, a doubling the length reduces the light intensity at the fiber output by about 50%. After being subjected to such intense deformations, the fibers did not show signs of permanent deterioration and return to their original size and shape. An experimental SR fiber was fabricated of 2 mm in diameter, which is substantially thicker than the conventional glass or polymer fibers. The large cross sectional area yields a high light transmission efficiency, however, substantially it is smaller than in the traditional fibers. The silicone rubber is reasonably transparent in the wavebands of 550-700, 750-820, and 940-970nm. Outside these relatively narrow bands, the material absorbs heavily. A change in the transmission is near 5% over the temperature range from 0 to 100°C. A transmittance is moisture dependent, if the fiber not coated with a waterproof jacket.

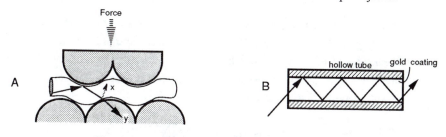

Fig. 3-16.13 Fiber-optic micro-bend strain gauge (A)
and a wave guide for the far infrared radiation (B)

Advantages of using the SR fibers for sensing is in their ability to modulate a light transmittance in response to an applied pressure and elongation. As a moisture sensor, an unprotected fiber can be exposed to water molecules to change the transmission coefficient. At the time of this writing, however, the SR fiber has demonstrated a strong moisture sensitivity only over 70%RH [37].

For operation in the spectral range where loss in fibers is too great, hollow tubes are generally used for the light channeling (Fig. 3-16.13B). The tubes are highly polished inside and coated with reflective metals. For instance, to channel thermal radiation, a tube may be fabricated of brass and coated inside by two layers: nickel as an under-layer to level the surface, and the optical quality gold having thickness in the range of 500-1000Å. Hollow waveguides may be bent to radii of 20 or more of their diameters. While fiber optics use the effect of the total internal reflection, tubular waveguides use a first surface mirror reflection, which is always less than 100%. As a result, loss in a hollow waveguide is a function of a number of reflections. That is, loss is higher for the smaller diameter and the longer length of a tube. At length/diameter ratios more than 20, hollow waveguides become quite inefficient.

3.16.6   Concentrators

Regarding optical sensors and their applications, there is an important issue of the increasing density of the photon flux impinging on the sensor's surface. In many cases, when only the energy factors are of importance, and a focusing or imaging is not required, special optical devices can be used quite effectively. These are the so-called nonimaging collectors, or concentrators [38]. They have some properties of the waveguides and some properties of the imaging optics (like lenses and curved mirrors). The most important characteristic of a concentrator is the ratio of the area of the input aperture divided by the area of the output aperture. The concentration ratio C is always more than unity. That is, the concentrator collects light from a larger area and directs it to a smaller area (Fig. 3-16.14A) where the sensing element is positioned. There is a theoretical maximum for C

$$C_{max} = \frac{1}{\sin^2 \Theta_i} , \tag{3.16.17}$$

where Θ_i is the maximum input semiangle. Under these conditions, the light rays emerge at all angles up to $\pi/2$ from the normal to the exit face. This means that the exit aperture diameter is smaller by $\sin\Theta_i$ times the input aperture. This gives an advantage in the sensor design as its linear dimensions can be reduced by that number while maintaining a near equal efficiency. The input rays entering at angle Θ will emerge as an output cone.

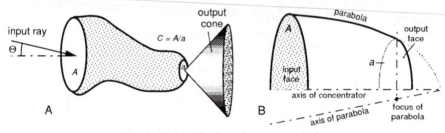

Fig. 3-16.14 Nonimaging concentrator
A: a general schematic; B: a concentrator having a parabolic profile

The concentrators can be fabricated with reflective surfaces (mirrors) or refractive bodies (lenses), or as combinations of both. A practical shape of the reflective parabolic concentrator is shown in Fig. 3-16.14B. It is interesting to note that the cone receptors of human retina have a shape similar to that shown in Fig. 3-16.14B [39]. The tilted parabolic concentrators have very high efficiency[1] - they can collect well over 90% of the incoming radiation. If a lesser efficiency is acceptable, a conical concentrator can be employed. Some of the incoming rays will be turned back

---

[1] This assumes that reflectivity of the inner surface of the concentrator is ideal.

after several reflections inside the cone, however, its overall efficiency is still near 80%. Clearly, the cones are much easier to fabricate than the paraboloids of revolution.

3.16.7 Electrooptic and acoustooptic modulators

One of the essential steps of a stimulus conversion in optical sensors is their ability to modify light in some way, for instance to alter its intensity by a control signal. This is called a modulation of light. The control signal can have different origins: temperature, chemical compounds with different refractive indices, electric filed, mechanical stress, etc. Here, we examine light modulation by electric signals and acoustic waves.

In some crystals, refractive index can be linked to an applied electric field [40]. The effect is characterized in the context of propagation of a light beam through a crystal. For an arbitrary propagation direction, light maintains constant linear polarization through a crystal for only those polarization directions allowed by the crystal symmetry. An external electric field applied to a crystal may change that symmetry, thus modulating the light intensity. Lithium niobate ($LiNbO_3$) is one of the most widely used materials for the electrooptic devices. A crystal is positioned between two polarizing filters which are oriented at 90° with respect to one another (Fig. 3-16.15). The input polarizer is oriented at 45° to the axis of the modulating crystal [41]. The crystal modulator has two electrodes attached to its surface. By changing the modulator voltage, the polarization of the light incident on the output polarizer is varied, which, in turn, leads to the intensity modulation.

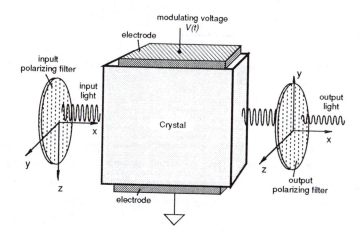

Fig. 3-16.15 Electrooptic modulator consists of two polarizing filters and a crystal

A similar effect can be observed when the crystal is subjected to mechanical effects, specifically - to acoustic waves [40, 42]. However, acoustooptic devices are used most often in fiber-optic applications as optical frequency shifters, and only to a lesser extent as intensity modulators. In the modulator, the light beam propagating through a crystal interacts with a traveling-wave index perturbation generated by an acoustic wave. The perturbation results from a photoelastic effect, whereby a mechanical strain produces a linear variation in refractive index. This resembles a traveling-wave diffraction grating, which under certain conditions can effectively deflect an optical beam (Fig. 3-16.16). Acoustooptic devices are often fabricated from lithium niobate and quartz, since acoustic waves can effectively propagate through these crystal over a frequency range from tens of megahertz to several gigahertz. The acoustic velocity in lithium niobate is about $6 \cdot 10^3$ m/s, thus 1GHz acoustic wave has a wavelength of about 6 µm, which is comparable to light in the infrared spectral range.

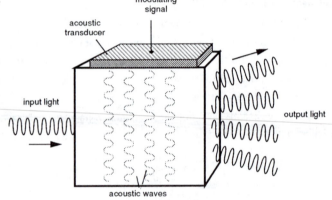

Fig. 3-16.16 Acoustooptic modulator produces multiple diffracted beams

3.16.8 Interferometric fiber-optic modulation

Light intensity in an optical fiber can be modulated to produce a useful output signal. Fig. 3-16.17 illustrates an optical waveguide which is split into two channels [43]. The waveguides are formed in a $LiNbO_3$ substrate doped with Ti to increase its refractive index. They are fabricated by a standard photolithographic liftoff technique. A substrate is patterned using a photomask. A Ti layer is electron-beam evaporated over the material, then the photoresist is removed by a solvent, leaving the Ti in the waveguide pattern. The Ti atoms later diffused into the substrate by baking [41]. This process results in a graded refractive index profile with a maximum difference at the surface of about 0.1% higher than the bulk value. Light is coupled to the waveguides through polished endfaces. Electrodes are positioned in parallel to the waveguides, which recombine at the output. Voltage across the electrodes produces a significant phase shift in the light waves.

The optical transmission of the assembly varies sinusoidally with the phase shift $\Delta\phi$ between two recombined signals, which is controlled by voltage $V(t)$:

$$\frac{P_{out}}{P_{in}} = \frac{1}{2}\left\{1 + \cos\left[\frac{\pi V(t)}{V_\pi} + \Theta_B\right]\right\} \quad , \tag{3.16.18}$$

where V_π is the voltage change required for the full on-off modulation, and Θ_B is the constant which can be adjusted for the optimum operating point. When the phase difference between the light in two waveguides before the recombination $\Delta\phi=0$, the output couples to the output waveguide. When the shift $\Delta\phi=\pi$, light propagates into the substrate. The well designed modulators can achieve a high extinction ratio on the order of 30 dB.

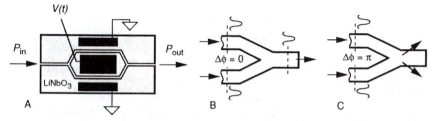

Fig. 3-16.17 Channel-waveguide interferometric intensity modulator (A).
Light recombines in the exit fiber when the phase shift is zero (B).
Light radiates into the substrate when the phase shift is π (C).
(Adapted from [41])

3.17 SILICON AS A SENSING MATERIAL

Silicon is present in the sun and stars and is a principle component of a class of meteorites known as *aerolites*. Silicon is the second most abundant material on Earth, being exceeded only by oxygen - it makes up to 25.7% of the earth's crust, by weight. Silicon is not found free in nature, but occurs chiefly as the oxide, and as silicates. Some oxides are sand, quartz, rock crystal, amethyst, clay, mica, etc. Silicon is prepared by heating silica and carbon in an electric furnace, using carbon electrodes. There are also several other methods for preparing the element.

Crystalline silicon has a metallic luster and grayish color[1]. The Czochralski process is commonly used to produce single crystals of silicon used for the solid-state semiconductors and micro-machined sensors. Silicon is a relatively inert element, but it is attacked by halogens and dilute alkali. Most acids, except hydrofluoric, do not affect it. Elemental silicon transmits infrared radiation and is commonly used as windows in far infrared sensors.

Silicon atomic weight is 28,0855±3, and its atomic number is 14. Its melting point is 1410°C and boiling point is 2355°C. Specific gravity at 25°C is 2.33 and valence is 4.

Properties of silicon are well studied and its applications to sensor designs have been extensively researched around the world. The material is inexpensive and can now be produced and processed controllably to unparalleled standards of purity and perfection. Silicon exhibits a number of physical effects which are quite useful for sensor applications.

Silicon based sensors can be responsive to [44]:

Radiant Stimuli	Photovoltaic effect, photoelectric effect, photoconductivity, photomagneto-electric effect.
Mechanical Stimuli	Piezoresistivity, lateral photoelectric effect, lateral photovoltaic effect
Thermal Stimuli	Seebeck effect, temperature dependence of conductivity and junction, Nernst effect
Magnetic Stimuli	Hall effect, magneto-resistance, Suhi effect
Chemical Stimuli	Ion-sensitivity

Unfortunately, silicon does not posses the piezoelectric effect. Most effects such as the Hall effect, the Seebeck effect, the piezo-resistance, etc. are quite large, however a major problem with silicon is that its response to many stimuli show substantial temperature sensitivity. For instance: strain, light, and magnetic field responses are temperature dependent. When silicon does not display the proper effect, it is possible to deposit layers of materials with the desired sensitivity on top of the silicon substrate. For instance, sputtering of ZnO thin films is used to form piezoelectric transducers which are useful for fabrication of SAW (surface acoustic waves) devices and accelerometers. In the later case, the strain at the support end of the an etched micromechanical cantilever is detected by a ZnO overlay.

---

[1] Silicon should not be confused with *silicone* which is made by hydrolyzing silicon organic chloride, such as dimethyl silicon chloride. Silicones are used as insulators, lubricants, and for production of silicone rubber.

Silicon itself exhibits very useful mechanical properties which nowadays are widely used to fabricate such devices as pressure transducers, temperature sensors, force and tactile detectors by employing the so-called *micromachining*. Thin film and photolithographic fabrication procedures make it possible to realize a great variety of extremely small, high precision mechanical structures using the same processes that have been developed for electronic circuits. High-volume batch-fabrication techniques can be utilized in the manufacture of complex, miniaturized mechanical components which may not be possible with other methods. Table 3-21 presents a comparative list of mechanical characteristics of silicon and other popular crystalline materials.

Table 3-21 Mechanical properties of some crystalline materials.
(From [45])

Material	Yield Strength 10^{10} dyne/cm^2	Knoop Hardness kg/mm^2	Young's Modulus 10^{12} dyne/cm^2	Density g/cm^3	Thermal Conductivity W/cm°C	Thermal Expansion 10^{-6}/°C
Diamond*	53	7000	10.35	3.5	20.0	1.0
SiC*	21	2480	7.0	3.2	3.5	3.3
TiC*	20	2470	4.97	4.9	3.3	6.4
Al$_2$O$_3$*	15.4	2100	5.3	4.0	0.5	5.4
Si$_3$N$_4$*	14	3486	3.85	3.1	0.19	0.8
Iron*	12.6	400	1.96	7.8	0.803	12.0
SiO$_2$ (fibers)	8.4	820	0.73	2.5	0.014	0.55
Si*	7.0	850	1.9	2.3	1.57	2.33
Steel (max. strength)	4.2	1500	2.1	7.9	0.97	12.0
W	4.0	485	4.1	19.3	1.78	4.5
Stainless Steel	2.1	660	2.0	7.9	0.329	17.3
Mo	2.1	275	3.43	10.3	1.38	5.0
Al	0.17	130	0.70	2.7	2.36	25.0

*Single crystal

Although single-crystal silicon (SCS) is a brittle material, yielding catastrophically (not unlike most oxide-based glasses) rather than deforming plastically (like most metals), it certainly is not as fragile as is often believed. The Young's modulus of silicon ($1.9 \cdot 10^{12}$ dyne/cm or $27 \cdot 10^6$ psi), for example, has a value of that approaching stainless steel and is well above that of quartz and most of glasses. The misconception that silicon is extremely fragile is based on the fact that it is often obtained in thin slices (5-13 cm diameter wafers) which are only 250 to 500µm thick. Even stainless steel at these dimensions is very easy to deform inelastically.

As mentioned above, many of the structural and mechanical disadvantages of SCS can be alleviated by deposition of thin films. Sputtered quartz, for example, is utilized routinely by industry to passivate integrated circuit chips against airborne impurities and mild atmospheric corrosion effects. Another example is a deposition of silicon nitrate (see Table 3-21) which has a hardness second only to diamond. Anisotropic etching is a key technology for the micromachining of miniature three dimensional structures in silicon. Two etching systems are of practical interest. One based on ethylenediamine and water with some additives. The other consists of purely inorganic alkaline solutions like KOH, NaOH, or LiOH.

Forming the so-called polysilicon (PS) materials allows us to develop sensors with unique characteristics. Polysilicon layers (on the order of 0.5 μm) may be formed by vacuum deposition onto oxided silicon wafer with an oxide thickness of about 0.1 μm [46]. Polysilicon structures are doped with boron by a technique known in the semiconductor industry as LPCVD (low-pressure chemical vapor deposition). The simple process of surface micromachining is illustrated in Fig. 3-17.1. Let us assume that we would like to form a certain structure over the silicon substrate surface. To do that, a sacrificial layer is deposited on the silicon substrate, which may have been coated first with an isolation layer. Windows are opened in the sacrificial layer and the microstructural thin film is deposited and etched. Selective etching of the sacrificial layer leaves a free-standing micro-mechanical structure.

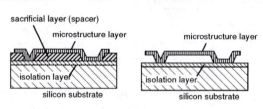

Fig. 3-17.1 Example of a micromachined structure
(adapted from [47])

Fig. 3-17.2A shows resistivity of boron doped LPCVD polysilicon in a comparison with SCS. Resistivity of PS layers is always higher than that of a single crystal material, even when the boron concentration is very high. At low doping concentrations the resistivity climbs rapidly, so that only the impurity concentration range is of interest to sensor fabrication. The resistance change of PS with temperature is not linear. The temperature coefficient of resistance may be selected over a wide range, both positive and negative, through selected doping (Fig. 3-17.2B). Generally, the temperature coefficient of resistance increases with decreased doping concentration. Resistance at any given temperature of a PS layer may be found from

$$R(T) = R_{20}e^{\alpha_R(T-T_o)} \ ,$$ \hfill (3.17.1)

where $\alpha_R = \dfrac{1}{R_{20}} \dfrac{dR(T_0)}{dT}$ is the temperature coefficient and R_{20} is the resistance at calibrating point ($T_0{=}20°C$). Fig. 3-17.3A shows that the temperature sensitivity of PS is substantially higher than that of SCS and can be controlled by doping. It is interesting to note that at a specific doping concentration, the resistance becomes insensitive to temperature variations (point Z).

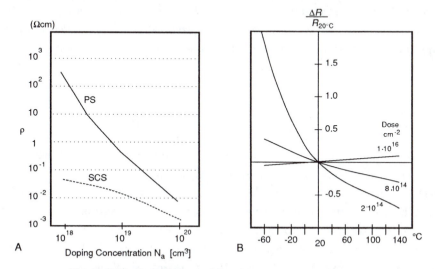

Fig. 3-17.2 Specific resistivity of boron doped silicon (A); temperature coefficient of resistivity of silicon for different doping concentrations (B)

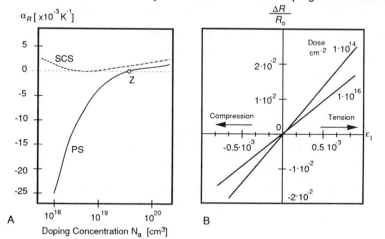

Fig. 3-17.3 Temperature coefficient as a function of doping (A) and piezoresistive sensitivity of silicon (B)

For the development of sensors for pressure, force, or acceleration, it is critical to know the strain sensitivity of PS resistors expressed through the gauge factor. Fig. 3-17.3B shows curves of the relative resistance change of boron dopes PS resistors, referenced to the resistance value R_0 under no-stress conditions, as a function of longitudinal strain ε_1. The parameter varies with the implantation dose. It can be seen that the resistance decreases with compression and increases under tension. It should be noted that the gauge factor (a slope of the line in Fig. 3-17.3B) is temperature dependent. PS resistors are capable of realizing at least as high a level of long term stability as any can be expected from resistors in SCS, since surface effects play only a secondary role in device characteristics.

3.18 DYNAMIC MODELS OF SENSOR ELEMENTS

To determine a sensor's dynamic response a variable stimulus should be applied to its input while observing the output values. Generally, a test stimulus may have any shape or form, which should be selected depending on a practical need. For instance, while determining a natural frequency of an accelerometer, sinusoidal vibrations of different frequencies are the best. On the other hand, for a thermistor probe, a step-function of temperature would be preferable. In many other cases, a step or square-pulse input stimulus is often employed. The reason for that is a theoretically infinite frequency spectrum of a step function. That is, the sensor simultaneously can be tested at all frequencies.

Mathematically, a sensor can be described by a differential equation whose order depends on the sensor's physical nature and design. There are three general types of relationships between the input s and the output S: a zero-order, a first-order and a second-order responses.

A *zero-order* is a static or time independent characteristic

$$S(t) = Gs(t) \ , \tag{3.18.1}$$

where G is a constant transfer function. This relationship may take any form, for instance, described by equations (2.1) through (2.4). The important point is that G is not a function of time. That is, a zero-order response to a step function is a step function.

A *first-order* response is characterized by a first-order differential equation

$$a_1 \frac{dS(t)}{dt} + a_0 S(t) = s(t) \ , \tag{3.18.2}$$

where a_1 and a_0 are constants. This equation characterizes a sensor which can store energy before dissipating it. An example of such a sensor is a temperature sensor which has a thermal capacity and is coupled to the environment through a thermal resistance. A first-order response to a step function is exponential:

$$S(t) = S_O(1 - e^{-t/\tau}) \ , \tag{3.18.3}$$

where S_O is a sensor's static response and τ is a time constant which is a measure of inertia. A typical first-order response is shown in Fig. 2-9B.

A second-order response is characterized by a second-order differential equation

$$a_2\frac{d^2S(t)}{dt^2} + a_1\frac{dS(t)}{dt} + a_0S(t) = s(t) \ . \tag{3.18.4}$$

This response is specific for a sensor or a system which contains two components which may store energy, for instance, an inductor and a capacitor, or a temperature sensor and a capacitor. A second-order response contains oscillating components and may lead to instability of the system. A typical shape of the response is shown in Fig. 2-10. A dynamic error of the second-order response depends on several factors, including its natural frequency ω_0 and damping coefficient b. A relationship between these values and the independent coefficients of (3.18.4) are the following

$$\omega_0 = \sqrt{\frac{a_0}{a_2}} \ , \tag{3.18.5}$$

$$b = \frac{a_1}{2\sqrt{a_0 a_2}} \ . \tag{3.18.6}$$

A critically damped response (see Fig. 2-10) is characterized by $b=1$. The overdamped response has $b>1$ and the underdamped has $b<1$.

For a more detailed description of dynamic responses the reader should refer to specialized texts, for instance [48].

Mathematical modeling of a sensor is a powerful tool in assessing its performance. The modeling may address two issues: static and dynamic. Static models usually deal with the sensor's transfer function as it is defined in Chapter 2. Here we briefly outline how sensors can be evaluated dynamically. The dynamic models may have several independent variables, however, one of them must be time. The resulting model is referred to as a lumped parameter model. In this section, mathematical models are formed by applying physical laws to some simple lumped parameter sensor elements. In other words, for the analysis, a sensor is separated into simple elements and each element is considered separately. However, once the equations describing the elements have been formulated, individual elements can be recombined to yield the mathematical model of the original sensor. The treatment is intended not to be exhaustive, but rather to introduce the topic.

3.18.1 Mechanical Elements

Dynamic mechanical elements are made of masses, or inertias, which have attached springs and dampers. Often the damping is viscous, and for the rectilinear motion the retaining force is proportional to velocity. Similarly, for the rotational motion, the retaining force is proportional to angular velocity. Also, the force, or torque, exerted by a spring, or shaft, is usually proportional to displacement. The various elements and their governing equations are summarized in Table 3-22.

One of the simplest methods of producing the equations of motion is to isolate each mass or inertia and to consider it as a free body. It is then assumed that each of the free bodies is displaced from the equilibrium position, and the forces or torques acting on the body then drive it back to its equilibrium position. Newton's second law of motion can then be applied to each body to yield the required equation of motion.

For a rectilinear system Newton's second law indicates that for a consistent system of units *the sum of forces equals to the mass times the acceleration.* In the SI system of units, force is measured in newtons (N), mass in kilograms (kg), and acceleration in meters per second squared (m/s^2).

For a rotational system, Newton's law becomes: *the sum of the moments equals the moment of inertia times the angular acceleration.* The moment, or torque, has units of newton-meters (Nm), the inertia units of kilogram per meter squared (kg/m^2) and the angular acceleration units of radians per second squared (rad/s^2).

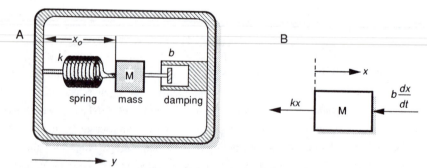

Fig. 3-18.1 Mechanical model of an accelerometer (A)
and a free-body diagram of mass (B)

Let us consider a monoaxial accelerometer which consists of an inertia element whose movement may be transformed into an electric signal. The mechanism of conversion may be, for instance, piezoelectric. Fig. 3-18.1A shows a general mechanical structure of such an accelerometer. Mass M is supported by a spring having stiffness k and the mass movement is damped by a damping element with a coefficient b. Mass may be displaced with respect to the accelerometer housing only in the horizontal direction. During operation, the accelerometer case is subjected to acceleration d^2y/dt^2, and the output signal is proportional to the deflection x_o of the mass M.

Since the accelerometer mass M is constrained to linear motion, the system has one degree of freedom. Giving the mass M a displacement x from its equilibrium position produces the free-body diagram shown in Fig. 3-18.1B. Note that x_o is equal to x plus some fixed displacement. Applying Newton's second law of motion gives

$$Mf = -kx - b\frac{dx}{dt} \; , \tag{3.18.7}$$

where f is the acceleration of the mass relative to the Earth and is given by

$$f = \frac{d^2x}{dt^2} - \frac{d^2y}{dt^2} \; . \tag{3.18.8}$$

Substituting for f gives the required equation of motion as:

$$M\frac{d^2x}{dt^2} + b\frac{dx}{dt} + kx = M\frac{d^2y}{dt^2} \; . \tag{3.18.9}$$

Note that each term in the above equation has units of newtons (N). The differential equation (3.18.9) is of a second order which means that the accelerometer output signal may have the oscillating shape. By selecting an appropriate damping coefficient b the output signal may be brought to a critically damped state which, in most cases, is a desirable response.

3.18.2 Thermal Elements

Thermal elements include such things as heat sinks, heating elements, insulators, heat reflectors, and absorbers. If heat is of a concern, a sensor should be regarded as a component of a larger device. In other words, heat conduction through the housing and the mounting elements, air convection and radiative heat exchange with other objects should not be discounted.

Heat may be transferred by three mechanisms: conduction, natural and forced convection, and thermal radiation (Section 3.14). For simple lumped parameter models, the first law of thermodynamics may be used to determine the temperature changes in a body. The rate of change of a body's internal energy is equal to the flow of heat into the body less the flow of heat out of the body, very much like fluid moves through pipes into and out of a tank. This balance may be expressed as

$$C\frac{dT}{dt} = \Delta Q \ ,$$ (3.18.10)

where $C=Mc$ is the thermal capacity of a body (J/K), T is the temperature (K), ΔQ is the heat flow rate (W), M is the mass of the body (kg), and c is the specific heat of the material (J/kg·K). The heat flow rate through a body is a function of the thermal resistance of the body. This is normally assumed to be linear, and therefore

$$\Delta Q = \frac{T_1 - T_2}{R} \ ,$$ (3.18.11)

where R is the thermal resistance (K/W) and $T_1–T_2$ is a temperature gradient across the element, where heat conduction is considered.

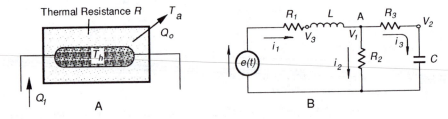

Fig. 3-18.2 Thermal model of a heating element (A); an electrical circuit diagram (B) with resistive, capacitive and inductive components

For the illustration, we analyze a heating element (Fig. 3-18.2A) having temperature T_h. The element is coated with insulation. The temperature of the surrounding air is T_a. Q_1 is the rate of heat supply to the element, and Q_O is the rate of heat loss. From Eq. (3.18.10)

$$C\frac{dT_h}{dt} = Q_1 - Q_0 \ ,$$ (3.18.12)

but, from Eq. (3.18.11)

$$Q_0 = \frac{T_h - T_a}{R} \ ,$$ (3.18.13)

and, in the result, we obtain a differential equation

$$\frac{dT_h}{dt} + \frac{T_h}{RC} = \frac{Q_1}{C} + \frac{T_a}{RC} \ ,$$ (3.18.14)

This is a first order differential equation which is typical for thermal systems. A thermal element, if not a part of a control system with a feedback loop, is inherently stable. A response of a simple thermal element may be characterized by a thermal time constant which is a product of thermal capacity and thermal resistance: $\tau_T = CR$. The time constant is measured in units of time (sec) and, for a passively cooling element, is equal to time which takes to reach about 37% of the initial temperature gradient.

3.18.3 Electrical Elements

There are three basic electrical elements: the capacitor, the inductor, and the resistor. Again, the governing equation describing the idealized elements are given in Table 3-22. For the idealized elements, the equations describing the sensor's behavior may be obtained from Kirchhoff's laws which directly follow from the law of conservation of energy:

Kirchhoff's 1st law: The total current flowing toward a junction is equal to the total current flowing from that junction, i.e., the algebraic sum of the currents flowing toward a junction is zero.

Kirchhoff's 2nd law: In a closed circuit, the algebraic sum of the voltages across each part of the circuit is equal to the applied e.m.f.

Let us assume that we have a sensor whose elements may be represented by a circuit shown in Fig. 3-18.2B. To find the circuit equation, we will use the 1st Kirchhoff's law, which sometimes is called Kirchhoff's current law. For the node, A

$$i_1 - i_2 - i_3 = 0 \; , \tag{3.18.15}$$

and for each current

$$i_1 = \frac{e - V_3}{R_1} = \frac{1}{L}\int (V_3 - V_1)dt$$

$$i_2 = \frac{V_1 - V_2}{R_3} = C\frac{dV_2}{dt}$$

$$i_3 = \frac{V_1}{R_2} \tag{3.18.16}$$

When these expressions are substituted into Eq. (3.18.15), the resulted equation becomes

$$\frac{V_3}{R_1} + \frac{V_1 - V_2}{R_3} + 2\frac{V_1}{R_2} + C\frac{dV_2}{dt} - \frac{1}{L}\int (V_3 - V_1)dt = \frac{e}{R_1} \; . \tag{3.18.17}$$

In the above equation, e/R_1 is the forcing input, and the measurable outputs are V_1, V_2, and V_3. To produce the above equation, three variables i_1, i_2, and i_3 have to be specified and three equations of motion derived. By applying the equation of constrain $i_1 - i_2 - i_3 = 0$ it has been possible to condense all three equations of motion into a single expression. Note that each element in this expression has a unit of current (A).

3.18.4 Analogies

Above, we considered mechanical, thermal, and electrical elements separately. However, the dynamic behavior of these systems is analogous. It is possible, for example, to take mechanical elements or thermal components, convert them into an equivalent electric circuit and analyze the circuit using Kirchhoff's laws. Table 3-22 gives the various lumped parameters for mechanical, thermal, and electrical circuits, together with their governing equations. For the mechanical components, Newton's second law was used and for thermal we apply Newton's law of cooling.

Table 3-22 Mechanical, thermal, and electrical analogies

MECHANICAL	THERMAL	ELECTRICAL	
MASS M $F = M\dfrac{d(v)}{dt}$	CAPACITANCE $\dashv\vdash$ C $Q = C\dfrac{dT}{dt}$	INDUCTOR $\sim$ L $V = L\dfrac{d(i)}{dt}$	CAPACITOR $\dashv\vdash$ C $i = C\dfrac{d(V)}{dt}$
SPRING $\sim$ k $F = k\int(v)dt$	CAPACITANCE $\dashv\vdash$ C $T = \dfrac{1}{C}\int(Q)dt$	CAPACITOR $\dashv\vdash$ c $V = \dfrac{1}{C}\int(i)dt$	INDUCTOR $\sim$ L $i = \dfrac{1}{L}\int(V)dt$
DAMPER b $F = bv$	RESISTANCE R $Q = \dfrac{1}{R}(T_2 - T_1)$	RESISTOR $\sim$ R $V = Ri$	RESISTOR $\sim$ R $i = \dfrac{1}{R}V$

The table gives various lumped parameters for mechanical, thermal, and electrical circuits, together with their governing equations. In the first column are the linear mechanical elements and their equations in terms of force (F). In the second column are the linear thermal elements and their equations in terms of heat (Q). In the third and fourth columns are electrical analogies (capacitor, inductor, and resistor) in terms of voltage and current (V and i). These analogies may be quite useful in a practical assessment of a sensor and for the analysis of its mechanical and thermal interface with the object and the environment.

REFERENCES

1. Halliday, D., Resnick, R. Fundamentals of physics. 2nd Ed., John Wiley & Sons, New York, 1986

2. Sprague, CN-207 Hall effect IC applications, 1986

3. Meissner, A. Über piezoelectrische Krystalle bei Hochfrequenz. *Z. tech. Phys.*, vol. 8, No. 74, 1927

4. Neubert, Herman K. P. Instrument transducers. An introduction to their performance and design. 2nd ed., Clarendon Press. Oxford, 1975

5. Radice, P. F. Corona Discharge poling process, *U.S. Patent* No. 4,365, 283; 1982

6. Southgate, P.D., *Appl. Phys. Lett.* 28, 250, 1976

7. Jaffe, B., Cook, W. R., Jaffe, H. Piezoelectric ceramics. Academic Press. London, 1971

8. Mason, W. P. Piezoelectric crystals and their application to ultrasonics. Van Nostrand, New York, 1950

9. Megaw, H. D. Ferroelectricity in crystals. Methuen, London, 1957

10. Tamura, M., Yamaguchi, T., Oyaba, T., Yoshimi, T. J. *Audio Eng. Soc.* vol. 23, No. 31, 1975

11. Elliason, S. Electronic properties of piezoelectric polymers. *Report TRITA-FYS 6665 from Dept. of Applied Physics*, The Royal Inst. of Techn., S-100 44 Stockholm, Sweden, 1984

12. Kawai, H. The Piezoelectricity of poly (vinylidene fluoride). *Jap. J. of Appl. Phys.* vol. 8, pp: 975-976, 1969

13. Meixner, H., Mader, G. and Kleinschmidt, P. Infrared sensors based on the pyroelectric polymer polyvinylidene fluoride (PVDF). *Siemens Forsch.-u. Entwicl. Ber. Bd.*, vol. 15, No. 3, pp: 105-114, 1986

14. Kleinschmidt, P. Piezo- und pyroelektrische Effekte. Heywang, W., ed. In: *Sensorik.* Kap. 6: Springer, 1984

15. Semiconductor Sensors. Data Handbook. Philips Export B.V, 1988

16. Ye, C., Tamagawa, T., and Polla, D.L. Pyroelectric $PbTiO_3$ thin films for microsensor applications. In: *Transducers'91. International conference on solid-state sensors and actuators. Digest of technical papers*, pp: 904-907, ©IEEE, 1991

17. Beer, A. C. Galvanomagnetic effect in semiconductors. *Suppl. to Solid State Physics.* F. Seitz and D. Turnbull, Eds. Academic Press, N.Y., 1963

18. Putlye, E. H. The Hall effect and related phenomena. Semiconductor monographs., Hogarth, ed., Butterwort, London, 1960

19. Sprague Hall Effect and Optoelectronic Sensors. Data Book SN-500, 1987

20. Williams, J. Thermocouple measurement, AN28, Linear applications handbook, © Linear Technology Corp., 1990

21. Seebeck, T., Dr. Magnetische Polarisation der Metalle und Erze durch Temperatur-Differenz. *Abhaandulgen der Preussischen Akademic der Wissenschaften*, pp: 265-373, 1822-1823

22. Benedict, R. P. Fundamentals of temperature, pressure, and flow measurements, 3rd ed., John Wiley & Sons, New York, 1984

23. MacDonald, D.K.C. Thermoelectricity: an introduction to the principles. John Wiley & Sons, New York, 1962

24. LeChatelier, *H. Copt. Tend.*, 102, 1886

25. Carter, E. F. ed., Dictionary of inventions and discoveries. Crane, Russak and Co., N.Y., © by F. Muller, 1966

26. Peltier, J.C.A. Investigation of the heat developed by electric currents in homogeneous materials and at the junction of two different conductors, *Ann. Phys. Chem.*, vol. 56 (2nd ser.), 1834

27. Thomson, W. On the thermal effects of electric currents in unequal heated conductors. *Proceedings of the Royal Soc.*, vol. VII, May 1854

28. Manual on the use of thermocouples in temperature measurement. ASTM Publication code number 04-470020-40, ©ASTM, Philadelphia, 1981

29. Raman, V. V. The second law of motion and Newton equations. The Physics Teacher, March 1, 1972

30. Shigley, J. E. and Mischke, C. R. Mechanical engineering design. 5th Ed., McGraw-Hill Book Co., 1989

31. Holman J. P. Heat transfer, 3rd edition, McGraw-Hill Book Co., New York, 1972

32. Bayley, F. J., Owen, J. M. and Turner, A. B. Heat transfer. Barnes & Noble, New York, 1972

33. Applications of phototransistors in electro-optic systems. AN-508. © Motorola, 1988

34. Begunov, B. N., Zakaznov, N. P., Kiryushin, S. I., Kuzichev, V. I. Optical instrumentation. Theory and design. Mir Publishers, Moscow, 1988

35. Giuliani, J. F. Optical waveguide chemical sensors. In: *Chemical sensors and microinstrumentation*. Chapt. 24, American Chemical Society, Washington, 1989

36. Mitchell G. L. Intensity-based and Fabry–Perot interferometer sensors. In: *Fiber optic sensors: an introduction for engineers and scientists*. Chapt. 6. E. Udd, ed., John Wiley & Sons, Inc., 1991

37. Muhs, J. D. Silicone rubber fiber optic sensors. In: *Photonics Spectra*, pp: 98-102, July 1992

38. Welford, W. T., Winston, R. High collection nonimaging optics. Academic Press, Inc., San Diego, 1989.

39. Winston, R., Enoch, J.M. Retinal cone receptor as an ideal light collector. *J. Opt. Soc. Am.* vol. 61, pp: 1120-21, 1971

40. Yariv, A. Optical electronics, 3rd ed., Holt, Reinhart and Winston. New York, 1985

41. Johnson, L. M. Optical modulators for fiber optic sensors. In: *Fiber optic sensors: Introduction for engineers and scientists*. E. Udd, ed., John Wiley & Sons, Inc., 1991

42. Haus, H. A. Waves and fields in optoelectronics. Prentice-Hall, Englewood Cliffs, NJ, 1984

43. Alferness, R.C. *IEEE Trans. Microwave Theory Technol.* MTT-30, p: 1121, 1982

44. Middelhoek, S. and Hoogerwerf A.C. Smart sensors: when and where? In: Sensors and actuators, vol. 8, No. 1, © by Elsevier Sequoya, pp: 39-48, 1985

45. Petersen, K. E. Silicon as a mechanical material. *Proc. IEEE*, vol. 70, No. 5, pp: 420-457, May 1, 1982

46. Obermier, E., Kopystynski, P. and Neißl, R. Characteristics of polysilicon layers and their application in sensors. In: *IEEE Solid-State Sensors Workshop*, 1986

47. Howe, R. T. Surface micromachining for microsensors and microactuators. *J. Vac. Sci. Technol. B*, vol. 6, No. 6, © American Institute of Physics, pp: 1809-1813, Nov.-Dec. 1988

48. Thompson, S. Control systems. engineering & design. Longman Scientific & Technical. Essex, England, 1989

4

Interface Electronic Circuits

4.1 INPUT CHARACTERISTICS OF INTERFACE CIRCUITS

A system designer is rarely able to connect a sensor directly to processing, monitoring or recording instruments. When a sensor generates an electric signal, that signal often is either too weak, or too noisy, or it contains undesirable components. Besides, the sensor output may be not compatible with the input requirements of a data acquisition system, that is, it may have a wrong format. Therefore, the signal usually has to be *conditioned* before it is fed into a processing device (a load). As its input signal, such a device usually requires either voltage or current. An interface or a signal conditioning circuit has a specific purpose - to bring the signal from the sensor up to the format which is compatible with the load device. Fig. 4-1.1 shows a stimulus that acts on a sensor which is connected to a load through an interface circuit. To do its job effectively, an interface circuit must be a faithful slave of two masters: the sensor's and the load device's. Its input characteristics must be matched to the output characteristics of the sensor and its output must be interfaceable with the load. This book, however, focuses on the sensors, therefore, below we will discuss only the front stages of the interface circuits.

The input part of an interface circuit may be specified through several standard numbers. These numbers are useful for calculating how accurately the circuit can process the sensor's signal and what is the circuit's contribution to a total error budget?

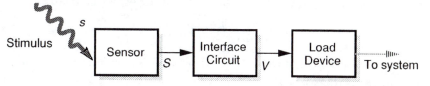

Fig. 4-1.1 Interface circuit matches the signal formats of a sensor and a load device

❖ *The input impedance* shows by how much the circuit loads the sensor. The impedance may be expressed in a complex form as:

$$Z = \frac{V}{I} ,$$ (4.1.1)

where *V* and *I* are complex notations for the voltage and the current across the input impedance. For example, if the input of a circuit is modeled as a parallel connection of input resistance *R* and input capacitance *C* (Fig. 4-1.2A), the complex input impedance may be represented as

$$Z = \frac{R}{1 + j\omega R C} ,$$ (4.1.2)

where ω is the circular frequency and j=√-1 is the imaginary unity. At very low frequencies, a circuit having a relatively low input capacitance and resistance has an input impedance which is almost equal to the input resistance: $Z \approx R$. Relatively low, here it means that the reactive part of the above equation becomes small, i.e., the following holds

$$RC << \frac{1}{\omega} .$$ (4.1.3)

Whenever an input impedance of a circuit is considered, the output impedance of the sensor must be taken into account. For example, if the sensor is of a capacitive nature, to define a frequency response of the input stage, its capacitance must be connected in parallel with the circuit's input capacitance. Formula (4.1.2) suggests that the input impedance is a function of the signal frequency. With an increase in the signal rate of change, the input impedance becomes lower.

Fig. 4-1.2B shows an equivalent circuit for a voltage generating sensor. The circuit is comprised of the sensor output, Z_{out}, and the circuit input, Z_{in}, impedances. The output signal from the sensor is represented by a voltage source *e* which is connected in series with the output impedance. Instead of a voltage source, for some sensors it is more convenient to represent the output signal as outgoing from a current source, which would be connected in parallel with the sensor output impedance. Both representations are equivalent to one another, so we will use voltage. Accounting for both impedances, the circuit input voltage V_{in} is represented as

$$V_{in} = e \frac{Z_{in}}{Z_{in} + Z_{out}} .$$ (4.1.4)

In any particular case, an equivalent circuit of a sensor should be defined. This helps to analyze the frequency response and the phase lag of the sensor-interface combination. For instance, a capacitive detector may be modeled as a pure capacitance connected in parallel with the input

impedance. Another example is a piezoelectric sensor which can be represented by a very high resistance (on the order of $10^{11}\Omega$) shunted by a capacitance (in the order of 10pF).

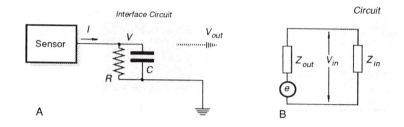

Fig. 4-1.2 Complex input impedance of an interface circuit (A)
and equivalent circuit of a voltage generating sensor (B)

To illustrate the importance of the input impedance characteristics, let us consider a purely
resistive sensor connected to the input impedance as shown in Fig. 4-1.2. The circuit's input voltage as function of frequency f can be expressed by a formula

$$V = \frac{e}{\sqrt{1 + \left(\dfrac{f}{f_c}\right)^2}} , \qquad (4.1.5)$$

where $f_c = (2\pi RC)^{-1}$ is the corner frequency, that is the frequency where the amplitude drops by
3 dB. If we assume that a 1% accuracy in the amplitude detection is required, then we can
calculate the maximum stimulus frequency which can be processed by the circuit:

$$f_{max} \approx 0.14 f_c , \qquad (4.1.6)$$

or $f_c \approx 7 f_{max}$, that is, the impedance must be selected in such a way as to assure a sufficiently high
corner frequency. For example, if the stimulus' highest frequency is 100 Hz, the corner frequency
must be selected at least at 700 Hz. In practice, f_c is selected even higher, because of the additional
frequency limitations in the subsequent circuits.

One should not overlook a speed response of the front stage of the interface circuit.
Operational amplifiers, which are the most often used building blocks of interface circuits, usually
have limited frequency bandwidths. There are so-called programmable operational amplifiers
which allow the user to control (to program) the bias current and, therefore, the first stage frequency response. The higher the current, the faster would be the response.

Fig. 4-1.3 is a more detailed equivalent circuit of the input properties of a passive electronic interface circuit[1], for instance, an amplifier or an A/D converter. The circuit is characterized by the input impedance, Z_{in}, and several generators. They represent voltages and currents which are generated by the circuit itself. These signals are spurious and may pose substantial problems if not handled properly. All these signals are temperature dependent. They discussed in detail in Section 4.9.

Input Stage of Interface Circuit

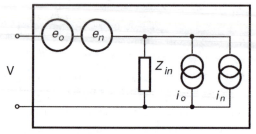

Fig. 4-1.3 Equivalent circuit of electrical noise sources at an input stage

❖ Voltage e_o is called the input *offset voltage*. If the input terminals of the circuit are shorted together, that voltage would simulate a presence of an input dc signal having a value of e_o. It should be noted that the offset voltage source is connected in series with the input and its resulting error is independent of the output impedance of the sensor.

❖ The input *bias current*, i_o, is also internally generated by the circuit. Its value is quite high for many bipolar transistors, much smaller for the JFETs, and even more lower for the CMOS circuits. This current may present a serious problem when a circuit or a sensor employs high impedance components. The bias current passes through the input impedance of the circuit and the output impedance of the sensor, resulting in a spurious voltage drop. This voltage may be of a significant magnitude. For instance, if a piezoelectric sensor is connected to a circuit having an input resistance of $1G\Omega$ $(10^9\Omega)$ and the input bias current of 1nA $(10^{-9}A)$, the voltage drop at the input becomes equal to $1G\Omega\cdot1nA=1V$ - a very high value indeed. In contrast to the offset voltage, the bias current resulting error is proportional to the output impedance of the sensor. This error is negligibly small for the sensors having low output resistances. For instance, an inductive detector is not sensitive to a magnitude or variations in the bias current.

❖ A circuit board *leakage current* may be a source of errors while working with high impedance circuits. This current may be the result of lower surface resistance in the printed circuit

---

[1] Here the word passive means that the circuit does not generate any excitation signal.

board. Fig. 4-1.4A shows that a power supply bus and the board resistance R_L may cause leakage current i_L through the sensor's output impedance. If the sensor is capacitive, its output capacitance will be very quickly charged by the leakage current. This will not only cause an error, but may even lead to the sensor's destruction.

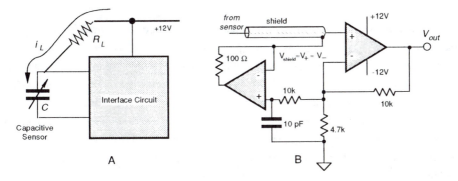

Fig. 4-1.4 Circuit board leakage affects input stage (A);
driven shield of the input stage (B)

There are several techniques known to minimize the board leakage current effect. One is a careful board layout to keep higher voltage conductors away from the high impedance components. A leakage through the board thickness in multilayer boards should not be overlooked. Another method is electrical guarding, which is an old trick. The so-called driven shield is also highly effective. Here, the input circuit is surrounded by a conductive trace that is connected to a low impedance point at the same potential as the input. The guard absorbs the leakage from other points on the board, drastically reducing currents that may reach the input terminal. To be completely effective, there should be guard rings on both sides of the printed circuit board. As an example, an amplifier is shown with a guard ring, driven by a relatively low impedance of the amplifier's inverting input.

It is highly advisable to locate the high impedance interface circuits as close as possible to the sensors. However, sometimes connecting lines can not be avoided. Coaxial shielded cables with good isolation are recommended [1]. Polyethylene or virgin (not reconstructed) Teflon is best for the critical applications. In addition to potential insulation problems, even short cable runs can reduce bandwidth unacceptably with high source resistances. These problems can be largely avoided by bootstrapping the cable's shield. Fig. 4-1.4B shows a voltage follower connected to the inverting input of an amplifier. The follower drives the shield of the cable, thus reducing the cable capacitance, the leakage and spurious voltages resulting from cable flexing. A small capacitance at the follower's noninverting input improves its stability.

4.2   AMPLIFIERS

Most passive sensors produce weak output signals. Magnitudes of these signals may be on the order of millivolts (mV) or picoamperes (pA). On the other hand, standard electronic data processors, such as A/D converters, frequency modulators, data recorders, etc. require input signals of sizable magnitudes - on the order of volts (V) and milliamperes (mA). Therefore, an amplification of the sensor output signals has to be made with a voltage gain up to 1,000 and a current gain up to 1 million. Amplification is a part of a signal conditioning. There are several standard configurations of amplifiers which might be useful for the amplifying signals from various sensors. These amplifiers may be built of discrete components, such as transistors, diodes, resistors, capacitors and inductors.

4.2.1   Operational amplifiers

Nowadays, one of the principle building blocks for the amplifiers is the so-called *operational amplifier* or OPAM, which is either an integrated (monolithic) or hybrid (a combination of monolithic and discrete parts) circuit. An integrated OPAM may contain hundreds of transistors, as well as resistors and capacitors. An analog circuit designer, by arranging around the OPAM discrete components (resistors, capacitors, inductors, etc.), may create an infinite number of useful circuits – not only the amplifiers, but many others circuits as well. Below, we will describe some circuits which are often used in conjunction with various sensors.

As a building block, a good operational amplifier has the following properties (a schematic representation of OPAM is shown in Fig. 4-2.1):

- ○   two inputs: one is inverting (−) and the other is noninverting (+);
- ○   a high input resistance (on the order of hundreds of MΩ or even GΩ);
- ○   a low output resistance (a fraction of Ω);
- ○   a low input offset voltage e_o (few mV or even μV);
- ○   a low input bias current i_o (few pA or even less);
- ○   a very high open loop gain (on the orders of 10^4 to 10^6) A_{OL}. That is, the OPAM must be able to magnify (amplify) a voltage difference V_{in}, between its two inputs by a factor of A_{OL};

○   a high common mode rejection ratio (CMRR). That is, the amplifier suppresses the in-phase equal magnitude input signals (common-mode signals) V_{CM} applied to its both inputs;

○   low intrinsic noise;

○   a broad operating frequency range;

○   a low sensitivity to variations in the power supply voltage.

○   a high environmental stability of its own characteristics.

As an example, specifications of some operational amplifiers which are useful for the sensor interfaces are given in Table 4-1. However, for the detailed information and the application guidance the user should refer to data books published by the respective manufacturers. Such books usually contain selection guides for the every important feature of an OPAM. For instance, OPAMs are grouped by such criteria as low offset voltages, low bias currents, low noise, a broad bandwidth, etc.

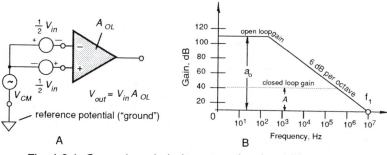

A   B

**Fig. 4-2.1 General symbol of an operational amplifier (A)
and gain/frequency characteristic of an OPAM (B)**

Fig. 4-2.1A depicts an operational amplifier without any feedback components. Therefore, it operates under the so-called *open-loop* conditions. An open loop gain A_{OL} of an OPAM is not constant over the frequency range. It may be approximated by a graph of Fig. 4-2.1B. The A_{OL} changes with the load resistance, temperature and the power supply fluctuations. Many amplifiers have an open loop gain temperature coefficient on the order of 0.2 to 1%/°C and the power supply gain sensitivity on the order of 1%/%. An OPAM is very rarely used with an open loop (without the feedback components) because the high open-loop gain may result in a circuit instability, a strong temperature drift, noise, etc. For instance, if the open-loop gain is 10^5, the input voltage drift of 10μV (ten microvolts) would cause the output drifts by about 1V.

The ability of an OPAM to amplify small magnitude, high frequency signals is specified by the gain-bandwidth product (GBW) which is equal to the frequency f_1 where the amplifier gain becomes equal to unity. In other words, above the f_1 frequency, the amplifier can not amplify. Fig. 4-2.2A depicts a non-inverting amplifier where resistors R_1 and R_2 define the feedback loop. The resulting gain $A=1+R_2/R_1$ is a closed-loop gain. It may be considered constant over a much broader frequency range (see Fig. 4-2.1B), however, f_1 is the frequency limiting factor regardless of the feedback. A linearity, gain stability, the output impedance and gain accuracy are all improved by the amount of feedback. As a general rule for moderate accuracy, the open loop gain of an OPAM should be at least 100 times greater than the closed loop gain at the highest frequency of interest. For even higher accuracy, the ratio of the open and closed loop gains should be 1000 or more.

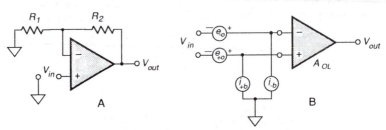

Fig. 4-2.2 Noninverting amplifier (A);
offset voltages and bias currents in an operational amplifier
are represented by generators connected to its inputs (B)

A typical data sheet for an OPAM specifies the bias and offset voltages. Due to limitations in manufacturing technologies, any OPAM acts not only as a pure amplifier, but as a generator of voltages and currents which may be related to its input (Fig. 4-1.3). Since these spurious signals are virtually applied to the input terminals, they are amplified along with the useful signals.

Because of offset voltages and bias currents, an interface circuit does not produce zero output when zero input signal is applied. In dc-coupled circuits, these undesirable input signals may be indistinguishable from the useful signal. If the input offset voltage is still too large for the desired accuracy, it can be trimmed out either directly at the amplifier (if the amplifier has dedicated trimming terminals) or in the independent offset compensation circuit, as is shown in Fig. 4-2.3A. It should be noted that the circuit produces an inverted output signal as compared with the input. The trimming potentiometer P_1 may adjust offset voltages originated in both the input and the trimming (U_2) amplifiers. The circuit, however, does not compensate for a temperature drift in the offset voltage. The offset voltage should be considered whenever an amplifier is used with either low or high impedance input device. It may be ignored if a dc component of the input signal is not important. For instance, in a temperature measurement, a dc component is important for overall accuracy, while in the thermal motion detectors, only the variable part of the input signal may be amplified. In the latter case, an amplifier with a relatively large offset voltage may be selected. Thermal drifts in the offset voltage may be compensated for by special networks containing temperature sensitive components, usually, thermistors.

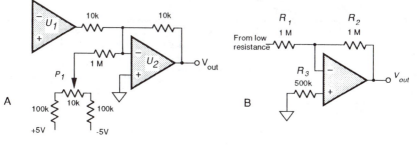

Fig. 4-2.3 Trimming of the offset voltage (A)
and compensation of the offset current by a balancing resistor R_3 (B)

The input *offset current* is the difference between the input bias currents of two inputs. In the CMOS amplifiers, the offset current may be as low as 1pA. If an amplifier is used with a sensor having a relatively low output resistance, the bias and offset currents of an OPAM may be ignored as they produce a very low voltage drop across the sensor's output resistance. On the other hand, if a sensor has a very high output resistance (for instance a piezoelectric detector), bias and offset currents must be taken into account. One useful technique to compensate for the offset current is shown in Fig. 4-2.3B. The inverting input of the amplifier is connected to two 1MΩ feedback resistors which form an equivalent network totaling 500kΩ. If the input bias current at the negative input is equal to 1nA (10^{-9}A) it will result in the voltage drop of 0.5mV at that input. This may be a source of an unacceptable error. To compensate for it, a ballast resistor of 500kΩ is connected in series with the positive input. A bias current of a noninverting input is usually very close to that of an inverting input. Hence, the ballast resistor results in an almost equal voltage drop. Since the amplifier rejects the in-phase signals (a common-mode rejection), the voltage drops at both inputs will be canceled out.

Quite often, cost considerations are the prime criteria for selecting an OPAM. If the financial constrains are not tight, there are plenty of operational amplifiers available for designing circuits with exceptionally high performances. In many cases, especially involving consumer electronics, the cost is a leading factor and the circuit designer is limited by choosing the OPAMs of lower grades. Generally, it is much easier to design a circuit when the high quality components are available. It is more challenging to achieve a sufficiently high performance with the inexpensive (e.g. lower grade) parts.

Table 4-1 Selected operational amplifiers for sensor interface

Abbreviations: $T_c(V_{os})$ is temperature coefficient of V_{os}
LTC - Linear Technologies Corp.
TI - Texas Instruments, Inc.
PMI - Precision Monolithics Inc.
AD - Analog Devices, Inc.

Part Number	Manufacturer	V_{os} max (μV)	$T_c(V_{os})$ (μV/°C)	I_B (nA) at 25°C	A_{OL} (V/mV) at dc	Noise (10Hz) $nV\sqrt{Hz}$	Note
TLC2652A	TI	1	0.003	4 pA typ	135 dB min	94	Chopper stabilized LinCMOS™
OP-27E	TI, PMI, LTC, Motorola	25	0.2	10 typ	1800	6	low noise, high speed
OP-07	PMI	25	0.2 typ	0.7 typ	500	10.3 typ	ultra stable
LT1012C	LTC	50	1.5	0.15	200	30	Low Vos; low power
LT1008C	LTC	120	1.5	0.1	200	30	low power
LT1022AC	LTC	250	5	0.05	150	50	High speed JFET
OP-41E	PMI	250		3 pA typ	5000	40	High stability JFET
LM11	Motorola, National	300	1	17 pA typ	1200	150	low drift low power
LM108A	Motorola, National	500	1	0.8 typ	300		low drift
TL051ACP	TI	800	8	4 pA typ	100	75	Precision JFET
AD549L	AD	1mV	10	60 fA	1000	90	Lowest bias current JFET
AD546	AD	1mV	20	1 pA	1000	90	Low Cost JFET
LF356A	LTC, National, Motorola, PMI	2mV	5	0.05	75	15	JFET inputs
TLC271BCP	TI Thomson	2mV	2	0.6 pA typ	23	100	Programmable LinCMOS™
MC34071	Motorola	3mV	10	100 typ	100	55	low drift
LM324	generic	3mV	7 typ	45 typ	100	10.3 typ	quad, low cost single supply
MB47833	Fujitsu	5mV		500 typ	110 dB	10	High speed low noise dual
MC1776	Motorola	5mV		2 typ	200	progr.	micro power programmable
MB42082	Fujitsu	15mV		30 pA typ	200	90	High speed JFET dual

4.2.2 Voltage follower

A voltage follower (Fig. 4-2.4) is a an electronic circuit that provides impedance conversion from a high to low level. A typical follower has high input impedance (the high input resistance and the low input capacitance) and low output resistance (the output capacitance makes no difference). A follower has a voltage gain very close to unity (typically, 0.999) and a high current gain. In essence, it is a current amplifier and impedance converter. That is, its high input impedance and low output impedance make it indispensable for the interfacing between many sensors and signal processing devices.

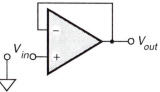

Fig. 4-2.4 Voltage follower with an operational amplifier

A follower, when connected to a sensor, makes very little effect on the latter's performance, thus providing a buffering function between the sensor and the load. When designing a follower, the following tips might be useful:

○ For the current generating sensors, the input bias current of the follower must be at least 100 times smaller than the sensor's current.

○ The input offset voltage must be either trimable or smaller than the required LSB.

○ The temperature coefficient of the bias current and the offset voltage should not result in errors of more than 1 LSB over an entire temperature range.

An application engineer should be concerned with, the output offset voltage, which can be derived from formula:

$$V_o = A(e_o + i_o R_{eqv})$$ (4.2.1)

where R_{eqv} is the equivalent resistance at the input (a combination of the sensor's output resistance and the input resistance of the amplifier), e_o is the input offset voltage and i_o is the input bias current. The offset is temperature dependent. In circuits where the amplifier has high gain, the output voltage offset may be a source of substantial error. There are several ways to handle this difficulty. Among them is selecting an amplifier with low bias current, high input resistance and low offset voltage. Chopper stabilized amplifiers are especially efficient (See Table 4-1).

4.2.3 Monopolar amplifiers

A monopolar amplifier is similar to a voltage follower, except that its gain may be either higher or lower than unity. Thus, it may serve as both voltage and current amplifier (or scaling circuit). A practical circuit is shown in Fig. 4-2.2A Resistors R_1 and R_2 define gain of the amplifier:

$$A = \left(1 + \frac{R_2}{R_1}\right) . \tag{4.2.2}$$

It can be seen that the minimum gain is 1 (when $R_1 = \infty$) and the amplifier becomes a voltage follower. A capacitor in parallel with R_2 may be used to limit the bandwidth, which is often important for noise reduction. A bandwidth of such a circuit at 3dB level may be estimated from a simple formula

$$f_u = \frac{0.159}{R_2 C} \text{ [Hz]} . \tag{4.2.3}$$

While the circuit of Fig. 4-2.2A shows a noninverting amplifier, Fig. 4-2.3B depicts an inverting amplifier. Its gain is defined as $A = \frac{R_2}{R_1}$, while the balance resistor[1] is $R_3 = R_1 \| R_2$. It should be noted, however, that such an inverting amplifier has a relatively low input resistance equal to R_1. This may result in the excessive loading of the sensor. The solution would be to use instead either a noninverting amplifier, or to employ a voltage follower between the sensor and the inverting amplifier.

4.2.4 Instrumentational amplifier

An instrumentational amplifier (IA) has two inputs and one output. It is distinguished from an operational amplifier in its finite gain (which is usually no more than 100) and the availability of both inputs for connecting to the signal sources. The latter feature means that all necessary feedback components are connected to other parts of the instrumentational amplifier, rather than to its noninverting and inverting inputs. The main function of the IA is to produce an output signal which is proportional to the difference in voltages between its two inputs:

$$V_{out} = A(V_+ - V_-) = A\Delta V , \tag{4.2.4}$$

where V_+ and V_- are the input voltages at noninverting and inverting inputs respectively, and A is the gain. An instrumentational amplifier can be either built from an OPAM or in a monolithic form.

---

[1] Resistor R_3 may not be required if bias current is small enough not to cause significant errors.

It is important to assure high input resistances for both inputs, so that the amplifier can be used in a true differential form. A differential input of the amplifier is very important for rejection of common mode interferences having an additive nature (see Section 4.9).

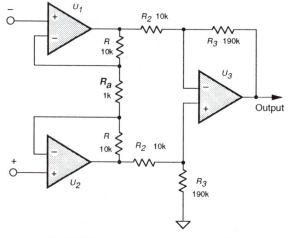

Fig. 4-2.5 Instrumentational amplifier
with three operational amplifiers and matched resistors

While several monolithic instrumentational amplifiers are presently available, quite often discrete component circuits prove to be more cost effective and often more efficient. A basic circuit of IA is shown in Fig. 4-2.5. The voltage across R_a is forced to become equal to the input voltage difference, ΔV. This sets the current through that resistor equal to $i = \Delta V / R_a$. The output voltages from the U_1 and U_2 OPAMs are equal to one another in amplitudes and opposite in the phases. Hence, the front stage (U_1 and U_2) has a differential input and a differential output configuration. The second stage (U_3) converts the differential output into a unipolar output and provides an additional gain. The overall gain of the IA is

$$A = \left(1 + \frac{2R_1}{R_a}\right)\frac{R_3}{R_2} \ . \tag{4.2.5}$$

A good and cost effective instrumentational amplifier can be built of two identical operational amplifiers and several precision resistors (Fig. 4-2.6A). The circuit uses the FET-input OPAMs to provide lower noise and lower input bias currents. The U_1 acts as a noninverting amplifier and U_2 is the inverting one. Each input has a high impedance and can be directly interfaced with a sensor. A feedback from each amplifier forces voltage across the gain-setting resistor R_a to become equal to ΔV. The gain of the amplifier is equal to

$$A \approx 2(1 + \frac{R}{R_a}) \ . \tag{4.2.6}$$

Hence, gain may vary from 2 (R_a is omitted) to a potentially open loop gain (R_a=0). With the components whose values are shown in Fig. 4-2.6, the gain is A=100. It should be remembered, however, that the input offset voltage will be amplified with the same gain. The CMRR primarily depends on matching values of resistors, R. At very low frequencies, it is the reciprocal of the net fractional resistor mismatch, i.e., CMRR=10,000 (-80 dB) for a 0.01% mismatch. At higher frequencies, the impedance mismatch must be considered, rather than the resistor mismatch. To balance the impedances, a trimpot and a capacitor C_1 may be used.

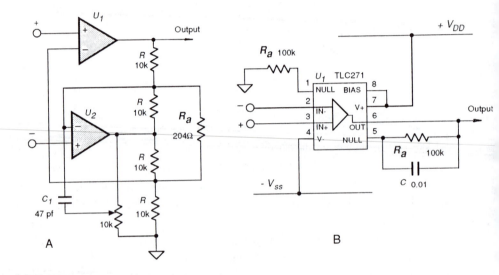

Fig. 4-2.6   A: Instrumentational amplifier with two operational amplifiers;
B: Low cost ac instrumentational amplifier with one operational amplifier

When cost is a really limiting factor and no high quality dc characteristics are required, a very simple IA can be designed with just one operational amplifier and two resistors (Fig. 4-2.6B). The feedback resistor R_a is connected to the null-balance terminal of the OPAM which is the output of the front stage of the monolithic circuit. The amount of the feedback through R_a depends on the actual circuit of an OPAM and somewhat varies from part to part. For the TLC271 operational amplifier (Texas Instruments), gain of the circuit may be found from

$$A \approx 1 + \frac{R_a}{2k\Omega} \qquad (R_a \text{ is in } k\Omega), \qquad\qquad (4.2.7)$$

which for values indicated in Fig. 4-2.6B gives gain of about 50.

4.2.5 Charge amplifiers

Charge amplifiers (CA) is a very special class of circuits which must have extremely low bias currents. These amplifiers are employed to convert to voltage signals from capacitive sensors, quantum detectors, pyroelectric sensors, and other devices which generate very low charges (on the order of pico-coulombs, pC) or currents (on the order of pico-amperes). A basic circuit of a charge-to-voltage converter is shown in Fig. 4-2.7A. A capacitor, C, is connected into a feedback network of an OPAM. Its leakage resistance, r, must be substantially larger than the impedance of the capacitor at the lowest operating frequency. A transfer function of the converter is:

$$V_{out} = -\frac{\Delta Q}{C} . \qquad\qquad (4.2.8)$$

Special hybrid charge sensitive preamplifiers are available for precision applications. One example is DN630 from Dawn Electronics, Inc.[1] The amplifier can operate with sources of less than 1pF capacitance. An internally connected 1pF capacitor sets the gain of the amplifier to 1 volt per 1pC (picocoulomb) sensitivity. The gain can be reduced by connecting one or a combination of the internal capacitor array to the input of the amplifier. It features low noise and has less than 5ns rise and fall times.

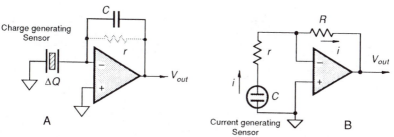

Fig. 4-2.7 Charge-to-voltage (A) and current-to-voltage (B) converters

Many sensors can be modeled by capacitors. Some capacitive sensors are active, that is, they require an excitation signal. Examples are the capacitive force and pressure transducers and humidity detectors. Other capacitive sensors are passive, that is they directly convert a stimulus into

[1] Carson City, Nevada. Tel. (702) 882-7721

an electric current. Examples are the piezoelectric and pyroelectric detectors. Ohm's law suggests that to convert an electric current into voltage, current should pass through an appropriate resistor and the voltage drop across that resistor is proportional to the magnitude of the current. Fig. 4-2.7B shows a basic current-to-voltage converter where the capacitive current generating sensor is connected to the inverting input of an OPAM which serves as a virtual ground. That is, voltage at the input is almost equal to that at the noninverting input which is grounded. The sensor operates at zero voltage across its terminals and its current is represented by the output voltage of the OPAM:

$$V_{out} = -iR \qquad\qquad (4.2.9)$$

Resistor $r \ll R$ is generally required for the circuit stability. At high frequencies, the OPAM would operate near the open loop gain which may result in oscillations. The advantage of the virtual ground is that the output signal does not depend on the sensor's capacitance. The circuit produces voltage whose phase is shifted by $180°$ with respect to the current. A noninverting circuit shown in Fig. 4-2.8A can convert and amplify the signal, however, its speed response depends on both the sensor's capacitance and the converting resistor. Thus, the response to a step function in a time domain can be described by:

$$V_{out} = iR_b \left(1 + \frac{R_2}{R_1} \right)(1 - e^{-t/RC}) \ . \qquad\qquad (4.2.10)$$

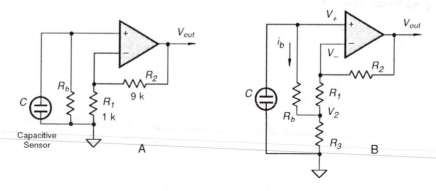

Fig. 4-2.8   Noninverting current-to-voltage converter (A)
and resistance multiplier (B)

When converting currents from such sensors as piezo and pyroelectrics, the resistor R_b (R in circuit 4-2.7B) may be required on the order of tens or even hundreds of gigohms. In many cases resistors of such high values may be not available or impractical to use due to poor stability. A high ohmic resistor can be simulated by a circuit which is known as a **resistance multiplier**. It is implemented by adding a positive feedback around the amplifier. Fig. 4-2.8B shows that R_1 and R_3 form a resistive divider. Due to a high open loop gain of the OPAM, voltages at noninverting and inverting inputs are almost equal to one another: $V_+ \approx V_-$. As a result, voltage V_2 at the divider is

$$V_2 = V_- \frac{R_3}{R_1+R_3} \approx V_+ \frac{R_3}{R_1+R_3} \quad , \qquad (4.2.11)$$

and current through the resistor is defined trough the voltage drop:

$$i_b = \frac{\Delta V}{R_b} = \frac{V_+ - V_2}{R_b} = \frac{V_+}{R_b} \frac{R_1}{R_1+R_3} \quad . \qquad (4.2.12)$$

From this equation, the input voltage can be found as a function of the input current and the resistive network:

$$V_+ = i_b R_b (1 + \frac{R_3}{R_1}) \quad . \qquad (4.2.13)$$

It is seen that the resistor R_b is multiplied by a factor of $(1+R_3/R_1)$. Resistance multiplication, while being a powerful trick should be used with some caution. Specifically, noise, bias current and offset voltage – all of them are also multiplied by the same factor $(1+R_3/R_1)$, which may be undesirable in some applications. Further, since the network forms a positive feedback, it may cause circuit instability. Therefore, in practical circuits, a resistance multiplication should be limited to a factor of 10.

4.3   EXCITATION CIRCUITS

External power is required for the operation of *active* sensors. Examples are: temperature sensors (thermistors and RTDs), pressure sensors (piezoresistive and capacitive), displacement (electromagnetic and optical). The power may be delivered to a sensor in different forms. It can be a constant voltage, constant current, sinusoidal or pulsing currents. It may even be delivered in the form of light or ionizing radiation. The name for that external power is an excitation signal. In many cases, stability and precision of the excitation signal directly relates to the sensor's accuracy and stability. Hence, it is imperative to generate the signal with such accuracy that the overall performance of the sensing system is not degraded. Below, we review several electronic circuits which feed sensors with appropriate excitation signals.

4.3.1 Current generators

Current generators are often used as excitation circuits to feed sensors with predetermined currents. In general terms, a current generator (current pump) is a device which produces electric current independent of the load impedance. That is, within the capabilities of the generator, characteristics of its output current (amplitude, phase, frequency) must remain substantially independent of any changes in the impedance of the load. A double circle is the symbol for a current generator. It is depicted in Fig. 4-3.1. According to Ohm's law, if current is constant, any change in voltage across the impedance, must be a linear function of that impedance. When a current generator is connected to a load, its output current i divides between its own output resistance R_{out} and the load impedance Z_L:

$$i = i_o + i_L ,$$
(4.3.1)

and current through the load is:

$$i_{out} = i - \frac{V}{R_{out}}$$
(4.3.2)

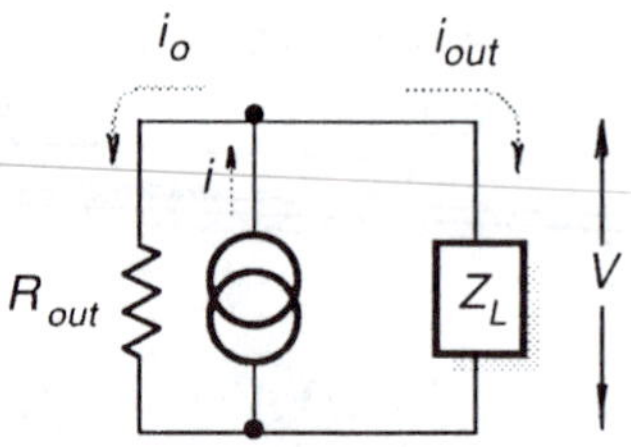

Fig. 4-3.1 Equivalent circuit of a current source

It is seen that the output current is not exactly equal to that produced by the current generator, because of the term which is dependent on R_{out}. However, if the output resistance is made very large ($R_{out} \to \infty$), the current generator puts all its output through the load. Good current generators must have a very large output resistance, practically, at least 3-4 orders of magnitude larger than that of the load.

Usefulness of the current generators for the sensor interfaces is in their ability to produce excitation currents of precisely controlled magnitude and shape. Hence, a current generator should not only produce current which is load independent, but it also must be controllable from an external signal source (a wave-form generator), which in most cases has a voltage output. A good current generator must produce current which follows the control signal with high fidelity and is independent of the load over a broad range of impedances.

There are two main characteristics of a current generator: the output resistance and the voltage compliance. A voltage compliance is the highest voltage which can be developed across the load without affecting the output current. For a high resistive load, according to Ohm's law, a higher voltage is required for a given current. For instance, if the required excitation current is $i = 10\text{mA}$ and the highest load impedance at any given frequency is $R_L = 10\text{k}\Omega$, a voltage compliance of at least $iR_L = 100\text{V}$ would be needed. Below, we cover some useful circuits with increased voltage compliance where the output currents can be controlled by external signals.

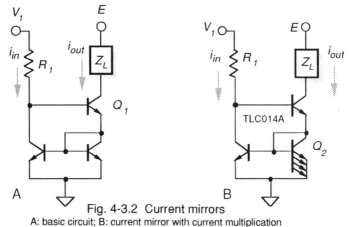

Fig. 4-3.2 Current mirrors
A: basic circuit; B: current mirror with current multiplication

A unipolar current generator is called either a current *source* (generates the out-flowing current), or a current *sink* (generates the in-flowing currents). Here, unipolar means that it can produce currents flowing in one direction only. Many of such generators utilize current-to-voltage characteristics of transistors. It is well known, that the transistor's collector current is very little dependent on collector voltages. This feature was employed by the so-called current mirrors. A current mirror has one current input and at least one (may be several) current output. Therefore, the output current is controlled by the input current. The input current is supplied from an external source and should be of a known value. Fig. 4-3.2A shows the so-called Wilson current mirror where voltage V_1 and resistance R_1 produce the input current i_{in}. The output transistor Q_1 acts as a current controlled resistor, thus regulating the output current i_{out} in such a manner as to maintain it equal to i_{in}. The output current may be multiplied several times if the transistor Q_2 (Fig. 4-3.2B) is fabricated with several emitters. Such a current sink is commercially available from Texas Instruments (part TLC014A). That current mirror has a voltage compliance of 35V and the output resistance ranging from 2 to 200MΩ (depending on the current). Texas Instruments also fabricates adjustable-ratio current mirrors (part TL010) with up to 33 distinct input-to-output ratios.

A voltage controlled current source or sink may include an operational amplifier (Fig. 4-3.3). In such a circuit, a precision and stable resistor R_1 defines the output current i_{out}. To deliver a higher current at a maximum voltage compliance, as little as possible voltage drop should be developed across the sensing resistor R_1. In effect, that current is equal to V_1/R_1. For better performance, the current through the base of the output transistor should be minimized, hence, a field-effect rather than bipolar transistor is often used as an output device.

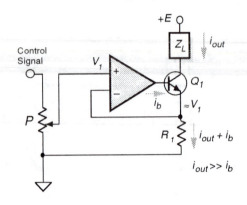

Fig. 4-3.3 Current sink with an operational amplifier and *npn* transistor

For many applications, *bipolar* current generators may be required. Such a generator provides a sensor with the excitation current which may flow in both directions (in- and out-flowing). Fig. 4-3.4A shows an operational amplifier where the load is connected as a feedback. Current though the load Z_L, is equal to V_1/R_1 which is load independent. The load current follows V_1 within the operating limits of the amplifier. An obvious limitation of the circuit is that the load is "floating", i.e., it is not connected to a ground bus or any other reference potential. For some applications, this is quite all right, however, many sensors need to be grounded or otherwise referenced. A circuit shown in Fig. 4-3.4B keeps one side of the load impedance near the ground potential, because a non-inverting input of the OPAM is a virtual ground. Nevertheless, even in this circuit, the load is still fully isolated from the ground. One negative implication of this isolation may be an increased pick up of various kinds of transmitted noise.

In cases where the sensor must be grounded, a current pump invented over 30 years ago by Brad Howland at MIT may be used (Fig. 4-3.5). The pump operation is based on utilizing both negative and positive feedbacks around the operational amplifier. The load is connected to the positive loop [2]. Current through the load is defined by:

$$i_{out} = \frac{R_2}{R_1} \frac{(V_1 - V_2)}{R_5}$$

(4.3.3)

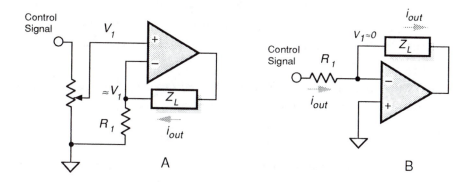

Fig. 4-3.4 Bipolar current generators with floating loads
A: a noninverting circuit; B: a circuit with a virtual ground

A trimming resistor P must be adjusted to assure that

$$R_3 = R_1 \frac{R_4 + R_5}{R_2} \tag{4.3.4}$$

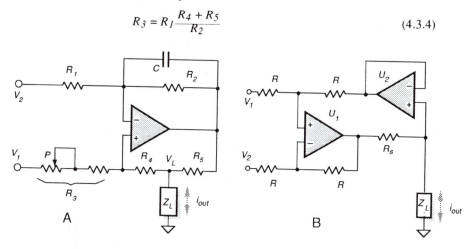

Fig. 4-3.5 Current generators with ground referenced loads
A: Howland current pump; B: current pump with two OPAMs

In that circuit, each resistor may have a relatively high value ($100k\Omega$ or higher), but the value for R_5 should be relatively small. This condition improves the efficiency of the Howland current pump, as smaller voltage is wasted across R_5 and smaller current is wasted through R_4 and R_3. The circuit is stable for most of the resistive loads, however, to insure stability, a few picofarad capacitor C may be added in a negative feedback or/and from the positive input of an operational amplifier to ground. When the load is inductive, an infinitely large compliance voltage would be required to deliver the set current when a fast transient control signal is applied. Therefore, the current pump will produce a limited rising slope of the output current. The flowing current will generate an inductive spike across the output terminal, which may be fatal to the operational amplifier. It is advisable, for the large inductive load to clamp the load with diodes to the power supply buses.

An efficient current pump with four matched resistors and two operational amplifiers is shown in Fig. 4-3.5B. Its output current is defined by the equation

$$i_{out} = \frac{(V_1 - V_2)}{R_s} \quad . \tag{4.3.5}$$

The advantage of this circuit is that resistors R may be selected with a relatively high value and housed in the same thermally homogeneous packaging for better thermal tracking.

For the generation of low level constant currents, a monolithic voltage reference shown in Fig. 4-3.6 may be found quite useful. The circuit contains a 2.5-volt reference from Analog Devices, Inc. which is powered by the output current from the voltage follower U_1. The voltage regulator keeps the voltage drop across R_s precisely equal to 2.5V, thus assuring that the current through that resistor and, subsequently, through the load is constant. The load current is defined as

$$i_{out} = \frac{2.5V}{R_s} \quad . \tag{4.3.6}$$

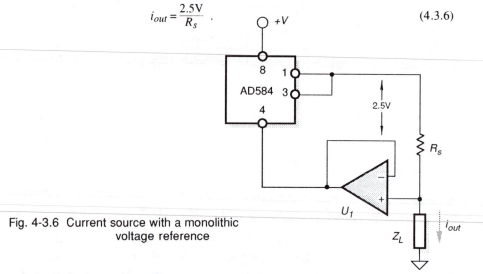

Fig. 4-3.6 Current source with a monolithic
voltage reference

4.3.2 Voltage references

A voltage reference is an electronic device which generates constant voltage that is affected little by variations in power supply, temperature, load, aging, and other external factors. There are several techniques known for generation of such voltages. Many voltage references are available in monolithic forms, however, in low cost applications, especially in consumer products, a simple device known as zener diode is often employed.

A zener diode has a constant voltage drop in a circuit when provided with a fairly constant current derived from a higher voltage elsewhere within the circuit. The active portion of a zener diode is a reverse-biased semiconductor pn-junction. When the diode is forward biased (the p-region is more positive), there is little resistance to current flow. Actually, a forward-biased zener diode looks very much like a normal semiconductor diode (Fig. 4-3.7A). When the diode is reverse biased (minus is at the anode and plus is at the cathode), very little current flows through it if the applied voltage is less than V_z. A reverse saturation current is a small leakage which is almost independent of the applied voltage. When the reverse voltage approaches the break-down voltage V_z, the reverse current increases rapidly and if not limited will result in the diode overheating and destruction. For that reason, zener diodes usually are used with current limiting components, such as resistors, PTC thermistors and others. The most common circuit with a zener diode is shown in Fig. 4-3.7B. In this circuit, the diode is connected in parallel with its load, thus implying the name a shunting reference. The zener voltage decreases as temperature of the junction rises.

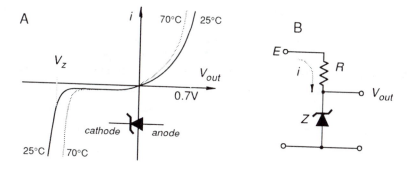

Fig. 4-3.7 Zener diode
A: a volt-ampere characteristic; B: a shunting-type voltage reference

Zener diodes (zeners) are available with voltages ranging from 2 to 200V and allow power dissipation from 1 to 50 watts. While being convenient and readily available, zeners suffer from many drawbacks, such as sensitivity to supply current i, temperature, part-to-part variations. To minimize the effect of the supply current, zeners should be selected with voltages below 3V and from 6 to 8V.

An efficient voltage reference (Fig. 4-3.8A) operating at very low currents (in a microampere range) can be implemented by using a regular bipolar transistor in a zener mode. The circuit uses a reverse breakdown of a base-emitter junction. When the junction voltage reaches V_r, which for most bipolar transistors is in the range between 6 and 8V, a further increase in voltage results in a dramatic increase in the reverse junction current (Fig. 4-3.8B). Thus, the junction acts as a zener diode. A combination of a transistor and a resistor forms a parallel (shunting) voltage regulator which may be used as a voltage reference in low cost circuits. It should be noted however, that such a voltage reference is temperature dependent and the breakdown voltage also depends on the manufacturing semiconductor process. Besides, part-to-part variations may be as large as 20%. Nevertheless, if a circuit, either does not need a precision reference point, or uses a ratiometric technique, this configuration is quite useful for applications where cost is a limiting factor. A significant advantage of this reference is that the breakdown curve is very sharp resulting in low output resistance at low currents.

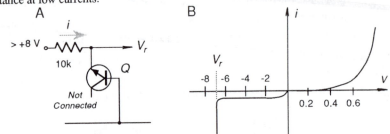

Fig. 4-3.8 Bipolar transistor as a micro-current zener diode
A: shunting-type circuit; B: volt-ampere characteristic

The so-called band gap references are often useful substitutes for zener diodes. They have typically ten times lower output impedance than low voltage zeners and can be obtained in a variety of nominal output voltages, ranging from 1.2 to 10V. Nowadays, a large variety of high quality voltage references with selectable outputs is available from many manufacturers.

4.3.3 Oscillators

Oscillators are generators of variable electrical signals. Any oscillator is essentially comprised of a circuit with a gain stage and some nonlinearity, and a certain amount of positive feedback. By definition, an oscillator is an unstable circuit (as opposed to an amplifier which better be stable) whose timing characteristics should be either stable or changeable according to a predetermined functional dependence. The latter is called a modulation. Generally, there are three types of electronic oscillators classified according to time-keeping components: the RC, the LC, and the crystal oscillators. In the RC-oscillators, the operating frequency is defined by capacitors and resistors, in the LC-oscillators - by the capacitive and inductive components. In the crystal oscillators, operating frequency is defined by a mechanical resonant in specific cuts of piezoelectric crystals, usually quartz.

There is a great variety of oscillation circuits, coverage of which is beyond the scope of this book. Below, we briefly describe some practical circuits which can be used for either direct interface with sensors or may generate excitation signals in an economical fashion.

Many various multivibrators can be built with logic circuits, for instance with NOR, NAND gates, or binary inverters. Also, many multivibrators can be designed with comparators or operational amplifiers having a high open loop gain. In all these oscillators, a capacitor is being charged, and the voltage across it is compared with another voltage, which is either constant or changing. The moment when both voltages are equal is detected by a comparator. A comparator is a two-input circuit which generates an output signal when its input signals are equal. The indication of a comparison is fed back to the RC-network to alter the capacitor charging in the opposite direction. Recharging in a new direction goes on until the next moment of comparison. This basic principle essentially requires the following minimum components: a capacitor, a charging circuit, and a threshold device (a comparator). Several relaxation oscillators are available from many manufacturers, for instance a very popular timer, type 555, which can operate in either monostable, or astable modes. For the illustration, below we describe just two discrete-component square-wave oscillators, however, there is a great variety of such circuits, which the reader can find in many books on operational amplifiers and digital systems, for instance [3].

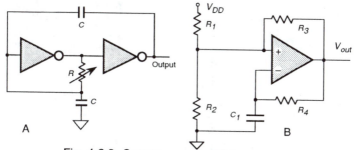

Fig. 4-3.9 Square-wave oscillators
A: with two logic inverters; B: with a comparator or OPAM

A simple square wave oscillator can be built with two logic inverters, for instance CMOS (Fig. 4-3.9A). A logic inverter has a threshold near a half of the power supply voltage. When voltage at its input crosses the threshold, the inverter generates output signal of the opposite direction. That is, if the input voltage is ramping up, at the moment when it reaches 1/2 of the power supply, the output voltage will be a negative-going transient. Timing properties of the oscillator are determined by the resistor R and the capacitor C. Both capacitors should be of the same value. Stability of the circuit primarily depends on the stabilities of the R and C.

A very popular square-wave oscillator (Fig. 4-3.9B) can be built with one OPAM or a voltage comparator[1]. The amplifier is surrounded by two feedback loops - one is negative (to an inverting input) and the other is positive (to a noninverting input). The positive feedback (R_3) controls the threshold level, while the negative loop charges and discharges timing capacitor C_1, through the resistor R_4. Frequency of this oscillator can be determined from

$$f = \frac{1}{R_4 C_1} \left[\ln(1 + \frac{R_1 \| R_2}{R_3}) \right]^{-1} , \qquad (4.3.7)$$

where $R1 \| R2$ is an equivalent resistance of parallel-connected R_1 and R_2.

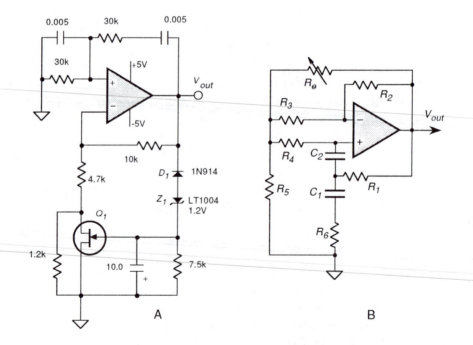

Fig. 4-3.10 Sine-wave oscillators
A: RC-circuit; B: RC-circuit for very low frequencies

---

[1] A voltage comparator differs from an operational amplifier by its faster speed response and the output circuit which is easier interfaceable with TTL and CMOS logic.

A sine wave oscillator is shown in Fig. 4-3.10A. The positive feedback is a Wein bridge, which for the shown component values is tuned to 1.5kHz. The Q_1, the zener diode Z_1, voltage reference and additional passive components form a negative feedback loop. The output voltage can be coupled to a sensor directly or through an additional buffer.

When generating low frequency sinusoidal signals, a circuit shown in Fig. 4-3.10B may be found quite useful. Its advantage is that the frequency can be independently controlled by a single variable resistance R_e. It can be shown [4] that the condition for oscillation is

$$R_6\frac{C_1}{C_2} + R_1\left(1 + \frac{C_1}{C_2}\right) = \frac{R_2R_4}{R_3} , \tag{4.3.8}$$

and the frequency of oscillation is given by

$$f = \frac{1}{2\pi\sqrt{C_1C_2R_1R_4}}\sqrt{\frac{R_e/R_5}{1 + (R_2/R_3)}} , \tag{4.3.9}$$

provided that

$$R_2R_4R_6 = R_1[R_4R_5 + R_3(R_4+R_5+R_6)] . \tag{4.3.10}$$

It follows from (4.3.9) that f is proportional to the square root of R_e and can be controlled by that resistor independently without conflicting with the conditions (4.3.8) and (4.3.10). In practice, it is convenient to select $C_1=C_2$ and $R_1=R_3=R_4=R_6$.

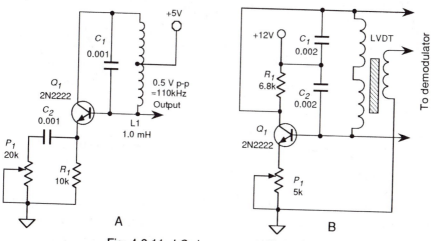

Fig. 4-3.11 LC sine-wave oscillators

Two other circuits (Fig. 4-3.11) also generate sine wave signals. They use npn-transistors as amplifiers and the LC-networks to set the oscillating frequency. The (B) circuit is especially useful for driving LVDT position sensors, as the sensor's transformer becomes a part of the oscillating circuit.

A radio-frequency oscillator can be used as a part of a capacitive occupancy detector to detect the presence of people in the vicinity of its antenna (Fig. 4-3.12)[1]. The antenna is a coil which together with the C_2 capacitors determines the oscillating frequency. The antenna is coupled to the environment through its distributed capacitance which somewhat reduces the frequency of the oscillator. When a person moves into the vicinity of the antenna, he/she brings in an additional capacitance which lowers the oscillator frequency even further. The output of the oscillator is coupled to a resonant tank (typically, an LC-network) which is tuned to a baseline frequency (near 30 MHz). A human intrusion lowers the frequency, thus substantially reducing the amplitude of the output voltage from the tank. The high frequency signal is rectified by a peak detector and a low frequency voltage is compared with a predetermined threshold by a comparator. This circuit employs an oscillator whose frequency is modulated by a sensing antenna. However, for other applications, the same circuit with a small inductor instead of antenna can produce stable sinusoidal oscillations.

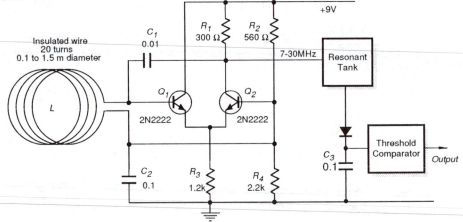

Fig. 4-3.12 *LC* radio-frequency oscillator as a capacitive occupancy detector

4.3.4 Drivers

As opposed to current generators, voltage drivers must produce output voltages which over broad ranges of the loads and operating frequencies are independent of the output currents. Sometimes, the drivers are called hard voltage sources. Usually, when the sensor which has to be

----

[1] See Section 6.3.

driven is purely resistive, a driver can be a simple output stage which can deliver sufficient current. However, when the load contains capacitances or inductances, that it, the load is reactive, the output stage becomes a more complex device.

In many instances, when the load is purely resistive, there still can be some capacitance associated with it. This may happen when the load is connected though lengthy wires or coaxial cables. A coaxial cable behaves as a capacitor connected from its central conductor to its shield if the length of the cable is less than 1/4 of the wavelength in the cable at the frequency of interest f. For a coaxial cable, this maximum length is given by

$$L \leq 0.0165 \frac{c}{f} , \qquad (4.3.11)$$

where c is the velocity of light in a coaxial cable dielectric.

For instance, if f=100 kHz, $L \leq 0.0165 \frac{3 \cdot 10^8}{10^5} = 49.5$, that is, a cable less than 49.5m (162.4 ft) long will behave as a capacitor connected in parallel with the load (Fig. 4-3.13A). For an R6-58A/U cable, the capacitance is 95pF/m. This capacitance must be considered for two reasons: for the speed and stability of the circuits. The instability results from the phase shift produced by the output resistance of the driver R_o and the loading capacitance C_L:

$$\varphi = arctang 2\pi f R_o C_L . \qquad (4.3.11)$$

For instance, for R_o=100Ω and C_L=1000pF, at f=1MHz, the phase shift $\varphi \approx 32°$. This shift significantly reduces the phase margin in a feedback network which may cause a substantial degradation of the response and a reduced ability to drive capacitive loads. The instability may be either overall, when an entire system oscillates, or localized when the driver alone becomes unstable. The local instabilities often can be cured by large by-pass capacitors (on the order of 10μF) across the power supply or the so-called Q-spoilers consisting of a serial connection of 3-10Ω resistor and a disc ceramic capacitor connected from the power supply pins of the driver chip to ground.

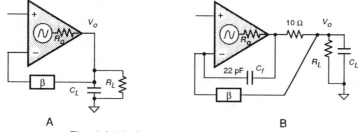

Fig. 4-3.13 Driving a capacitive load
A: a load capacitor is coupled to the driver's input through a feedback;
B: decoupling of a capacitive load

To make a driver stage more tolerant to capacitive loads, it can be isolated by a small serial resistor as it is shown in Fig. 4-3.13B. A small capacitive feedback (C_f) to the inverting input of the amplifier, and a 10Ω resistor may allow to drive loads as large as $0.5\mu F$. However, in any particular case it is recommended to find the best values for the resistor and the capacitor experimentally.

4.4   ANALOG-TO-DIGITAL CONVERTERS

4.4.1   Basic concepts

The analog-to-digital (A/D) converters range from discrete circuits, to monolithic ICs (integrated circuits), to high-performance hybrid circuits, modules, and even boxes. Also, the converters are available as standard cells for custom and semicustom application specific integrated circuits (ASIC). The A/D converters transform analog data – usually voltage – into an equivalent digital form, compatible with digital data processing devices. Key characteristics of A/D converters include absolute and relative accuracy, linearity, no-missing codes, resolution, conversion speed, stability and price. Quite often, when price is of a major concern, discrete component or monolithic IC versions are the most efficient. The most popular A/D converters are based on a successive-approximation technique because of an inherently good compromise between speed and accuracy. However, other popular techniques are used in a large variety of applications, especially when no high conversion speed is required. These include dual-ramp, quad-slope, and voltage-to-frequency (V/F) converters. The art of an A/D conversion is well developed. Here, we briefly review some popular architectures of the converters, however, for detailed descriptions the reader should refer to specialized texts, such as [5].

The best known digital code is *binary* (base 2). Binary codes are most familiar in representing integers, i.e., in a natural binary integer code having n bits, the LSB (least significant bit) has a weight of 2^0 (i.e., 1), the next bit has a weight of 2^1 (i.e., 2), and so on up to MSB (most significant bit), which has a weight of 2^{n-1} (i.e., $2^n/2$). The value of a binary number is obtained by adding up the weights of all nonzero bits. When the weighted bits are added up, they form a unique number having any value from 0 to 2^n-1. Each additional trailing zero bit, if present, essentially doubles the size of the number.

When converting signals from analog sensors, because full scale is independent of the number of bits of resolution, a more useful coding is *fractional* binary [5], which is always normalized to full scale. Integer binary can be interpreted as fractional binary if all integer values are divided by

2^n. For example, the MSB has a weight of 1/2 (i.e., $2^{n-1}/2^n = 2^{-1}$), the next bit has a weight of 1/4 (i.e., 2^{-2}), and so forth down to the LSB, which has a weight of $1/2^n$ (i.e., 2^{-n}). When the weighted bits are added up, they form a number with any of 2^n values, from 0 to $(1-2^{-n})$ of full scale. Additional bits simply provide more fine structure without affecting the full-scale range. To illustrate these relationships, Table 4-2 lists 16 permutations of 4-bit's worth of 1's and 0's, with their binary weights, and the equivalent numbers expressed as both decimal and binary integers and fractions.

Table 4-2 Integer and fractional binary codes
(adapted from [5])

Decimal Fraction	Binary fraction	MSB x1/2	Bit2 x1/4	Bit3 x1/6	Bit4 x1/16	Binary integer	Decimal integer
0	0.0000	0	0	0	0	0000	0
1/16=2-4(LSB)	0.0001	0	0	0	1	0001	1
2/16=1/6	0.0010	0	0	1	0	0010	2
3/16=1/8+1/16	0.0011	0	0	1	1	0011	3
4/16=1/4	0.0100	0	1	0	0	0100	4
5/16=1/4+1/16	0.0101	0	1	0	1	0101	5
6/16=1/4+1/8	0.0110	0	1	1	0	0110	6
7/16=1/4+1/8+1/16	0.0111	0	1	1	1	0111	7
8/16=1/2 (MSB)	0.1000	1	0	0	0	1000	8
9/16=1/2+1/16	0.1001	1	0	0	1	1001	9
10/16=1/2+1/8	0.1010	1	0	1	0	1010	10
11/16=1/2+1/8+1/16	0.1011	1	0	1	1	1011	11
12/16=1/2+1/4	0.1100	1	1	0	0	1100	12
13/16=1/2+1/4+1/16	0.1101	1	1	0	1	1101	13
14/16=1/2+1/4+1/8	0.1110	1	1	1	0	1110	14
15/16=1/2+1/4+1/8+1/16	0.1111	1	1	1	1	1111	15

When all bits are "1" in natural binary, the fractional number value is $1-2^{-n}$, or normalized full-scale less 1 LSB (1-1/16=15/16 in the example). Strictly speaking, the number that is represented, written with an "integer point", is 0.1111 (=1-0.0001). However, it is almost universal practice to write the code simply as the integer 1111 (i.e., "15") with the fractional nature of the corresponding number understood: "1111" $\rightarrow$ 1111/(1111+1), or 15/16.

For convenience, Table 4-3 lists bit weights in binary for numbers having up to 20 bits. However, the practical range for the vast majority of sensors rarely exceeds 16 bits.

The weight assigned to the LSB is the resolution of numbers having n bits. The dB column represents the logarithm (base 10) of the ratio of the LSB value to unity (full scale), multiplied by 20. Each successive power of 2 represents a change of 6.02dB [i.e., $20 \log_{10}(2)$] or "6dB/octave".

Table 4-3 Binary bit weights and resolutions

BIT	2^{-n}	$1/2^n$ fraction	dB	$1/2^n$ decimal	%	ppm
FS	2^0	1	0	1.0	100	1,000,000
MSB	2^{-1}	1/2	-6	0.5	50	500,000
2	2^{-2}	1/4	-12	0.25	25	250,000
3	2^{-3}	1/8	-18.1	0.125	12.5	125,000
4	2^{-4}	1/16	-24.1	0.0625	6.2	62,500
5	2^{-5}	1/32	-30.1	0.03125	3.1	31,250
6	2^{-6}	1/64	-36.1	0.015625	1.6	15,625
7	2^{-7}	1/128	-42.1	0.007812	0.8	7,812
8	2^{-8}	1/256	-48.2	0.003906	0.4	3,906
9	2^{-9}	1/512	-54.2	0.001953	0.2	1,953
10	2^{-10}	1/1,024	-60.2	0.0009766	0.1	977
11	2^{-11}	1/2,048	-66.2	0.00048828	0.05	488
12	2^{-12}	1/4,096	-72.2	0.00024414	0.024	244
13	2^{-13}	1/8,192	-78.3	0.00012207	0.012	122
14	2^{-14}	1/16,384	-84.3	0.000061035	0.006	61
15	2^{-15}	1/32,768	-90.3	0.0000305176	0.003	31
16	2^{-16}	1/65,536	-96.3	0.0000152588	0.0015	15
17	2^{-17}	1/131,072	-102.3	0.00000762939	0.0008	7.6
18	2^{-18}	1/262,144	-108.4	0.000003814697	0.0004	3.8
19	2^{-19}	1/524,288	-114.4	0.000001907349	0.0002	1.9
20	2^{-20}	1/1,048,576	-120.4	0.0000009536743	0.0001	0.95

4.4.2 V/F converters

The voltage-to-frequency (V/F) converters can provide a high-resolution conversion, and such useful for sensors special features as a long-term integration (from seconds to years), a digital-to-frequency conversion (together with a D/A converter), a frequency modulation, a voltage isolation, and an arbitrary frequency division and multiplication. The converter accepts an analog output from the sensor which can be either voltage or current. In some cases, a sensor may become a part of an A/D converter as it is illustrated in Section 4.5. Here, however, we will discuss only the conversion of voltage to frequency, or, in other words, to a number of square pulses per unit of time. For a given input value, the number of pulses per second, or *frequency*, is proportional to the *average* value of the input value.

By using a V/F converter, an A/D can be performed in the most simple and economical manner. The time required to convert an analog voltage into a digital number is related to the full-scale frequency of the V/F converter and the required resolution. Generally, the V/F converters are relatively slow, as compared with successive approximation devices, however, they are quite appropriate for the vast majority of sensor applications. When acting as an A/D converter, the V/F converter is coupled to a counter which is clocked with the required sampling rate. For instance, if a full-scale frequency of the converter is 32kHz, and the counter is clocked 8 times per second, the highest number of pulses which can be accumulated every counting cycle is 4,000 which approximately corresponds to a resolution of 12 bit (see table 4-3). By using the same combination of components: the V/F converter and the counter, an integrator can be build for the applications, where the stimulus needs to be integrated over a certain time. The counter accumulates pulses over the gated interval rather than as an average number of pulses per counting cycle.

Another useful feature of a V/F converter is that its pulses can be easily transmitted through communication lines. The pulsed signal is much less susceptible to noisy environment than a high resolution analog signal.

In the ideal case, the output frequency f_{out} of the converter is proportional to the input voltage V_{in}:

$$\frac{f_{out}}{f_{FS}} = \frac{V_{in}}{V_{FS}} ,$$

(4.4.1)

where f_{FS} and V_{FS} are the full-scale frequency and input voltage, respectively. For a given converter, ratio $f_{FS}/V_{FS}=G$ is constant and is called a conversion factor, then:

$$f_{out} = GV_{in} .$$

(4.4.2)

There are several known types of V/F converters. The most popular of them are the multivibrator and the charge-balance configurations.

A *multivibrator V/F converter* employs a free-running square-wave oscillator where charge-discharge currents of a timing capacitor are controlled by the input signal (Fig. 4-4.1). Input voltage V_{in} is amplified by a differential amplifier whose output signal controls two voltage-to-current converters (transistors Q_1 and Q_2). A precision multivibrator alternatively connects timing capacitor C to both current converters. The capacitor is charged for a half of period through transistor Q_1 by the current i_a. During the second half of the timing period, it is discharged by the current i_b through transistor Q_2. Since currents i_a and i_b are controlled by the input signal, the capacitor charging and discharging slopes vary accordingly, thus changing the oscillating frequency. An apparent advantage of this circuit is its simplicity and potentially very low power consumption, however, its ability to reject high frequency noise in the input signal is not as good as in the charge-balance architecture.

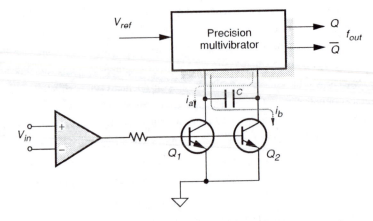

Fig. 4-4.1 Multivibrator type of a voltage-to-frequency converter

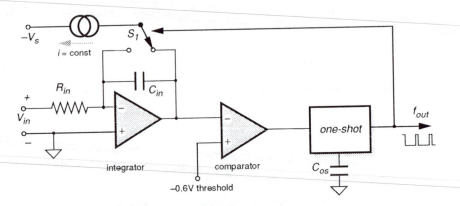

Fig. 4-4.2 Charge-balance V/F converter

The *charge-balance* type of converter employs an analog integrator and a voltage comparator as shown in Fig. 4-4.2. This circuit has such advantages as high speed, high linearity, and good noise rejection. The circuit is available in an integral form from several manufacturers, for instance, ADVFC32 and AD650 from Analog Devices, LM331 from National Semiconductors. The converter operates as follows. Input voltage V_{in} is applied to an integrator through the input resistor

R_{in}. The integrating capacitor is connected as a negative feedback loop to the operational amplifier whose output voltage is compared with a small negative threshold of -0.6V. The integrator generates a saw-tooth voltage (Fig. 4-4.4) which at the moment of comparison with the threshold results in a transient at the comparator's output. That transient enables a one-shot generator which produces a square pulse of a fixed duration t_{os}. A precision current source generates constant current i which is alternatively applied either to the summing node of the integrator, or to its output. The switch S_1 is controlled by the one-shot pulses. When the current source is connected to the summing node, it delivers a precisely defined packet of charge $\Delta Q = i t_{os}$ to the integrating capacitor. The same summing node also receives an input charge through the resistor R_{in}, thus the net charge is accumulated on the integrating capacitor C_{in}.

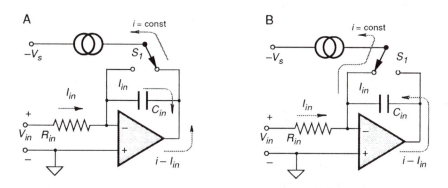

Fig. 4-4.3 Integrate and deintegrate phases in a charge-balance converter

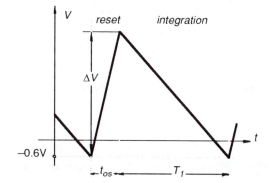

Fig. 4-4.4 Integrator output in a charge-balance converter

When the threshold is reached, the one-shot is triggered and the switch, S_1, changes its state to high, thus initiating a reset period (Fig. 4-4.3B). During the reset period, the current source delivers its current to the summing node of the integrator. The input current charges the integrating capacitor upward. The total voltage between the threshold value and the end of the de-integration is determined by the duration of a one-shot pulse:

$$\Delta V = t_{os}\frac{dV}{dt} = t_{os}\frac{i - I_{in}}{C_{in}} \quad . \tag{4.4.1}$$

When the output signal of the one-shot circuit goes low, switch S_1 diverts current i to the output terminal of an integrator, which makes no effect on the state of the integrating capacitor C_{in}. That is, the current source sinks a portion of the output current from the operational amplifier. This time is called the integration period (Figs. 4-4.3A and 4-4.4). During the integration, the positive input voltage delivers current $I_{in}=V_{in}/R_{in}$ to the capacitor C_{in}. This causes the integrator to ramp down from its positive voltage with the rate proportional to V_{in}. The amount of time required to reach the comparator's threshold is:

$$T_1 = \frac{\Delta V}{dV/dt} = t_{os}\frac{i - I_{in}}{C_{in}}\frac{1}{I_{in}/C_{in}} = t_{os}\frac{i - I_{in}}{I_{in}} \quad . \tag{4.4.2}$$

It is seen that the capacitor value does not effect duration of the integration period. The output frequency is determined by:

$$f_{out} = \frac{1}{t_{os} + T_1} = \frac{I_{in}}{t_{os}\cdot i} = \frac{V_{in}}{R_{in}}\frac{1}{t_{os}\cdot i} \quad . \tag{4.4.3}$$

Therefore, the frequency of one-shot pulses is proportional to the input voltage. It depends also on quality of the integrating resistor, stability of the current generator, and a one-shot circuit. With a careful design, this type of a V/F converter may reach nonlinearity error below 100ppm and can generate frequencies from 1 Hz to 1 MHz.

A major advantage of the integrating-type converters, such as a charge-balanced V/F converter, is the ability to reject large amounts of additive noise. By integrating of the measurement noise is reduced or even totally eliminated. Pulses from the converter are accumulated for a gated period T in a counter. Then, the counter behaves like a filter having a transfer function in the form

$$H(f) = \frac{\sin\pi f T}{\pi f T} \quad , \tag{4.4.4}$$

where f is the frequency of pulses. For low frequencies, this transfer function is close to unity, meaning that the converter and the counter make correct measurements. However, for a frequency $1/T$ the transfer function is zero, meaning that these frequencies are completely rejected. For example, if gating time $T=16.67$ ms which corresponds to a frequency of 60Hz – the power line frequency which is a source of substantial noise in many sensors, then the 60Hz noise will be rejected. Moreover, the multiple frequencies (120Hz, 180Hz, 240Hz, and so on) will also be rejected.

4.4.3 Dual-slope converter

This type of converter performs indirect conversion of the input voltage. First, it converts V_{in} into a function of time, then the time function is converted into a digital number by a pulse counter. Dual slope converters are quite slow, however, for stimuli which don't exhibit fast changes, they are often the converters of choice, due to their simplicity, cost effectiveness, noise immunity, and potentially high resolution. The operating principle of the converter is as follows (Fig. 4-4.5). Like in a charge-balanced converter, there is an integrator and a threshold comparator. The threshold level is set at zero (ground) or any other suitable constant voltage. The integrator can be selectively connected through the analog selector S_1 either to the input voltage or to the reference voltage. In this simplified schematic, the input voltage is negative, while the reference voltage is positive. However, by shifting a dc level of the input signal (with the help of an additional OPAM), the circuit will be able to convert bipolar input signals as well. The output of the comparator sends a signal to the control logic when the integrator's output voltage crosses zero. The logic controls both the selector S_1 and the reset switch S_2, which serves for discharging the integrating capacitor, C_{in}.

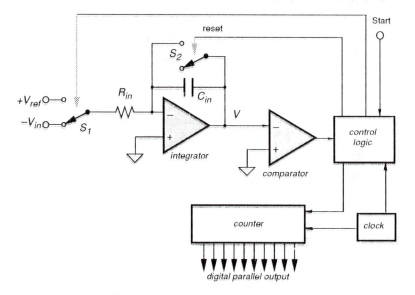

Fig. 4-4.5 Dual-slope A/D converter

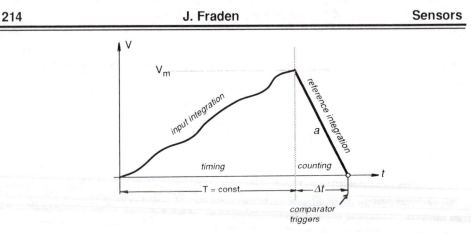

Fig. 4-4.6 Integrator output in a dual-slope A/D converter

When the start input is enabled, S_1 connects the integrator to the input signal and the logic starts a timer. The timer is preset for a fixed time interval T. During that time, the integrator generates a positive-going ramp (Fig. 4-4.6) which changes according to the input signal. It should be noted that the input signal does not have to be constant. Any variations in the signal are averaged during the integration process. Upon elapsing time T, the integrator output voltage reaches the level

$$V_m = \overline{V}_{in} \frac{T}{R_{in}C_{in}} , \qquad (4.4.5)$$

where $\overline{V}_{in}$ is an average input signal during time T. At that moment, S_1 switches to the reference voltage which is of the opposite polarity with respect to the input signal, thus setting the deintegrate phase (a reference voltage integration), during which the integrator's voltage ramps downward until it crosses a zero threshold. The integral of the reference has slope

$$a = -\frac{V_{ref}}{R_{in}C_{in}} . \qquad (4.4.6)$$

During the deintegrate phase, the counter counts clock pulses. When the comparator indicates a zero-crossing, the count is stopped and the analog integrator is reset by discharging its capacitor through S_2. The charge at the capacitor gained during the input signal integrate phase is precisely equal to the charge lost during the reference deintegrate phase. Therefore, the following holds

$$\overline{V}_{in} \frac{T}{R_{in}C_{in}} = V_{ref} \frac{\Delta t}{R_{in}C_{in}} , \qquad (4.4.7)$$

which leads to

$$\frac{\overline{V}_{in}}{V_{ref}} = \frac{\Delta t}{T} . \qquad (4.4.8)$$

Therefore, the ratio of the average input voltage and the reference voltage is replaced by the ratio of two time intervals. Then, the counter does the next step - it converts the time interval Δt into a digital form by counting the clock pulses during Δt. The total count is the measure of $\overline{V}_{in}$ (V_{ref} and T are constants).

The dual slope has the same advantage as the charge-balanced converter – they both reject frequencies $1/T$ corresponding to the integrate timing. It should be noted that selecting time $T=200$ms will reject noise produced by both 50 and 60Hz, thus making the converter immune to the power line noise originated at either standard frequency. Further, the conversion accuracy is independent on component values and the clock frequency stability, because the same clock sets timing T and the counter. The resolution of the conversion is limited only by the analog resolution, hence, the excellent fine structure of the signal may be represented by more bits than would be needed to maintain a given level of scale-factor accuracy. The integration provides rejection of high frequency noise and averages all signal instabilities during the interval T. The throughput of a dual-slope conversion is limited to somewhat less than $1/2T$ conversions per second.

To minimize errors produced in the analog portion of the circuit (the integrator and the comparator) a third timing phase is usually introduced. It is called an auto-zero phase because during that phase, the capacitor C_{in} is charged with zero-drift errors, which are then introduced in the opposite sense during the integration, in order to nullify them. An alternative way to reduce the static error is to store the auto-zero counts and then digitally subtract them.

Dual-slope converters are often implemented as a combination of analog components (OPAMs, switches, resistors, and capacitors) and a microprocessor, which handles functions of timing, control logic, and counting. Sometimes, the analog portion is packaged in a separate integrated circuit. An example is the TS500 module from Texas Instruments.

4.4.4 Successive approximation converter

These converters are widely used in a monolithic form thanks to their high speed (to 1MHz throughput rates) and high resolution (to 16 bit). Conversion time is fixed and independent of the input signal. Each conversion is unique, as the internal logic and registers are cleared after each conversion, thus making these A/D converters suitable for the multichannel multiplexing. The converter (Fig. 4-4.7A) consists of a precision voltage comparator, a module comprising shifter registers and a control logic, and a digital-to-analog converter (D/A) which serves as a feedback from the digital outputs to the input analog comparator.

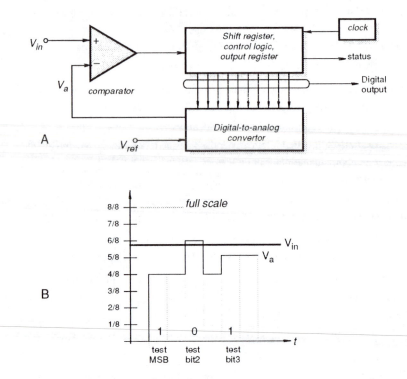

Fig. 4-4.7 Successive-approximation A/D converter
A: block diagram; B: 3-bit weighing

The conversion technique consists of comparing the unknown input, V_{in}, against a precise voltage V_a or current generated by a D/A converter. The conversion technique is similar to a weighing process using a balance with a set of n binary weights (for instance, 1/2 kg, 1/4 kg, 1/8 kg, 1/16 kg, etc. up to total of 1 kg). Before the conversion cycles, all the registers must be cleared and the comparator's output is HIGH. The D/A converter has MSB (1/2 scale) at its inputs and generates an appropriate analog voltage V_a equal to 1/2 of a full scale input signal. If the input is still greater than the D/A voltage (Fig. 4-4.8), the comparator remains HIGH, causing "1" at the register's output. Then, the next bit (2/8=1/4 of FS) is tried. If the second bit does not add enough weight to exceed the input, the comparator remains HIGH ("1" at the output), and the third bit is tried. However, if the second bit tips the scale too far, the comparator goes LOW resulting in "0" in the register, and the third bit is tried. The process continues in order of descending bit weight until the last bit has been tried. After the completion, the status line indicates the end of conversion and data can be read from the register as a valid number corresponding to the input signal.

To make the conversion valid, the input signal V_{in} must not change until all the bits are tried, otherwise, the digital reading may be erroneous. To avoid any problems with the changing input, a successive approximation converter usually is supplied with a sample-and-hold (S&H) circuit. This circuit is a short-time analog memory which samples the input signal and stores it as a dc voltage during an entire conversion cycle.

4.4.5 Resolution extension

In many data acquisition systems built with discrete components or ASIC[1] circuits, especially when cost is a serious consideration, the maximum resolution of an available A/D converter is often limited to 8 bits. However, this may be not nearly enough for the correct representation of a stimulus. One method of achieving higher resolution is use a dual-slope A/D converter whose resolution limited only by the available counter rate and the speed response of a comparator[2]. Another method is to use an 8-bit A/D converter (for instance, of a successive approximation type) with a resolution extension circuit. Such a circuit can boost the resolution by several bits, for instance from 8 to 12. A block-diagram of the circuit is shown in Fig. 4-4.8. In addition to a conventional 8-bit A/D converter, it includes a digital-to-analog (D/A) converter, a subtraction circuit and an amplifier having gain A. In the ASIC or discrete circuits, a D/A converter may be shared with an A/D part (See Fig. 4-4.7A).

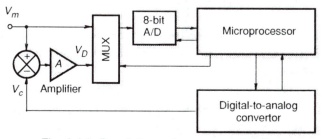

Fig. 4-4.8 Resolution enhancement circuit

The input signal V_m has a full-scale value E, thus for an 8-bit converter, the initial resolution will be

$$R_o = E/(2^8 - 1) = E/255 , \qquad (4.4.9)$$

---

[1] Application Specific Integrated Circuit (ASIC)
[2] A resolution should not be confused with accuracy.

which is expressed in volts per bit. For instance, for a 5-volt full scale, the 8-bit resolution is 19.6 mV/bit. Initially, the multiplexer (MUX) connects the input signal to the A/D converter which produces the output digital value M which is expressed in bits. Then, the microprocessor outputs that value to a D/A converter which produces output analog voltage V_c, which is an approximation of the input signal. This voltage is subtracted from the input signal and amplified by the amplifier to value

$$V_D = (V_m - V_c) \cdot A \qquad (4.4.10)$$

The voltage V_D is an amplified error between the actual and digitally represented input signals. For a full scale input signal, the maximum error $(V_m - V_c)$ is equal to a resolution of an A/D converter, therefore, for an 8-bit conversion V_D=19.6A mV. The multiplexer connects that voltage to the A/D converter which converts V_D to a digital value C:

$$C = \frac{V_D}{R_o} = (V_m - V_c)\frac{A}{R_o} \qquad (4.4.11)$$

As a result, the microprocessor combines two digital values: M and C, where C represents the high resolution bits. If A=255, then for the 5 volt full scale, LSB$\approx$77μV which corresponds to a total resolution of 16 bit. In practice, it is hard to achieve such a high resolution because of the errors originated in the D/A converter, reference voltage, amplifier's drift, noise, etc. Nevertheless, the method is quite efficient when a modest resolution of 10 or 12 bit is deemed to be sufficient.

4.5   DIRECT DIGITIZATION

Almost all sensors produce low level signals. To bring these signals to levels compatible with data processing devices, amplifiers are generally required. Unfortunately, amplifiers and connecting cables and wires may introduce additional errors, add cost to the instrument and increase complexity. Some emerging trends in the sensor-based systems are causing the use of signal conditioning amplifiers to be reevaluated (at least for some transducers) [6]. In particular, many industrial sensor-fed systems are employing digital transmission and processing equipment. These trends point toward direct digitization of sensor outputs - a difficult task. It is especially true when a sensor-circuit integration on a single chip is considered.

Classical A/D conversion techniques emphasize high level input ranges. This allows LSB step size to be as large as possible, minimizing offset and noise error. For this reason, a minimum LSB signal is always selected to be at least 100-200μV. Therefore, a direct connection of many sensors, for instance, RTD temperature transducers or piezoresistive strain gauges is unrealistic.

Such transducer's full scale (FS) output may be limited by several millivolts, meaning that a 10-bit A/D converter must have about $1\mu V$ LSB.

Direct digitization of transducers eliminates a dc gain stage and may yield a better performance without sacrificing accuracy. The main idea behind a direct digitization is to incorporate a sensor into a signal converter, for instance, an A/D converter or an impedance-to-frequency converter. All such converters perform a modulation process and, therefore, are nonlinear devices. Hence, they have some kind of nonlinear circuit, often a threshold comparator. Shifting the threshold level, for instance, may modulate the output signal, which is a desirable effect.

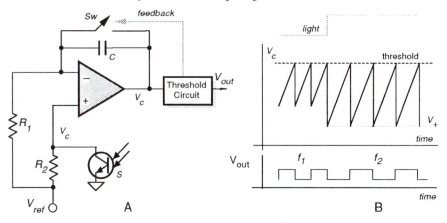

Fig. 4-5.1 Simplified schematic (A) and voltages (B) of a light-modulated oscillator

Fig. 4-5.1 shows a simplified circuit diagram of a modulating oscillator. It is comprised of an integrator built with an operational amplifier and a threshold circuit. The voltage across capacitor C is an integral of the current whose value is proportional to voltage in the noninverting input of the operational amplifier. When that voltage reaches the threshold, switch SW closes, thus fully discharging the capacitor. The capacitor starts integrating the current again until the cycle repeats. The operating point of the amplifier is defined by the resistor R_2, a phototransistor S, and the reference voltage V_{ref}. A change in light flux which is incident on the base of the transistor, changes its collector current, thus shifting the operation point. A similar circuit may be used for direct conversion of a resistive transducer, for instance, a thermistor. The circuit can be further modified for the accuracy enhancement, such as for the compensation of the amplifier's offset voltage or bias current, temperature drift, etc.

Capacitive sensors are very popular in many applications. Nowadays, micromachining technology allows us to fabricate small monolithic capacitive sensors. Capacitive pressure transducers employ a thin silicon diaphragm as a movable plate of the variable-gap capacitor which is com-

pleted by a metal electrode on the opposing plate. The principle problem in these capacitors is a
relatively low capacitance value per unit area (about $2pF/mm^2$) and resulting large die sizes. A typical device offers a zero pressure capacitance on the order of few picofarads, so that an 8 bit resolution requires the detection of capacitive shifts on the order of 50fF or less (1 femtofarad=10^{-15}F).
It is obvious that any external measurement circuit will be totally impractical, as parasitic capacitance of connecting conductors can be on the order of 1pF - too much with respect to the capacitance of the sensor. Therefore, the only way to make such a sensor practical is to build an interface
circuit as an integral part of the sensor itself. One quite effective way to design such a circuit is to
use a switched capacitor technique. The technique is based on charge transfer from one capacitor to
another by means of solid-state analog switches.

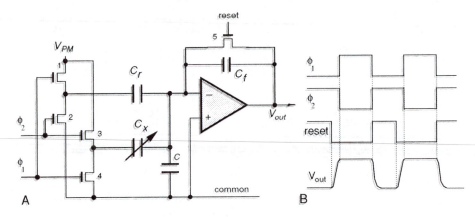

Fig. 4-5.2 Simplified schematic (A)
and timing diagrams (B) of a differential capacitance-to-voltage converter

Fig. 4-5.2A shows a simplified circuit diagram of a switched-capacitor converter [7], where
variable capacitance C_x and reference capacitance C_r are parts of a symmetrical silicon pressure
sensor. Monolithic MOS switches (1-4) are driven by opposite phase clock pulses ϕ_1 and ϕ_2.
When the clocks switch, a charge appears at the common capacitance node. The charge is provided
by the constant voltage source V_{PM} and is proportional to (C_x-C_r) and, therefore to applied
pressure to the sensor. This charge is applied to a charge-to-voltage converter which includes an
operational amplifier, integrating capacitor C_f, and MOS discharge (reset) switch 5. The output
signal is variable-amplitude pulses (Fig. 4-5.2B) which can be transmitted through the
communication line and either demodulated to produce linear signal or can be further converted into
digital data. So long as the open loop gain of the integrating OPAM is high, the output voltage is
insensitive to stray input capacitance C, offset voltage, and temperature drift. The minimum

detectable signal (noise floor) is determined by the component noise and temperature drifts of the components. The circuit analysis shows that the minimum noise power occurs when the integration capacitor C_f is approximately equal to frequency compensation capacitor of the OPAM.

When the MOS reset switch goes from the on-state to the off-state, the switching signal injects some charge from the gate of the reset transistor to the input summing node of the OPAM (inverting input). This charge propagated through the gate-to-channel capacitance of the MOS transistor 5. An injection charge results in an offset voltage at the output. This error can be compensated for by a charge-canceling device [8] which can improve the signal-to-noise ratio by two orders of magnitude of the uncompensated charge. Temperature drift of the circuit can be expressed as:

$$\frac{dV_{out}}{dT} = V_{PM} \frac{(C_x - C_r)}{C_f} (T_{Cr} - T_{Cf}) \; , \tag{4.5.1}$$

where T_{Cr} is the nominal temperature coefficient of C_x and C_r, and T_{Cf} is the temperature coefficient of integrating capacitor C_f. This equation suggests that the temperature drift primarily depends on the mismatch of the capacitances in the sensor. Typical transfer functions of the circuit for two different integrating capacitors C_f are shown in Fig. 4-5.3. An experimental circuit similar to the above was built in a silicon die having dimensions 0.68x0.9 mm [9] using the standard CMOS process. The circuit operates with clock frequencies in the range from 10 to 100 kHz.

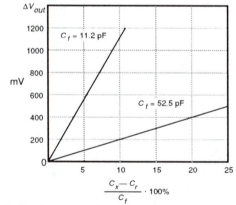

Fig. 4-5.3 Transfer function of a capacitance-to-voltage converter
for two values of integrating capacitor
(adapted from [7])

Presently, various manufacturers supply several monolithic integrated circuits which internally perform essential steps to interface sensors with peripheral digital devices. An example is SSC8830 Prosensor™ (TLSI, Inc., Farmingdale, N.Y.) which directly accepts a low voltage signal from a sensor, amplifies and digitizes it, and serially outputs a digital pulse stream to a processor or microcontroller. The analog portion of the device consists of a differential input, chopper-stabilized amplifier with an adjustable gain and a 5V regulator. A chopper amplifier allows to eliminate the amplifier's offset as a source of error. The chopping process is a multiplication of an input analog signal and a sampling frequency signal. This results in shifting a spectral characteristic of the input signal from the low frequency into a high frequency domain where amplification is free from the offset influence and $1/f$-type noise. After the amplification, the signal is converted back into a low frequency domain and turned into a digital format.

4.6   RATIOMETRIC CIRCUITS

A powerful method of improving the accuracy of a sensor is a ratiometric technique, which is one of the ways of signal conditioning. It should be emphasized, however, that the method is useful only if a source of error is of a multiplicative nature but not additive. That is, the technique would be useless to reduce, for instance, thermal noise. On the other hand, it is quite potent to solve such problems as dependence of a sensor's sensitivity to such factors as power supply instability, ambient temperature, humidity, pressure, effects of aging, etc. The technique essentially requires the use of two sensors where one is the acting sensor which responds to an external stimulus and the other is a reference sensor which is either shielded from that stimulus or is insensitive to it. Both sensors must be exposed to all other external effects which may multiplicatively change their sensitivity. The second sensor, which is called *reference*, must be subjected to a reference stimulus, which is ultimately stable during the life time of the product. In many practical systems, the reference sensor must not necessarily be exactly similar to the acting sensor, however its physical properties which are subject to instabilities should be the same. For example, Fig. 4-6.1A shows a simple temperature detector where the acting sensor is a negative temperature coefficient (NTC) thermistor R_T. A stable reference resistor R_o has a value equal to the resistance of the thermistor at some reference temperature, for instance at 25°C. Both are connected via an analog multiplexer to an amplifier with a feedback resistor R. The output signals produced by the sensor and the reference resistor respectively are:

$$V_N = - \frac{ER}{R_T} \, , \qquad\qquad\qquad (4.6.1)$$

$$V_D = - \frac{ER}{R_o} \, .$$

It is seen that both voltages are functions of a power supply voltage E and the circuit gain, which is defined by resistor R. That resistor as well as the power supply may be the sources of error. If two output voltages are fed into a divider circuit, the resulting signal may be expressed as $V_o = kV_N/V_D = kR_o/R_T$, where k is the divider's gain. The output signal is not a subject of neither power supply voltage nor the amplifier gain. It depends only on the sensor and its reference resistor. This is true only if spurious variables (like the power supply or amplifier's gain) do not change rapidly. That is, they must not change appreciably within the multiplexing period. This requirement determines the rate of multiplexing.

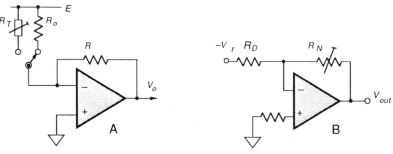

Fig. 4-6.1 Ratiometric temperature detector (A)
and analog divider of resistive values (B)

A ratiometric technique essentially requires the use of a division It can be performed by two standard methods: digital and analog. In a digital form, output signals from both the acting and the reference sensors are multiplexed and converted into binary codes in an analog-to-digital (A/D) converter. Subsequently, a computer (for instance a microprocessor or a state-machine) performs an operation of a division. In an analog form, a divider may be a part of a signal conditioner or the interface circuit. An analog "divider" circuit (Fig. 4-6.2A) produces an output voltage or current proportional to the ratio of two input voltages or currents:

$$V_o = k\frac{V_N}{V_D} , \tag{4.6.2}$$

where the numerator is denoted as V_N, the denominator V_D and k is equal to the output voltage, when $V_N = V_D$. The operating ranges of the variables (quadrants of operation) is defined by the polarity and magnitude ranges of the numerator and denominator inputs, and of the output. For instance, if V_N and V_D are both either positive or negative, the divider is of a 1-quadrant type. If the numerator is bipolar, the divider is 2-quadrant. Generally, the denominator is restricted to a single polarity, since the transition from one polarity to another would require the denominator to pass through zero, which would call for an infinite output (unless the numerator is also zero). In practice, the denominator is a signal from a reference sensor which usually is of a constant value.

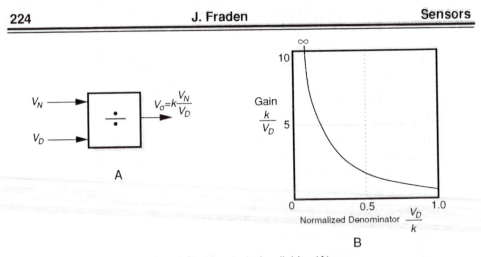

Fig. 4-6.2 Symbol of a divider (A)
and gain of a divider as a function of a denominator (B)

Division has long been the most difficult of the four arithmetic functions to implement with analog circuits. This difficulty stems primarily from the nature of division: the magnitude of a ratio becomes quite large, approaching infinity, for a denominator that is approaching zero (and a nonzero numerator). Thus, an ideal divider must have a potentially infinite gain and infinite dynamic range. For a real divider, both of these factors are limited by the magnification of drift and noise at low values of V_D. That is, the gain of a divider for a numerator is inversely dependent on the value of the denominator (Fig. 4-6.2B). Thus, the overall error is the net effect of several factors, such as gain dependence of denominator, numerator and denominator input errors, like offsets, noise and drift (which must be much smaller than the smallest values of input signals). Besides, the output of the divider must be constant for constant ratios of numerator and denominator, independent of their magnitudes. For example: $10/10=0.01/0.01=1$ and $1/10=0.001/0.01=0.1$.

There are several ways to implement a linear ratio circuit [10]: Below we describe several circuits which were proven to be quite effective for the sensor applications. The simplest circuit shown in Fig. 4-6.1B uses only one OPAM and a reference voltage source. It is useful only for resistive sensors, such as thermistor. The acting sensor R_N is connected in a feedback between the output and the inverting input, while a reference sensor R_D is connected to the negative voltage reference. A negative reference is required for the positive output voltage which is defined as:

$$V_o = -V_r \frac{R_N}{R_D} .$$

(4.6.3)

It should be noted that the acting sensor must be floating, i.e., it can not be connected to the ground or any other potential, except as shown in the schematic.

An analog divider which uses currents as input signals is shown in Fig. 4-6.3 A dual JFET contains a matched pair of two transistors. The circuit employs two OPAMs connected as virtual grounds. A concept of the virtual ground is very powerful in the art of the analog circuit design. The idea is based on the fact that because of a high open loop gain an inverting input of an OPAM has almost the same potential as the noninverting input, which is grounded. Thus, the inverting input is at the ground potential while being physically isolated from the actual ground. That implies the name the *virtual ground*.

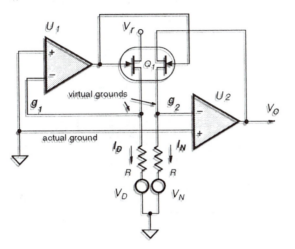

Fig. 4-6.3 Analog divider

The circuit of Fig. 4-6.3 has two current inputs I_D and I_N, meaning that input signals must come from current sources rather than from voltage sources. If current sources are not available, they can be very easily formed by voltage sources and serially connected resistors R_D and R_N. Sources of both transistors are connected to virtual grounds (g_1 and g_2). For better accuracy, the output signal must be maintained within 100 mV, thus a voltage gain stage may be required before further signal processing. The circuit operates as follows. OPAM $U1$ sets a channel resistance of the left JFET transistor to conduct current I_D from the voltage reference V_r which determines the JFET channel resistance

$$r_1 = \frac{V_r}{I_D} \ .$$

(4.6.4)

The same amplifier $U1$ sets the channel resistance for the right JFET transistor to conduct current I_N from the output voltage V_o:

$$r_2 = \frac{V_o}{I_N} \ .$$

(4.6.5)

Since sources of the transistors are at virtual grounds, and the gates are controlled by the same voltage, the channel resistances of both transistors become equal ($r_1 = r_2$). Hence:

$$V_o = \frac{I_N}{I_D} V_r \ , \tag{4.6.6}$$

which is proportional to the quotient of the two input currents. The full scale is determined by V_r whose value for better accuracy should be selected relatively low, on the order of 100 mV. The circuit operates from dc to several hundred kilohertz and has demonstrated accuracy of about 0.2% FS providing that $I_N > I_D$. Low bias current and low offset voltage OPAMs improve the circuit performance.

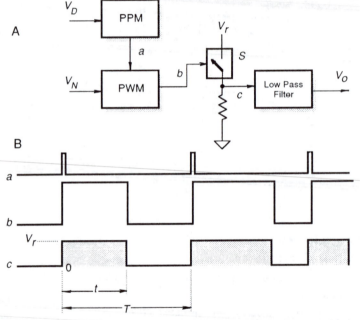

Fig. 4-6.4 Block diagram of a modulator-demodulator type divider (A) and timing diagrams (B)

A more complex circuit incorporates a modulation/demodulation technique. In a block-diagram form it is shown in Fig. 4-6.4A. Signal V_D which goes into the denominator, modulates period T of the pulse-position modulator (PPM):

$$T = k_1 V_D \ . \tag{4.6.7}$$

Each pulse a from the modulator sets a pulse b in the pulse-width modulator (PWM). The duration t of the pulse b is controlled by voltage V_N:

$$t = k_2 V_N \qquad (4.6.8)$$

Pulses from the PWM are fed into the clamping circuit which makes their amplitude constant and equal to a reference voltage V_r. A possible implementation of the clamping circuit is shown in Fig. 4-6.4A where the analog switch S is controlled by pulses b and connects a voltage reference to the resistor R. Pulses c across that resistor have amplitude V_r, period T, and width t. Therefore, the frequency spectrum of these pulses contain a low frequency component whose amplitude is proportional to the input ratio:

$$V_o = V_r \frac{t}{T} = V_r \frac{k_2}{k_1} \frac{V_N}{V_D} \ . \qquad (4.6.8)$$

To extract that component, pulses are fed into a low-pass filter whose cutoff frequency is matched with the upper frequency of the input signals. Naturally, this divider is useful only for relatively slow changing input signals. For better filtering, the lowest frequency generated by the PPM must be at least 100 times higher than the highest frequency of the input signals.

4.7   BRIDGE CIRCUITS

Wheatstone bridge circuits are popular and very effective implementations of the ratiometric technique. A basic circuit is shown in Fig. 4-7.1. Impedances Z may be either active or reactive, that is they may be either simple resistances, like in piezo-resistive gauges, or capacitors, or inductors. For the resistor, the impedance is R, for the ideal capacitor, the magnitude of its impedance is equal to $1/2\pi f C$ and for the inductor, it is $2\pi f L$, where f is the frequency of the current passing through the element. The bridge output voltage is represented by:

$$V_{out} = \left(\frac{Z_1}{Z_1 + Z_2} - \frac{Z_3}{Z_3 + Z_4} \right) V_{ref} \ , \qquad (4.7.1)$$

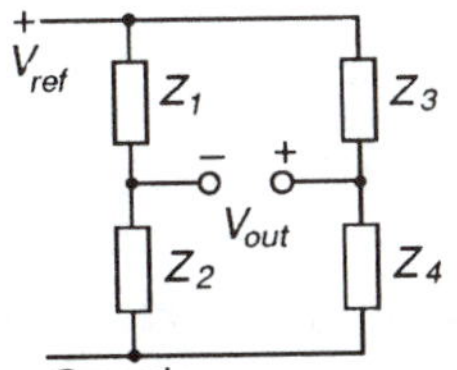

Fig. 4-7.1 General circuit of Wheatstone bridge

The bridge is considered to be in a balanced state when the following condition is met:

$$\frac{Z_1}{Z_2} = \frac{Z_3}{Z_4} \ . \qquad (4.7.2)$$

Under the balanced condition, the output voltage is zero. When at least one impedance changes, the bridge becomes imbalanced and the output voltage goes either in a positive or negative direction, depending on the direction of the impedance change. To determine the bridge sensitivity with respect to each impedance (calibration constant) partial derivatives may be obtained from Eq. (4.7.1):

$$\frac{\partial V_{out}}{\partial Z_1} = \frac{Z_2}{\left(Z_1 + Z_2\right)^2}V_{ref}$$

$$\frac{\partial V_{out}}{\partial Z_2} = -\frac{Z_1}{\left(Z_1 + Z_2\right)^2}V_{ref}$$

$$\frac{\partial V_{out}}{\partial Z_3} = -\frac{Z_4}{\left(Z_3 + Z_4\right)^2}V_{ref} \tag{4.7.3}$$

$$\frac{\partial V_{out}}{\partial Z_4} = \frac{Z_3}{\left(Z_3 + Z_4\right)^2}V_{ref}$$

By summing these equations, we obtain the bridge sensitivity:

$$\frac{\delta V_{out}}{V_{ref}} = \frac{Z_2\delta Z_1 - Z_1\delta Z_2}{\left(Z_1 + Z_2\right)^2} - \frac{Z_4\delta Z_3 - Z_3\delta Z_4}{\left(Z_3 + Z_4\right)^2}, \tag{4.7.4}$$

A closer examination of Eq. (4.7.4) shows that only the adjacent pairs of impedances (i.e., Z_1 and Z_2, Z_3 and Z_4) have to be identical in order to achieve the ratiometric compensation (such as the temperature stability, drift, etc.). It should be noted that impedances in the balanced bridge do not have to be equal, as long as a balance of the ratio (4.7.2) is satisfied. In many practical circuits, only one impedance is used as a sensor, thus for Z_1, the bridge sensitivity becomes:

$$\frac{\delta V_{out}}{V_{ref}} = \frac{\delta Z_1}{4Z_1}. \tag{4.7.5}$$

The resistive bridge circuits are commonly used with strain gauges, piezo-resistive pressure transducers, thermistor thermometers, and other sensors when immunity against environmental factors is required. Similar arrangements are used with the capacitive and magnetic sensors for measuring force, displacement, moisture, etc.

4.7.1 Disbalanced bridge

A basic Wheatstone bridge circuit (Fig. 4-7.2A) generally operates with a disbalanced bridge. This is called the *deflection* method of measurement. It is based on a detecting the voltage across the bridge diagonal. The bridge output voltage is a nonlinear function of a disbalance, Δ, however, for small changes ($\Delta<0.05$) it may be considered quasi-linear. The bridge maximum sensitivity is obtained when $R_1=R_2$ and $R_3=R$. When $R_1>>R_2$ or $R_2>>R_1$, the bridge output voltage is decreased. Assuming that $k=R_1/R_2$, the bridge sensitivity may be expressed as:

$$\alpha = \frac{V}{R}\frac{k}{(k+1)^2} \qquad (4.7.6)$$

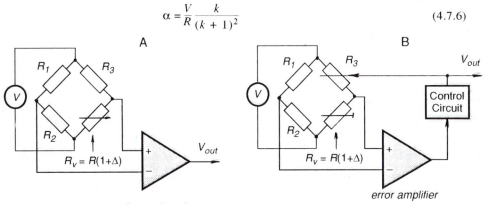

Fig. 4-7.2 Two methods of using a bridge circuit
A: disbalanced bridge and B: balanced bridge with a feedback control

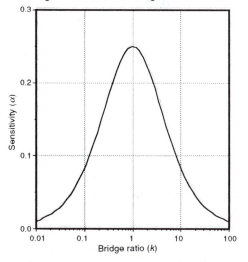

Fig. 4-7.3 Sensitivity of a disbalanced bridge as a function of impedance ratio

A normalized graph calculated according to this equation is shown in Fig. 4-7.3. It indicates that the maximum sensitivity is achieved at $k=1$, however, the sensitivity drops relatively little for the range where $0.5<k<2$. If the bridge is fed by a current source, rather by a voltage source, its output voltage for small Δ and a single variable component is represented by:

$$V_{out} \approx I\,\frac{k\Delta}{2(k+1)} \; , \qquad\qquad (4.7.7)$$

where I is the excitation current.

4.7.2 Null-balanced bridge

Another method of using a bridge circuit is called a *null-balance*. The method overcomes the limitation of small changes (Δ) in the bridge arm to achieve a good linearity. The null-balance essentially requires that the bridge is *always* maintained at the balanced state. To satisfy the requirement for a bridge balance (4.7.2), another arm of the bridge should vary along with the arm which is used as a sensor. Fig. 4-7.2B illustrates this concept. A control circuit modifies the value of R_3 on the command from the error amplifier. The output voltage may be obtained from the control signal of the balancing arm R_3. For example, both R_v and R_3 may be photoresistors. The R_3-photoresistor could be interfaced with a light emitting diode (LED) which is controlled by the error amplifier. Current through the LED becomes a measure of resistance R_v, and, subsequently, of the light intensity detected by the sensor.

4.7.3 Temperature compensation of resistive bridge

The connection of four resistive components in a Wheatstone bridge configuration is used quite extensively in measurements of temperature, force, pressure, magnetic fields, etc. In many of these applications, sensing resistors exhibit temperature sensitivity. This results in temperature sensitivity of a transfer function, which by using a linear approximation may be expressed by Eq. (2.1). In any detector, except of that intended for temperature measurements, this temperature dependence has a highly undesirable effect which usually must be compensated for. One way to do a compensation is to couple a detector with a temperature sensitive device, which can generate a temperature related signal for the electronic correction. Another way to do a temperature compensation is to incorporate it directly into the bridge circuit. Let us analyze the Wheatstone bridge output signal with respect to its excitation signal V_e. We consider all four arms in the bridge being responsive to a stimulus with the sensitivity coefficient α, so that each resistor has value

$$R_i = R(1 \pm \alpha s) \ , \tag{4.7.8}$$

where R is the nominal resistance, s is the stimulus (for instance, pressure or force), and sensitivity α is defined as

$$\alpha = \frac{1}{R} \frac{dR}{ds} \ . \tag{4.7.9}$$

An output voltage from the bridge is

$$V_{out} = V_e \cdot \alpha s + V_o \ , \tag{4.7.10}$$

where V_o is the offset voltage resulting from the initial bridge imbalance. If the bridge is not properly balanced, the offset voltage may be a source of error. However, an appropriate trimming of the bridge sensor during either its fabrication or in application apparatus may reduce this error to an acceptable level. In any event, even if the offset voltage is not properly compensated for, its temperature variations usually are several order of magnitude smaller than that of the sensor's transfer function. In this discussion, we consider V_o temperature independent ($dV_o/dT=0$), however, for a broad temperature range (wider than $\pm 15°C$) V_o should not be discounted.

In the Eq. (4.7.8) sensitivity α generally is temperature dependent for many sensors and is a major source of inaccuracy. It follows from equation (4.7.9) that α may vary if R is temperature dependent or when dR/ds is temperature dependent. If the bridge has a positive temperature coefficient of resistivity, coefficient α decreases with temperature, or it is said, it has a negative TCS (temperature coefficient of sensitivity). Taking a partial derivative with respect to temperature T, from Eq. (4.7.10) we arrive at:

$$\frac{\partial V_{out}}{\partial T} = s \left(\alpha \frac{\partial V_e}{\partial T} + \frac{\partial \alpha}{\partial T} V_e \right) . \tag{4.7.11}$$

A solution of this equation is the case when the output signal does not vary with temperature: $\partial V_{out}/\partial T = 0$. Then, the following holds:

$$\alpha \frac{\partial V_e}{\partial T} = - \frac{\partial \alpha}{\partial T} V_e \ , \tag{4.7.12}$$

and finally,

$$\frac{1}{V_e} \frac{\partial V_e}{\partial T} = - \frac{1}{\alpha} \frac{\partial \alpha}{\partial T} = - \beta \ , \tag{4.7.13}$$

where β is the TCS of the bridge arm.

The above is a condition for an *ideal* temperature compensation of a fully symmetrical Wheatstone bridge. That is, to compensate for temperature variations in α, the excitation voltage, V_e, must change with temperature at the same rate and with the opposite sign. To control V_e, several circuits were proven to be useful [11]. Fig. 4-7.4 shows a general circuit which incorporates a temperature compensation network to control voltage V_e across the bridge according to a predetermined function of temperature. Several options of the temperature compensation network are possible.

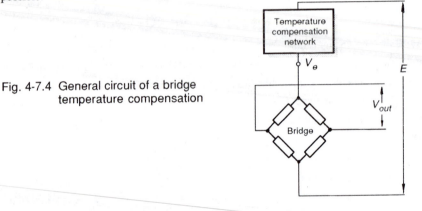

Fig. 4-7.4 General circuit of a bridge temperature compensation

Option 1: Use of a temperature sensor as a part of the compensation network. Such a network may be represented by an equivalent impedance R_t, while an entire bridge can be represented by its equivalent resistance R_B. Then, the voltage across the bridge is

$$V_e = E \frac{R_B}{R_B + R_t} \cdot \qquad (4.7.14)$$

Taking a derivative with respect to temperature we get

$$\frac{\partial V_e}{\partial T} = E\left(\frac{1}{R_B + R_t}\frac{\partial R_B}{\partial T} - \frac{R_B}{(R_B + R_t)^2}\left(\frac{\partial R_B}{\partial T} + \frac{\partial R_t}{\partial T}\right)\right), \qquad (4.7.15)$$

and substituting (4.7.14) into (4.7.15) we arrive to the compensation condition

$$\frac{1}{V_e}\frac{\partial V_e}{\partial T} = \frac{1}{R_B}\frac{\partial R_B}{\partial T} - \frac{1}{R_B + R_t}\left(\frac{\partial R_B}{\partial T} + \frac{\partial R_t}{\partial T}\right). \qquad (4.7.16)$$

Since for a bridge with four sensitive arms $R_B=R$, and $\frac{1}{R}\frac{\partial R}{\partial T} = \gamma$ which is a temperature coefficient of bridge arm resistance, R (TCR), the above equation according to (4.7.13) must be equal to negative TCS:

$$-\beta = \gamma - \frac{1}{R + R_t}\left(\frac{\partial R}{\partial T} + \frac{\partial R_t}{\partial T}\right) . \qquad (4.7.17)$$

This states that such a compensation is useful over a broad range of excitation voltages since E is not a part of the equation[1]. To make it work, the resistive network R_t must incorporate a temperature sensitive resistor, for instance a thermistor. When R, β and $\partial R/\partial T$ are known, Eq. (4.7.17) can be solved to select R_t. This method requires a trimming of the compensating network to compensate not only for TCS and TCR, but for V_e as well. The method, while being somewhat complex, allows for a broad range temperature compensation from -20 to + 70°C and with somewhat reduced performance or with a more complex compensating network, from -40 to 100°C. Fig. 4-7.5A shows an example of the compensating network which incorporates an NTC thermistor $R°$ and several trimming resistors. An example of such a compensation is a Motorola pressure sensor, PMX2000, which contains a diffused into silicon thermistor, and 8 thin film resistors which calibrate offset and compensate for temperature variations. The resistors are laser trimmed, on-chip, during the calibrating process to assure high stability over the range from -40 to +125°C.

Option 2: The compensation network is a fixed resistor. This is the most popular temperature compensation of a resistive Wheatstone bridge. A fixed resistor (Fig. 4-7.5B) $R_t=R_c$ must have low temperature sensitivity (50 ppm or less). It can be stated that:

$$\frac{1}{R_c}\frac{\partial R_c}{\partial T} = 0 , \qquad (4.7.18)$$

and Eq. (4.7.17) is simplified to

$$-\beta = \frac{\partial R}{\partial T}\left(\frac{1}{R} - \frac{1}{R + R_c}\right) . \qquad (4.7.19)$$

It can be solved for the temperature compensating resistor

$$R_c = -\frac{\beta R}{\frac{\partial R}{\partial T} + \beta R} . \qquad (4.7.20)$$

Then, the compensating resistor is

$$R_c = -R\frac{\beta}{\gamma + \beta} . \qquad (4.7.21)$$

The minus sign indicates that the equation is true for negative TCS β. Thus, when TCR, TCS and a nominal resistance of the bridge arm is known, a simple stable resistor in series with

---

[1] This demands that the compensation network contains no active components, such as diodes, transistors, etc.

voltage excitation E can provide a quite satisfactory compensation. It should be noted, however, that according to Eq. (4.7.19), to use this method, TCS of the bridge arm must be smaller than its TCR ($|\beta|<\gamma$). As option 1, this circuit is ratiometric with respect to the operating voltage, E, meaning that the compensation works over a broad range of power supply voltages. Selecting R_c according to Eq. (4.7.21) may result in a very large compensating resistance which may be quite inconvenient in many applications. The sensor's TCR can be effectively reduced by adding a resistor in parallel with the bridge. When a large resistor R_c is used, this method of compensation becomes similar to option 4 as described below, since a large resistor operates as a "quasi" current source.

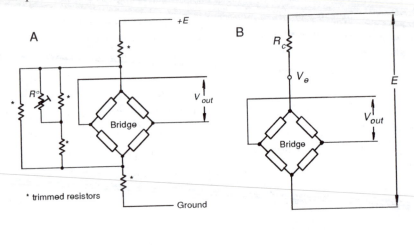

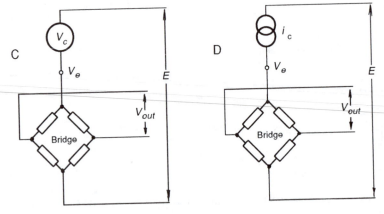

Fig. 4-7.5 Temperature compensation of a bridge circuit
A: with NTC thermistor; B: with a fixed resistor;
C: with a temperature controlled voltage source; D: with a current source

A first impression from this option is that it seems like a perfect solution - just one resistor ideally compensates for a temperature drift. However, this option does not yield satisfactory results for broad temperature ranges or precision applications. For instance, to select an appropriate R_c, γ and β must be precisely known. That is, each actual bridge must be characterized, which is not acceptable in low-cost applications. Using typical rather than actual values may result in span errors on the order of 100 ppm/°C. Further, large resistors R_c cause lower output voltages and a reduced signal-to-noise ratio. Practically, usefulness of this compensation is limited to the range of 25±15°C.

Option 3: A compensating network contains a temperature controlled voltage source, for instance a diode or transistor (Fig. 4-7.5C). For this circuit, to satisfy a condition for the best compensation of TCS β, temperature sensitivity β_c of the voltage source V_c must be

$$\beta_c = \beta \left(\frac{E}{V_c} - 1\right) . \tag{4.7.22}$$

Since β is a parameter of the bridge, by manipulating E and V_c one can select the optimum compensation. Since the compensating circuit contains a voltage source, it is not ratiometric with respect to the power supply. For the operation, this option requires a regulated source of E. An obvious advantage of the circuit is simplicity since diodes and transistors with predicable temperature characteristics are readily available. An obvious disadvantage of this method is a need to operate the sensor at a fixed specified voltage. A useful temperature range for this method is about 25±25°C.

Option 4: A current source is employed as an excitation circuit (Fig. 4-7.5D). This circuit requires that the bridge possesses a particular property. Its TCR (β) must be equal to TCS (α) with the opposite sign:

$$\alpha = -\beta . \tag{4.7.23}$$

The voltage across the bridge is equal to

$$V_e = i_c R_B . \tag{4.7.24}$$

Since the current source is temperature independent and for a bridge with four identical arms $R_B = R$, then

$$\frac{\partial V_e}{\partial T} = i_c \frac{\partial R}{\partial T} , \tag{4.7.25}$$

and dividing (4.7.25) by (4.7.24) we arrive at

$$\frac{1}{V_c} \frac{\partial V_e}{\partial T} = \frac{1}{R} \frac{\partial R}{\partial T} . \tag{4.7.26}$$

If condition (4.7.23) is fulfilled, we receive a provision of an ideal compensation as defined by equation (4.7.12). Unfortunately, this method of compensation has a similar limitation as option 2, specifically a reduced output voltage and a need for individual sensor characterization if used in a broad temperature range. Nevertheless, this method is acceptable when the accuracy of 1% to 2% of FS over 50°C is acceptable.

The above options provide a framework of the compensating techniques. While designing a practical circuit, many variables must be accounted for: temperature range, allowable temperature error, environmental conditions, size, cost, etc. Therefore, we can not recommend a universal solution - the choice of the most appropriate option must be a result of a typical engineering compromise.

4.7.4 Bridge amplifiers

The bridge amplifiers for resistive sensors are probably the most frequently used sensor interface circuits. They may be of several configurations, depending on the required bridge grounding and availability of either grounded or floating reference voltages. Fig. 4-7.6A shows the so-called active bridge, where a variable resistor (the sensor) is floating, i.e., isolated from ground, and is connected into a feedback of the OPAM. If a resistive sensor can be modeled by a first order function:

$$R_x \approx R_o(1 + \alpha) \; , \tag{4.7.27}$$

then, a transfer function of this circuit is:

$$V_{out} = -\frac{1}{2}\alpha V \; . \tag{4.2.28}$$

A circuit with a floating bridge and floating reference voltage source V is shown in Fig. 4-7.6B. This circuit may provide gain which is determined by a feedback resistor whose value is nR_o:

$$V_{out} = (1+n)\alpha\frac{V}{4}\ \frac{1}{1+\dfrac{\alpha}{2}} \approx (1+n)\alpha\frac{V}{4} \; . \tag{4.2.29}$$

A bridge with the asymmetrical resistors ($R \neq R_o$) may be used with the circuit shown in Fig. 4-7.6C. It requires a floating reference voltage source V:

$$V_{out} = n\alpha\frac{V}{4}\frac{1}{1+\dfrac{\alpha}{2}} \approx n\alpha\frac{V}{4} \quad .$$ (4.2.30)

When a resistive sensor is grounded, a circuit shown in Fig. 4-7.6D may be employed. Its transfer function is determined from

$$V_{out} = -\frac{n}{2}\frac{V}{1+\dfrac{1}{2n}}\frac{\alpha}{1+\alpha} \approx -\frac{n}{2}\frac{V}{1+\dfrac{1}{2n}}\alpha \quad .$$ (4.2.31)

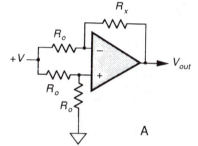
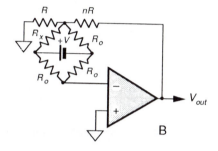

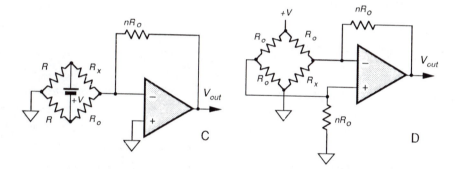

Fig. 4-7.6 Connection of operational amplifiers
to resistive bridge circuits (disbalanced mode)

Fig. 4-7.7 shows a circuit diagram of a differential amplifier for a piezoresistive pressure transducer. The amplifier incorporates a KTY-temperature sensor (Philips) with a positive temperature coefficient. The sensor must be in an intimate thermal coupling with the bridge, i.e., it should be incorporated into the same housing as the pressure transducer. The temperature sensor affects the gain of the first stage of a differential amplifier providing compensation for the TCR of the silicon bridge resistors.

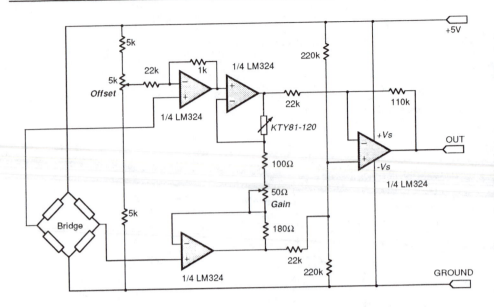

Fig 4-7.7 Temperature compensated amplifier for a piezoresistive bridge

4.8   DATA TRANSMISSION

4.8.1 Two-wire transmission

Two-wire transmitters are used to couple sensors to control and monitoring devices in the process industry [12]. When, for example, a temperature measurements is taken within a process, a 2-wire transmitter relays that measurement to the control room or interfaces the measurement directly to a process controller. Two wires can be used to transmit either voltage or current, however, current was accepted as an industry standard. It varies in the range from 4 to 20mA which represents an entire span of the input stimulus. Zero stimulus corresponds to 4mA while the maximum is at 20mA. There are two advantages of using current rather than voltage as it is illustrated in Fig. 4-8.1. A sensor is connected to the so-called *two-wire transmitter* which converts its output signal into current. In essence, the transmitter is a current source (current generator) that has a very high output resistance. The two wires form a current loop, which at the sensor's side has the sensor and a two-wire transmitter, while at the controller side it has a load resistor and a power supply which are connected in series. The first advantage of this arrangement is that the wire resistance has no effect on the current flowing in the loop, because the current source output resistance is high. In a voltage transmittance circuit, the wire resistance would pose a significant problem, especially if the transmission distance is long. Another advantage is that the same current which carries information is also used by the transmitter and the sensor to provide their operating power. Obviously, even for the lowest output signal which produces 4 mA current, that 4 mA must be sufficient to power the transmitting side of the loop. The loop current causes a voltage drop across the load resistor at the controller side. This voltage is a received signal which is suitable for further processing by electronic circuits.

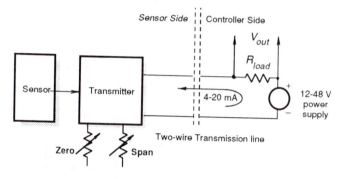

Fig. 4-8.1 Two-wire 20-mA analog data transmission

4.8.2 Four-wire sensing

Sometimes, it is desirable to connect a resistive sensor to a remotely located interface circuit. When such a sensor has a relatively low resistance (for instance, it is normal for the piezo-resistors or RTDs to have resistances in the order of 100Ω), connecting wire resistance pose a serious problem. The problem can be solved by using the so-called *4-wire method* (Fig. 4-8.2A). It allows us to measure the resistance of a remote resistor without measuring the resistance of connecting conductors. A resistor which is the subject of measurement is connected to the interface circuit through four rather than through two wires. Two wires are connected to a current source and two others to the voltmeter. A constant current source (current pump) has a very high output resistance, therefore the current which it pushes through the loop is almost independent of any resistances r in that loop. An input impedance of a voltmeter is very high, hence no current is diverted from the current loop to the voltmeter. The voltage drop across the resistor R_x is

$$V_x = R_x i_o \ , \tag{4.8.1}$$

which is independent of any resistances r of the connecting wires. The 4-wire method is a very powerful means of measuring resistances of remote detectors and is used in industry and science quite extensively.

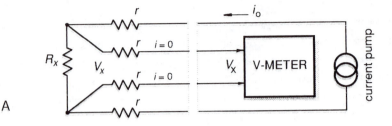

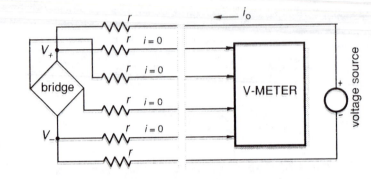

Fig. 4-8.2 Remote measurements of resistances
A: four-wire method; B: six-wire measurement of a bridge

4.8.3 Six-wire sensing

When a Wheatstone bridge circuit is remotely located, voltage across the bridge plays an important role in the bridge temperature stability, as it was shown in Section 4.7. That voltage often should be either measured or controlled. Long transmitting wires may introduce unacceptably high resistance in series with the bridge excitation voltage which interfere with the temperature compensation. The problem may be solved by providing two additional wires to feed the bridge with voltage and to dedicate two wires to measure the voltage across the bridge (Fig. 4-8.2B). The actual excitation voltage across the bridge and the bridge differential output voltage are measured by a high-input impedance voltmeter with negligibly small input currents. Thus, the accurate bridge voltages are available at the data processing site without being affected by the long transmission lines.

4.8.4 Telemetry

The term telemetry refers to data transmission using a high frequency carrier with some kind of modulation. A block diagram of a multichannel telemetry system (Fig. 4-8.3A) is similar to a general data acquisition system (Fig. 1-2) except that it is linked to the processing and recording part of the system via a radio-channel. The outputs from the sensors and/or signal conditioners are fed into a multiplexer (MUX) which combines them into a composite signal. This is done by a time sharing method: a multiplexer is a combination of analog gates. It has several inputs and one output. A composite signal consists of a sequence of steps where each one represents a single sensor. All sensors are queried in series with a frequency which is defined by a formula based on Shannon's theorem:

$$F > 2f_{max} \; , \tag{4.8.2}$$

where f_{max} is the highest frequency of the sensor signals and F is the multiplexing frequency. In practice, F is selected at least 4 times higher than f_{max}. A further increase in F does not improve the fidelity of the signal reproduction, however it may present some difficulty in data transmission. If a sampling period $T=1/F$ is too short, the time may be not sufficient enough for the data transmission and processing. Signals from the multiplexer may go to RF-modulator, whose function is to modify radio-frequency from RF-oscillator with a multiplexed composite signal. Fig. 4-8.3B shows that the RF-modulator changes the frequency of the oscillator in proportion with each step of the composite signal. The modified (modulated) RF-signal is transmitted by a radio transmitter. To make data transmission more noise-resistant, an additional sub-modulation step may be added before the RF-modulation. There is a large variety of modulations which may be used to transmit data through a radio-channel. The most popular modulations are:

Table 4-4 Some popular modulation combinations
for transmission information through radio-channel

AM/FM	amplitude/frequency	no sub-modulation
PW/FM	pulse-width/frequency	sub-modulation
PP/FM	pulse-position/frequency	sub-modulation
DM/FM	digital/frequency	sub-modulation

A modulated signal is received by a radio receiver and demodulated to reconstruct the original signals, so that each measurement can be displayed and evaluated individually. Fig. 4-8.4 shows different methods of sub-modulations which are commonly used for the multi-channel systems.

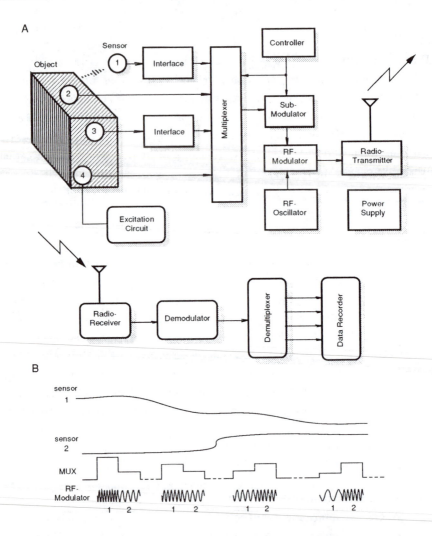

Fig. 4-8.3 Block diagram (A) and some signals (B)
in a radio-telemetry system with AM-FM modulation

Besides the radio-frequency carrier, other methods of data transmission are possible. For instance, an underwater telemetry may use acoustic carriers with the same sub-modulations as shown in Fig. 4-8.4. At relatively short distances, when a direct vision between the transmitter and the receiver can be established, an infrared beam can be used to carry information. However, optical transmittance may be susceptible to air pollutants, fog and obstructive opaque objects.

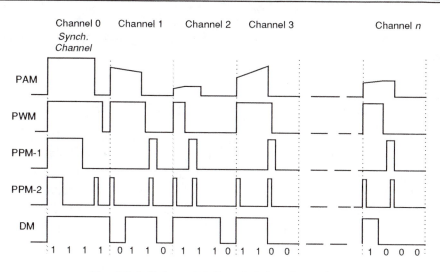

Fig. 4-8.4 Sub-modulations in telemetry systems

4.9 NOISE IN SENSORS AND CIRCUITS

Noise in sensors and circuits may present a substantial source of errors and should be seriously considered. "Like diseases, noise is never eliminated, just prevented, cured, or endured, depending on its nature, seriousness, and the cost/difficulty of treating"[13]. There are two basic classifications of noise for a given circuit: they are *inherent* noise, which is noise arising within the circuit, and *interference* (*transmitted*) noise, which is noise picked up from outside the circuit.

Any sensor, no matter how well it was designed, never produces an electric signal which is an ideal representation of the input stimulus. Often, it is a matter of judgment to define the goodness of the signal. The criteria for this are based on the specific requirements to accuracy and reliability. Distortions of the output signal can be either systematic or stochastic. The former are related to the sensor's transfer function, its linearity, dynamic characteristics, etc. They all are the result of the sensor's design, manufacturing tolerances, material quality, and calibration. During a reasonably short time, these factors either do not change or drift relatively slowly. They can be well defined, characterized and specified (see Chapter 2). In many applications, such a determination may be used as a factor in the error budget and can be accounted for. Stochastic disturbances, on the other hand, often are irregular, unpredictable to some degree and may change rapidly. Generally, they are termed *noise*, regardless of their nature and statistical properties. It should be noted that word *noise*, in association with audio equipment noise, is often mistaken for an irregular, somewhat fast changing signal. We use this word in a much broader sense for all disturbances, either in stimuli, environment, or in the components of sensors and circuits from dc to upper operating frequencies.

4.9.1 Inherent noise

A signal which is amplified and converted from a sensor into a digital form should be regarded not just by its magnitude and spectral characteristics, but also in terms of a digital resolution. When a conversion system employs an increased digital resolution, the value of the least-significant bit (LSB) decreases. For example, the LSB of a 10-bit system with a 5V full scale is about 5mV, the LSB of 16 bits is 77μV. This by itself poses a significant problem. It makes no sense to employ, say a 16-bit resolution system, if a combined noise is, for example, 300μV. In the real world, the situation is usually much worse. There are almost no sensors which are capable of producing a 5V full scale output signals. Most of them, require an amplification. For instance, if a sensor produces a full scale output of 5mV, at a 16-bit conversion it would correspond to a LSB of 77nV - an extremely small signal which makes amplification an enormous task by itself. Whenever a high resolution of a conversion is required, all sources of noise must be seriously considered. In the circuits, noise can be produced by the monolithic amplifiers and other components which are required for the feedback, biasing, bandwidth limiting, etc.

Input offset voltages and bias currents may drift. In dc circuits, they are indistinguishable from low magnitude signals produced by a sensor. These drifts are usually slow (having bandwidth of tenths and hundredths of a Hz), therefore they are often called ultralow frequency noise. To distinguish them from the higher frequency noise, the equivalent circuit (Fig. 4-1.3) contains two additional generators. One is a *voltage noise* generator e_n and the other is a *current noise* generator i_n [15] The noise signals (voltage and current) result from physical mechanisms within the resistors and transistors that are used to fabricate the circuits. There are several sources of noise whose combined effect is represented by the noise voltage and current generators. One cause for noise is a discrete nature of electric current because current flow is made up of moving charges, and each charge carrier transports a definite value of charge (charge of an electron is $1.6 \cdot 10^{-19}$ coulombs). At the atomic level, current flow is very erratic. The motion of the current carriers resembles popcorn popping. This was chosen as a good analogy for current flow and has nothing to do with the "popcorn noise" which we will discuss below. As popcorn, the electron movement may be described in statistical terms. Therefore, one never can be sure about very minute details of current flow. The movement of carriers are temperature related and noise power, in turn, is also temperature related. In a resistor, these thermal motions cause Johnson noise to result [15]. The mean-square value of noise voltage (which is representative of noise power) can be calculated from

$$\overline{e_n^2} = 4kTR\Delta f \ [\text{V}^2/\text{Hz}] \ , \qquad\qquad (4.9.1)$$

where
$k = 1.38 \times 10^{-23}$ J/K (Boltzmann constant)
T = temperature in K
R = resistance in Ω
Δf = bandwidth over which the measurement is made, in Hz.

For practical purposes, noise density per $\sqrt{\text{Hz}}$ generated by a resistor at room temperature may be estimated from a simplified formula: $\overline{e_n} \approx 0.13\sqrt{R}$ in nV/$\sqrt{\text{Hz}}$. For example, if noise bandwidth is 100Hz and the resistance of concern is 10MΩ ($10^7\Omega$), the average noise voltage is estimated as $\overline{e_n} \approx 0.13\cdot\sqrt{10^7}\cdot\sqrt{100} = 4{,}111\text{nV} \approx 4\mu\text{V}$.

Even a simple resistor is a source of noise. It behaves as a perpetual generator of electric signal. Naturally, relatively small resistors generate extremely small noise, however, in some sensors Johnson must be taken into account. For instance, a pyroelectric detector uses a bias resistor on the order of 50GΩ. If the sensor is used at room temperature within a bandwidth of 100Hz, one may expect the average noise voltage across the resistor to be on the order of 0.3mV - a pretty high value. To keep noise at bay, bandwidths of the interface circuits must be maintained small, just wide enough to pass the minimum required signal. It should be noted, that noise voltage is proportional to square root of the bandwidth. It implies that if we reduce the bandwidth 100 times, noise voltage will be reduced by a factor of 10. Johnson noise magnitude is constant over a broad range of frequencies. Hence, it is often called *white noise* because of the similarity to white light, which is composed of all the frequencies in visible spectrum.

The ac current noise source (i_n) represents the ac noise that results because of dc current flow in semiconductors. It is called *shot noise* - the name which was suggested by Schottky not in association with his own name but rather because this noise sounded like "a hail of shot striking the target" (nevertheless, shot noise is often called *Schottky noise*). Shot noise is also white noise. Its value becomes higher with the increase in the bias current. This is the reason why in FET and CMOS semiconductors current noise is quite small. For a bias current of 50pA, it is equal to about 4fA/$\sqrt{\text{Hz}}$ (1fA=10^{-15}A is extremely small current which is equivalent to movement of about 6000 electrons per second). A convenient equation for shot noise is

$$i_{sn} = 5.7\cdot10^{-4}\sqrt{I\Delta f} \ , \qquad\qquad (4.9.2)$$

where I is a semiconductor junction current in picoamperes and Δf is a bandwidth of interest in Hz.

An additional ac noise mechanism exists at low frequencies (Fig. 4-9.1). Both the noise voltage and noise current sources have a spectral density roughly proportional to $1/f$, which is called the *pink noise*, because of the higher noise contents at lower frequencies (lower frequencies are also at red side of the visible spectrum). This $1/f$ noise occurs in all conductive materials, therefore it is also associated with resistors. At extremely low frequencies it is impossible to separate the $1/f$ noise from dc drift effects. The $1/f$ noise is sometimes called a flicker noise. Mostly it is pronounced at frequencies below 100 Hz, where many sensors operate. It may dominate Johnson and Schottky noise and becomes a chief source of errors at these frequencies. The magnitude of pink noise depends on current passing through the resistive or semiconductive material. Nowadays progress in semiconductor technology resulted in significant reduction of $1/f$ noise in semiconductors, however, when designing a circuit, it is a good engineering practice to use metal film or wirewound resistors in sensors and the front stages of interface circuits wherever significant currents flow through the resistor and low noise at low frequencies is a definite requirement.

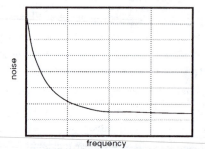

Fig. 4-9.1 Spectral distribution of 1/f noise

A peculiar ac noise mechanism is sometimes seen on the screen of an oscilloscope when observing the output of an operational amplifier - a principal building block of many sensor interface circuits. It looks like a digital signals transmitter from outer Space - noise has a shape of square pulses having variable duration of many milliseconds. This abrupt type of noise is called *popcorn noise* because of the sound it makes coming from a loudspeaker. Popcorn noise is caused by defects that are dependent on the integrated circuits manufacturing techniques. Thanks to advances fabricating technologies, this type of noise is drastically reduced in modern semiconductor devices.

A combined noise from all voltage and current sources is given by sum of squares of individual noise voltages:

$$e_E = \sqrt{e_{n1}^2 + e_{n2}^2 + \cdots + \left(R_1 i_{n1}\right)^2 + \left(R_1 i_{n2}\right)^2 + \cdots} \quad . \tag{4.9.3}$$

A combined random noise may be presented by its *r.m.s* value which is

$$E_{rms} = \sqrt{\frac{1}{T}\int_0^T e^2 dt} \tag{4.9.4}$$

where T is time of observation, e is noise voltage and t is time.

Also, noise may be characterized in terms of the peak values which are the differences between the largest positive and negative peak excursions observed during an arbitrary interval. For some applications, in which peak-to-peak (*p-p*) noise may limit the overall performance, *p-p* measurement may be essential. Yet, due to a generally Gaussian distribution of noise signal, *p-p* magnitude is very difficult to measure in practice. Because *r.m.s.* values are so much easier to measure repeatedly, and they are the most usual form for presenting noise data noncontroversially, the Table 4-5 should be useful for estimating the probabilities of exceeding various peak values given by the *r.m.s.* values. The casually-observed *p-p* noise varies between 3x*r.m.s.* and 8x*r.m.s.*, depending on the patience of observer and amount of data available.

Table 4-5 Peak-to-peak value vs. *r.m.s.* (for Gaussian distribution)

Nominal p-p voltage	% of time that noise will exceed nominal p-p value
2x*r.m.s.*	32.0%
3x*r.m.s.*	13.0%
4x*r.m.s.*	4.6%
5x*r.m.s.*	1.2%
6x*r.m.s.*	0.27%
7x*r.m.s.*	0.046%
8x*r.m.s.*	0.006%

4.9.2 Transmitted noise

A large portion of environmental stability is attributed to the resistance of a sensor and an interface circuit to noise which is originated in external sources. Fig. 4-9.2 shows a block diagram of the transmitted noise propagation. Noise comes from a source which often can be identified. Examples of the sources are: voltage surges in power lines, lightnings, change in ambient temperature, sun activity, etc. These interferences propagate toward the sensor and the interface circuit, and to present a problem eventually must appear at the output. However, before that, they somehow must affect the sensing element inside the sensor, its output terminals or the electronic components in a circuit. The sensor and the circuit function as receivers of the interferences.

There can be several classifications of transmitted noise, depending on how it affects the output signal, how it enters the sensor or circuit, etc. With respect to its relation to the output signals, noise can be either *additive* or *multiplicative*.

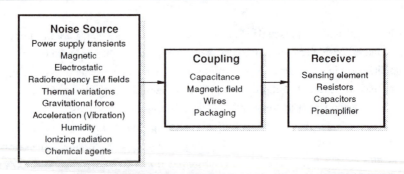

Fig. 4-9.2 Sources and coupling of transmitted noise

Additive noise e_n is added to the useful signal V_s and mixed with it as a fully independent voltage (or current)

$$V_{out} = V_s + e_n \ . \tag{4.9.5}$$

An example of such a disturbance is depicted in Fig. 4-9.3B. It can be seen, that the noise magnitude does not change when the actual (useful) signal changes. As long as the sensor and interface electronics can be considered linear, the additive noise magnitude is totally independent of the signal magnitude and, if the signal is equal to zero, the output noise still will be present.

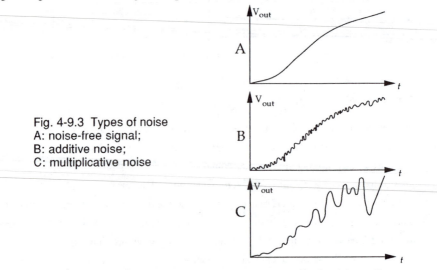

Fig. 4-9.3 Types of noise
A: noise-free signal;
B: additive noise;
C: multiplicative noise

Multiplicative noise affects the sensor's transfer function or the circuit's nonlinear components in such a manner as V_s signal's value becomes altered or *modulated* by the noise:

$$V_{out} = [1+N(t)]V_s \ , \tag{4.9.6}$$

where $N(t)$ is a function of noise. An example of such noise is shown in Fig. 4-9.3C. Multiplicative noise at the output disappears or becomes small (it also becomes additive) when the signal magnitude nears zero. Multiplicative noise grows together with the signal's V_s magnitude. As its name implies, multiplicative noise is a result of multiplication (which is essentially a nonlinear operation) of two values where one is a useful signal and the other is a noise dependent value.

To improve noise stability against transmitted additive noise, quite often sensors are combined in pairs, that is, they are fabricated in a dual form whose output signals are subtracted from one another (Fig. 4-9.4). This method is called a *differential* technique. One sensor of the pair (it is called the main sensor) is subjected to a stimulus of interest s_1, while the other (reference) is shielded from stimulus perception.

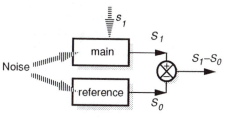

4-9.4 Differential technique

Since additive noise is specific for the linear or quasi-linear sensors and circuits, the reference sensor does not have to be subjected to any particular stimulus. Often, it may be equal to zero. It is anticipated that both sensors are subjected to identical transmitted noise, which it is said, is a common-mode noise. This means that noisy effects at both sensors are in-phase and have the same magnitude. If both sensors are identically influenced by common-mode spurious stimuli, the subtraction removes the noise component. Such a sensor is often called either a dual or a *differential* sensor. The quality of noise rejection is described by a number which is called the *common-mode rejection ratio* (*CMRR*):

$$CMRR = 0.5\frac{S_1 + S_0}{S_1 - S_0} ,$$ (4.9.7)

where S_1 and S_0 are output signals from the main and reference sensors, respectively. *CMRR* may depend on magnitude of stimuli and usually becomes smaller at greater input signals. The ratio shows how many times stronger the actual stimulus will be represented at the output, with respect to a common mode noise having the same magnitude. The value of the *CMRR* is a measure of the sensor's symmetry. To be an effective means of noise reduction, both sensors must be positioned as close as possible to each other, they must be very identical and subjected to the same environmental conditions. Also, it is very important that the reference sensor is reliably shielded from the actual stimulus, otherwise the combined differential response will be diminished.

To reduce transmitted multiplicative noise, a ratiometric technique is quite powerful (see section 4.6 for circuits description). Its principle is quite simple. The sensor is fabricated in a dual form where one part is subjected to the stimulus of interest and both parts are subjected to the same environmental conditions which may cause transmitted multiplicative noise. The second sensor is called reference because a constant environmentally stable reference stimulus s_o is applied to its input. For example, the output voltage of a sensor in a narrow temperature range may be approximated by equation

$$V_1 \approx [1+\alpha(T-T_o)]f(s_1) \ , \tag{4.9.8}$$

where α is the temperature coefficient of the sensor's transfer function, T is the temperature and T_o is the temperature at calibration. The reference sensor whose reference input is s_o generates voltage

$$V_o \approx [1+\alpha(T-T_o)]f(s_o) \ . \tag{4.9.9}$$

Table 4-6 Typical sources of transmitted noise
(adapted from [16])

External Source	Typical Magnitude	Typical Cure
60/50 Hz power	100 pA	Shielding; attention to ground loops; isolated power supply
120/100 Hz supply ripple	3 μV	Supply filtering
180/150 Hz magnetic pickup from saturated 60/50 Hz transformers	0.5 μV	Reorientation of components
Radio broadcast stations	1 mV	Shielding
Switch-arcing	1 mV	Filtering of 5 to 100 MHz components; attention to ground loops and shielding
Vibration	10 pA (10-100 Hz)	Proper attention to mechanical coupling; elimination of leads with large voltages near input terminals and sensors
Cable vibration	100 pA	Use a low noise (carbon coated dielectric) cable
Circuit boards	0.01-10 pA/√Hz below 10 Hz	Clean board thoroughly; use Teflon insulation where needed and guard well

We consider ambient temperature as a transmitted multiplicative noise which affects both sensors in the same way. Taking the $\dfrac{V_1}{V_0} = \dfrac{1}{f(s_o)}f(s_1)$ ratio of the above equations we arrive at

$$\frac{V_1}{V_0} = \frac{1}{f(s_o)}f(s_1) \ . \tag{4.9.10}$$

Since $f(s_o)$ is constant, the ratio is not temperature dependent. It should be emphasized however, that the ratiometric technique is useful only when the anticipated noise has a multiplicative nature, while a differential technique works only for additive noise. Neither technique is useful for inherent noise which is generated internally in sensors and circuits.

While inherent noise is mostly Gaussian, the transmitted noise is usually less suitable for conventional statistical description. Transmitted noise may be periodic, irregularly recurring, or essentially random, and it ordinarily may be reduced substantially by taking precautions to minimize electrostatic and electromagnetic pickup from power sources at line frequencies and their harmonics, radio broadcast stations, arcing of mechanical switches, and current and voltage spikes resulting from switching in reactive (having inductance and capacitance) circuits. Such precautions may include filtering, decoupling, shielding of leads and components, use of guarding potentials, elimination of ground loops, physical reorientation of leads, components and wires, use of damping diodes across relay coils and electric motors, choice of low impedances where possible, and choice of power supply and references having low noise. Transmitted noise from vibration may be reduced by proper mechanical design. A table outlining some of the sources of transmitted noise, their typical magnitudes, and some ways of dealing with them is shown in Table 4-6.

The most frequent channel for the coupling of electrical noise is a "parasitic" capacitance. Such a coupling exists everywhere. Any object is capacitively coupled to another object. For instance, a human standing on isolated earth develops a capacitance to ground on the order of 700pF, electrical connectors have a pin-to-pin capacitance of about 2pF, an optoisolator has an emitter-detector capacitance of about 2pF. Fig. 4-9.5A shows that an electrical noise source is connected to the sensor's internal impedance Z through a coupling capacitance C_s. That impedance may be a simple resistance or a combination of resistors, capacitors, inductors, and nonlinear elements, like diodes. Voltage across the impedance is a direct result of the change rate in the noise signal, the value of coupling capacitance C_s and impedance Z. For instance, a pyroelectric detector may have an internal impedance which is equivalent to a parallel connection of a 30pf capacitor and a 50GΩ resistor. The sensor may be coupled through just 1pf to a moving person who has the surface electrostatic charge on the body resulting in static voltage of 1000V. If we assume that the main frequency of human movement is 1Hz, the sensor would pickup the electrostatic interference of about 30V! This is 3 to 5 orders of magnitude higher than the sensor would normally produce in response to thermal radiation received from the human body. Since some sensors and virtually all electronic circuits have nonlinearities, high frequency interference signals, generally called RFI (radio-frequency interference) or EMI (electromagnetic interferences), may be rectified and appear at the output as a dc or slow changing voltage.

4.9.3 Electric shielding

Interferences attributed to electric fields can be significantly reduced by appropriate shielding of the sensor and circuit, especially of high impedance and nonlinear components. Each shielding problem must be analyzed separately and carefully. It is very important to identify the noise source and how it is coupled to the circuit. Improper shielding and guarding may only make matters worse or create a new problem.

A shielding serves two purposes [17]. First, it confines noise to a small region. This will prevent noise from getting into nearby circuits. However, the problem with such shields is that the noise captured by the shield can still cause problems if the return path that the noise takes is not carefully planned and implemented by an understanding of the ground system and making the connections correctly.

Second, if noise is present in the circuit, shields can be placed around critical parts to prevent the noise from getting into sensitive portions of the detectors and circuits. These shields may consist of metal boxes around circuit regions or cables with shields around the center conductors.

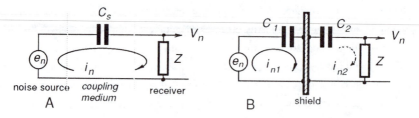

noise source   coupling   receiver
      A   medium   B   shield

Fig. 4-9.5 Capacitive coupling (A) and electric shield (B)

As it was shown in Section 3.1, the noise that resulted from the electric fields can be well controlled by metal enclosures because charge q cannot exist on the interior of a closed conductive surface. Coupling by a mutual, or stray, capacitance can be modeled by a circuit shown in Fig. 4-9.5. Here e_n is a noise source. It may be some kind of a part or component whose electric potential varies. C_s is the stray capacitance (having impedance Z_s at a particular frequency) between the noise source and the circuit impedance Z, which acts as a receiver of the noise. Voltage V_n is a result of the capacitive coupling. A noise current is defined as

$$i_n = \frac{V_n}{Z + Z_s} \, ,$$
(4.9.11)

and actually produces noise voltage

$$V_n = \frac{e_n}{(1+\frac{Z_C}{Z})} \quad .$$

(4.9.12)

For example, if C_s=2.5pf, Z=10kΩ (resistor) and e_n=100 mV, at 1.3MHz, the output noise will be 20mV (for a full scale of 5V, it is 16 LSB of 12 bits).

One might think that 1.3MHz noise is relatively easy to filter out from low frequency signals produced by a sensor. In reality, it cannot be done, because many sensors and, especially the front stages of the amplifiers, contain nonlinear components (*pn*-semiconductor junctions) which act as rectifiers. As a result, the spectrum of high frequency noise shifts into a low frequency region making the noise signal similar to voltage produced by a sensor.

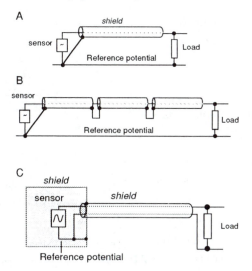

Fig. 4-9.6 Connections of an input cable to a reference potential

When a shield is added, the change to the situation is shown in Fig. 4-9.5B. With the assumption that the shield has zero impedance, the noise current at the left side will be $i_{n1}= e_n/Z_{c1}$. On the other side of the shield, current i_{n2} will be essentially zero since there is no driving source at the right side of the circuit. Subsequently, the noise voltage over the receiving impedance will also be zero and the sensitive circuit becomes effectively shielded from the noise source. There are several practical rules that must be observed when applying electrostatic shields:

O An electrostatic shield, to be effective, should be connected to the reference potential of any circuitry contained within the shield. If the signal is connected to a ground (chassis of the frame or to earth), the shield must be connected to that ground. Grounding of shield is useless if the signal is not returned to the ground.

○ If a shielding cable is used, its shield must be connected to the signal referenced node at the signal source side (Fig. 4-9.6A).

○ If the shield is split into sections, as might occur if connectors are used, the shield for each segment must be tied to those for the adjoining segments, and ultimately connected only to the signal referenced node (Fig. 4-9.6B).

○ The number of separate shields required in a data acquisition system is equal to the number of independent signals that are being measured. Each signal should have its own shield, with no connection to other shields in the system, unless they share a common reference potential (signal "ground"). In that case all connections must be made by a separate jumping wire connected to each shield at a single point.

Fig. 4-9.7 Cable shield is erroneously grounded at both ends

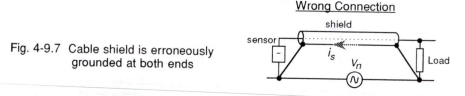

Wrong Connection

○ A shield must be grounded only at one point - preferably next to the sensor. A shielded cable must never be grounded at both ends (Fig. 4-9.7). The potential difference (V_n) between two "grounds" will cause shield current i_s to flow which may induce a noise voltage into the center conductor via magnetic coupling.

○ If a sensor is enclosed into a shield box and data are transmitted via a shielded cable (Fig. 4-9.6C), the cable shield must be connected to the box. It is a good practice to use a separate conductor for the reference potential ("ground") inside the shield, and not to use the shield for any other purposes except shielding: do not allow shield current to exist.

○ Never allow the shield to be at any potential with respect to the reference potential (except in case of driven shields as shown in Fig. 4-1.4). The shield voltage couples to the center conductor (or conductors) via a cable capacitance.

○ Connect shields to a ground via short wires to minimize inductance. This is especially important when both analog and digital signals are transmitted.

4.9.4 Bypass capacitors

The bypass capacitors are used to maintain a low power supply impedance at the point of load. Parasitic resistance and inductance in supply lines mean that the power supply impedance can be quite high. As the frequency goes up, the inductive parasitic becomes troublesome and may result in the circuit oscillation or ringing effects. Even if the circuit operates at lower frequencies, the bypass capacitors are still important as high frequency noise may be transmitted to the circuit and power supply conductors from external sources, for instance radio stations. At high frequencies no power supply or regulator has zero output impedance. What type of capacitor to use is determined by the application, frequency range of the circuit, cost, board space, and some other considerations. To select a bypass capacitor one must remember that a practical capacitor at high frequencies may be far away from the idealized capacitor which is described in textbooks.

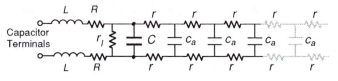

Fig. 4-9.8 Equivalent circuit of a capacitor

A generalized equivalent circuit of a capacitor is shown in Fig. 4-9.8. It is comprised of a nominal capacitance C, leakage resistance r_l, lead inductances L, and resistances R. Further, it includes dielectric absorption terms r and c_a, which are manifested in capacitor's "memory". In bypass applications, r_l and dielectric absorption are second order terms but series R and L are of importance. They limit the capacitor's ability to damp transients and maintain a low power supply output impedance. Often, bypass capacitors must be of large values ($10\mu F$ or more) so they can absorb longer transients, thus electrolytic capacitors are often employed. Unfortunately, these capacitors have large series R and L. Usually, tantalum capacitors offer better results, however, a combination of aluminum electrolytic with nonpolarized (ceramic or film) capacitors may offer even further improvement. A combination of wrong types of bypass capacitors may lead to ringing, oscillation and crosstalk between data communication channels. The best way to specify a correct combination of bypass capacitors is to first try them on a breadboard.

4.9.5 Magnetic shielding

Proper shielding may dramatically reduce noise resulting from electrostatic and electrical fields. Unfortunately, it is much more difficult to shield against magnetic fields because it pene-

trates conducting materials. A typical shield placed around a conductor and grounded at one end
has little if any effect on the magnetically induced voltage in that conductor. As a magnetic field
penetrates a shield, its amplitude decreases exponentially. An effective magnetic shielding can be
accomplished with thick steel shields at higher frequencies. Since magnetic shielding is very difficult, the most effective approach at low frequencies is to minimize the strength of magnetic fields,
minimize the magnetic loop area at the receiving end, and selecting the optimal geometry of
conductors. Some useful practical guidelines are as follows:

○ Locate the receiving circuit as far as possible from the source of the magnetic field.

○ Avoid running wires parallel to the magnetic field; instead, cross the magnetic field at
right angles.

○ Shield the magnetic field with an appropriate material for the frequency and strength.

○ Use a twisted pair of wires for conductors carrying the high-level current that is the
source of the magnetic field. If the currents in the two wires are equal and opposite, the net field in
any direction over each cycle of twist will be zero. For this arrangement to work, none of the current can be shared with another conductor, for example, a ground plane, which may result in
ground loops.

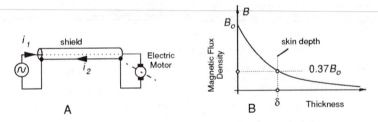

Fig. 4-9.9 Reduction of a transmitted magnetic noise
by powering a load device through a coaxial cable (A);
Magnetic shielding improves with the thickness of the shield (B)

○ Use a shielded cable with the high-level source circuit's return current carried by the
shield (Fig. 4-9.9A). If the shield current i_2 is equal and opposite to that of the center conductor i_1,
the center conductor field and the shield field will cancel, producing a zero net field. This case
seems a violation of a rule *"no shield currents"* for the receiver's circuit, however, the shielded cable here is not used to electrostatically shield the center conductor. Instead, the geometry produces
a cancellation of the magnetic field which is generated by a current supplied to a "current-hungry"
device (an electric motor in this example)

○ Since magnetically induced noise depends on the area of the receiver loop, the induced voltage due to magnetic coupling can be reduced by making the loop's area smaller.

What is the receiver's loop? Fig. 4-9.10 shows a sensor which is connected to the load circuit via two conductors having length L and separated by distance D. The rectangular circuit forms a loop area $a = L \cdot D$. The voltage induced in series with the loop is proportional to the area and cosine of its angle to the field. Thus, to minimize noise, the loop should be oriented at right angles to the field, and its area should be minimized.

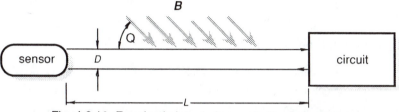

Fig. 4-9.10 Receiver's loop is formed by long conductors

The area can be decreased by reducing the length of the conductors and/or decreasing the distance between the conductors. This is easily accomplished with a twisted pair, or at least with a tightly cabled pair of conductors. It is a good practice to pair the conductors so that the circuit wire and its return path will always be together. This requirement shall not be overlooked. For instance, if wires are correctly positioned by a designer, a service technician may reposition them during the repair work. A new wire location may create a disastrous noise level. Hence, a general rule is - know the area and orientation of the wires and permanently secure the wiring.

Table 4-7 Skin depth δ in mm versus frequency
(adapted from [18])

Frequency	Copper	Aluminum	Steel
60 Hz	8.5	10.9	0.86
100 Hz	6.6	8.5	0.66
1 kHz	2.1	2.7	0.20
10 kHz	0.66	0.84	0.08
100 kHz	0.2	0.3	0.02
1 MHz	0.08	0.08	0.008

Magnetic fields are much more difficult to shield against than electric fields because they can penetrate conductive materials. A typical shield placed around a conductor and grounded at one end has little if any effect on the magnetically induced voltage in that conductor. As magnetic field B_o

penetrates the shield, its amplitude drops exponentially (Fig. 4-9.9B). The skin depth δ of the shield is the depth required for the field attenuation by 37% of that in the air. Table 4-7 lists typical values of δ for several materials at different frequencies. At high frequencies, any material may be used for effective shielding, however at a lower range steel yields a much better performance.

For improving low-frequency magnetic field shielding, a shield consisting of a high-permeability magnetic material (e.g., mumetal) should be considered. However, mumetal effectiveness drops at higher frequencies and strong magnetic fields.

4.9.6 Mechanical noise

Vibration and *acceleration effects* are also sources of transmitted noise in sensors which otherwise should be immune to them. These effects may alter transfer characteristics (multiplicative noise), or they may result in generation by a sensor of spurious signals (additive noise). If a sensor incorporates certain mechanical elements, vibration along some axes with a given frequency and amplitude may cause resonant effects. For some sensors acceleration is a source of noise. For instance, most of pyroelectric detectors also possess piezoelectric properties. The main function of the detector is to respond to thermal gradients. However, such environmental mechanical factors such as fast changing air pressure, strong wind or structural vibration cause the sensor to respond with output signals which often are indistinguishable from responses to normal stimuli.

4.9.7 Ground planes

For many years ground planes have been known to electronic engineers and printed circuit designers as a "mystical and ill-defined" cure for spurious circuit operation [19]. Ground planes are primarily useful for minimizing circuit inductance. They do this by utilizing the basic magnetic theory. Current flowing in a wire produces an associated magnetic field (Section 3.3). The field's strength is proportional to the current i and inversely related to the distance r from the conductor (equation 3.3.6):

$$B = \frac{\mu_0 i}{2\pi r} \quad .$$

(4.9.13)

Thus, we can imagine a current carrying wire surrounded by a magnetic field. Wire inductance is defined as energy stored in the field setup by the wire's current. To compute the wire's inductance requires integrating the field over the wire's length and the total area of the field. This implies integrating on the radius from the wire surface to infinity. However, if two wires carrying the same current in opposite directions are in close proximity, their magnetic fields are canceled. In this case, the virtual wire inductance is much smaller. An opposite flowing current is called *return current*. This is the underlying reason for ground planes. A ground plane provides a return path directly under the signal carrying conductor through which return current can flow. Return current has a direct path to ground, regardless of the number of branches associated with the conductor. Currents will always flow through the return path of the lowest impedance. In a properly designed ground plane this path is directly under the signal conductor. In practical circuits, a ground plane is one side of the board and the signal conductors are on the other. In the multilayer boards, a ground plane is usually sandwiched between two or more conductor planes. Aside from minimizing parasitic inductance, ground planes have additional benefits. Their flat surface minimizes resistive losses due to "skin effect" (ac current travel along a conductor's surface). Additionally, they aid the circuit's high frequency stability by referring stray capacitance to the ground. Some practical suggestions:

○ Make ground planes of as much area as possible on the components side (or inside for the multilayer boards). Maximize the area especially under traces that operate with high frequency or digital signals.

○ Mount components that conduct fast transient currents (terminal resistors, ICs, transistors, decoupling capacitors, etc.) as close to the board as possible.

○ Wherever a common ground reference potential is required use separate conductors for the reference potential and connect them all to the ground plane at a common point to avoid voltage drops due to ground currents.

○ Keep the trace length short. Inductance varies directly with length and no ground plane will achieve perfect cancellation.

4.9.8 Ground loops and ground isolation

When a circuit is used for low-level input signals, a circuit itself may generate enough noise to present a substantial problem for accuracy. Sometimes, a circuit is correctly designed on paper, a bench bread board shows quite a satisfactory performance, however, when a production prototype with the printed circuit board is tested, the accuracy requirement is not met. A difference between a bread board and PC-board prototypes may be in the physical layout of conductors. Usually, conductors between electronic components are quite specific — they may connect a capacitor to a resistor, a gate of a JFET transistor to the output of an operational amplifier, etc. However, there are at least two conductors which, in most cases, are common for the majority of the electronic circuit. These are the power supply bus and the ground bus. Both of them may carry undesirable signals from one part of the circuit to another, specifically, they may couple strong output signals to the sensitive input stages.

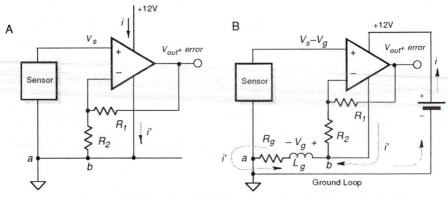

Fig. 4-9.11 Wrong connection of a ground terminal to a circuit (A);
Path of a supply current through the ground conductors (B)

A power supply bus carries supply currents to all stages. A ground bus also carries supply currents, but, in addition, it is often used to establish a reference base for an electrical signal. Interaction of these two functions may lead to a problem which is known as ground loop. We illustrate it in Fig. 4-9.11A where a sensor is connected to a positive input of an amplifier which may have a substantial gain. The amplifier is connected to the power supply and draws current i which is returned to the ground bus as i'. A sensor generates voltage V_s which is fed to the positive input of the amplifier. A ground wire is connected to the circuit in point a – right next to the sensor's terminal. A circuit has no visible error sources, nevertheless, the output voltage contain substantial error. A noise source is developed in a wrong connection of ground wires. Fig. 4-9.11B shows that the ground conductor is not ideal. It may have some finite resistance R_g and inductance L_g. In this example, supply current while returning to the battery from the amplifier, passes through the ground bus between points b and a resulting in voltage drop V_g. This drop, however small, may be comparable with the signal produced by the sensor. It should be noted that voltage V_g is serially connected with the sensor and is directly applied to the amplifier's input. Ground currents may also contain high frequency components, then the bus inductance will produce quite strong spurious high frequency signals which not only add noise to the sensor, but may cause circuit instability as well. For example, let us consider a thermopile sensor which produces voltage corresponding to $100\mu V/°C$ of the object's temperature. A low noise amplifier has quiescent current, $i=5mA$, which passes through the ground loop having resistance $R_g=0.2\Omega$. Ground loop voltage $V_g = iR_g = 1mV$

corresponds to an error of -10°C! The cure is usually quite simple - ground loops must be broken. A circuit designer should always separate a reference ground from current carrying grounds, especially serving digital devices. Fig. 4-9.12 shows that moving the ground connection from sensor's point a to the power terminal point c prevents formation of spurious voltage across the ground conductor connected to the sensor and a feedback resistor R_2. A rule of thumb is to connect the ground to the circuit board only at one point. Grounding at two or more spots may form ground loops which often is very difficult to diagnose.

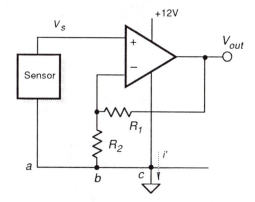

Fig. 4-9.12 Correct grounding of a sensor and interface circuit

4.9.9 Seebeck noise

This noise is a result of the Seebeck effect (Section 3.9) which is manifested as the generation of an electromotive force (e.m.f.) when two dissimilar metals are joined together. The Seebeck *e.m.f.* is small and for many sensors may be simply ignored. However, when absolute accuracy on the order of 10-100μV is required, that noise must be taken into account. The connection of two dissimilar metals produces a temperature sensor. However, when temperature sensing is not a desired function, a thermally induced e.m.f. is a spurious signal. In electronic circuits, connection of dissimilar metals can be found everywhere: connectors, switches, relay contacts, sockets, wires, etc. For instance, the copper PC board cladding connected to kovar[TM][1] input pins of an integrated circuit creates an offset voltage of $40\mu V \cdot \Delta T$ where ΔT is the temperature gradient in °C between two dissimilar metal contacts. The common lead-tin solder, when used with the copper cladding creates a thermoelectric voltage between 1 and 3μV/°C. There are special cadmium-tin solders available to reduce these spurious signals down to 0.3μV/°C. Fig. 4-9.13 shows Seebeck *e.m.f.*

---

[1] Trademark of Westinghouse Electric Corp.

for two types of solder. Connection of two identical wires fabricated by different manufacturers may result in voltage having slope on the order of 200nV/°C.

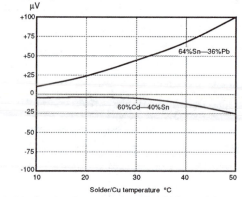

Fig. 4-9.13 Seebeck *e.m.f.* developed by solder-copper joints
(adapted from [20])

In many cases, Seebeck *e.m.f.* may be eliminated by a proper circuit layout and thermal balancing. It is a good practice to limit the number of junctions between the sensor and the front stage of the interface circuit. Avoid connectors, sockets, switches and other potential sources of *e.m.f.* to the extent possible. In some cases this will not be possible. In these instances, attempt to balance the number and type of junctions in the circuit's front stage so that differential cancellations occur. Doing this may involve deliberately creating and introducing junctions to offset necessary junctions. Junctions which intent to produce cancellations must be maintained at the same temperature. Fig. 4-9.14 shows a remote sensor connection to an amplifier where the sensor junctions, input terminal junctions and amplifier components junctions are all maintained while at different but properly arranged temperatures. Such thermally balanced junctions must be maintained at a close physical proximity and preferably on common heat sinks. Air drafts and temperature gradients in the circuit boards and sensor enclosures must be avoided.

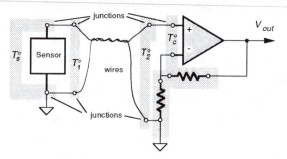

Fig. 4-9.14 Maintaining joints at the same temperature reduces Seebeck noise

REFERENCES

1.   Widlar, R. J. Working with high impedance Op Amps, AN24, *Linear Application Handbook*. National Semiconductor, 1980

2.   Pease, R. A. Improve circuit performance with a 1-op-amp current pump. *EDN*, pp: 85-90, Jan. 20,1983

3.   Bell, D.A. Solid state pulse circuits. 2nd ed. Reston Publishing Company, Inc., Reston, VA, 1981

4.   Senani, R. and Bhaskar, D. R. Single Op-Amp sinusoidal oscillators suitable for generation of very low frequencies. *IEEE Trans. on Instr. and Meas.* Vol. 40, No. 40, pp:777-79, Aug. 1991

5.   Sheingold, D. H., ed. Analog-digital conversion handbook. 3rd ed., Prentice-Hall, Englewood Cliffs, NJ 07632, 1986

6.   Williams, J. Some techniques for direct digitization of transducer outputs, AN7, *Linear Technology Application Handbook*, 1990

7.   Park, Y. E., Wise, K. D. An MOS switched-capacitor readout amplifier for capacitive pressure sensors. *IEEE Custom IC Conf.*, pp: 380-384, 1983

8.   Stafford, K.R., Gray, P. R., and Blanchard, R.A. A complete monolithic sample/hold amplifier. *IEEE J. of Solid-State Circuits*, 9, pp: 381-387, Dec. 1974

9.   Cho S. T., Wise, K. D. A self-testing ultrasensitive silicon microflow sensor. *Sensor Expo Proceedings*, 208B-1, 1991

10.   Sheingold, D. H., Ed. Nonlinear circuits handbook. Analog Devices, Inc. Northwood, Mass., 1974

11.   Weatherwax, S. Understanding constant voltage and constant current excitation for pressure sensors. *SenSym solid-state sensor handbook*. ©Sensym, Inc., 1991

12.   Coats, M. R. New technology two-wire transmitters. *Sensors*, Vol. 8, No. 1., 1/1/1991

13.   Sheingold, D. H., Ed. Analog-digital handbook. Analog Devices, Inc. Norwood, Mass, 1972

14.   Frederiksen, T. M. Intuitive IC OP AMPS. National Semiconductors Corp., Santa Clara, CA, 1984

15.   Johnson, J. B. Thermal agitation of electricity in conductors. *Phys. Rev.* July, 1928

16.   The Best of Analog Dialogue. © Analog Devices, Inc. 1991

17.   Rich, A. Shielding and guarding. *Best of Analog Dialogue*, ©Analog Devices, Inc., 1991

18.   Ott, H. W. Noise reduction techniques in electronic systems. John Wiley & Sons, New York, 1976

19.   Williams, J. High speed comparator techniques, AN13. *Linear applications handbook*. ©Linear Technology Corp., 1990

20.   Pascoe, G. The choice of solders for high-gain devices. *New Electronics (U.K.)*, Febr. 6, 1977

5

Position, Level, and Displacement

The measurement of position and displacement of physical objects is essential for many applications: process feedback control, performance evaluation, transportation traffic control, robotics, security systems - just to name the few. By *position*, we mean the determination of the object's coordinates (linear or angular) with respect to a selected reference. *Displacement* means moving from one position to another for a specific distance or angle. A critical distance is measured by *proximity* sensors. In effect, a proximity sensor is a threshold version of a position detector. In other words, a position sensor is a linear device whose output signal represents the distance to the object from a reference point. A proximity sensor may be a somewhat simpler device which generates the output signal when the distance to the object becomes essential for an indication. For instance, many moving mechanisms in process control and robotics use a very simple but highly reliable proximity sensor - the end switch. It is an electrical switch having normally open or normally closed contacts. When a moving object activates the switch by a physical contact, the latter sends a signal to a control circuit. The signal is an indication that the object has reached the end position. Obviously, such contact switches have many drawbacks, for example, high mechanical load on the moving object and hysteresis.

A displacement sensor may be part of a more complex sensor where the detection of movement is one of the steps in a signal conversion. An example is a capacitive pressure sensor where pressure is translated into a displacement of a diaphragm, and the diaphragm displacement is subsequently converted into an electrical signal representing pressure. Therefore, the positions sensors, some of which are described in this chapter, are essential for designs of many other sensors which are covered in the following chapters of this book.

Position and displacement sensors are static devices whose speed response usually is not critical for the performance[1]. In this Section, we do not cover any sensors whose response is a

---

[1] Nevertheless, the maximum rate of response is usually specified by a manufacturer.

function of time, which by definition, are dynamic sensors. They will be covered in the following chapters.

When designing or selecting position and displacement detectors, the following questions should be answered:

1. How big is the displacement and of what type (linear, circular)?

2. What resolution and accuracy are required?

3. What the measured object is made from (metal, plastic, ferromagnetic, etc.)?

4. How much space is available for mounting the detector?

5. How much play is there in the moving assembly and what is the required detection range?

6. What are the environmental conditions (humidity, temperature, sources of interference, vibration, corrosive materials, etc.)?

7. How much power is available for the sensor?

8. How much mechanical wear can be expected over the life time of the machine?

9. What is the production quantity of the sensing assembly (limited number, medium volume, mass production)?

10. What is the target cost of the detecting assembly?

A careful analysis will pay big dividends in the long term.

5.1   POTENTIOMETRIC SENSORS

A position or displacement transducer may be built with a linear or rotary potentiometer, or a *pot* for short. The operating principle of this sensor is based on a formula (3.5.16) for a wire resistance. From the formula it follows that a resistance linearly relates to the wire length. Thus, by making an object to control the length of the wire, as it is done in a pot, a displacement measurement can be performed. Since a resistance measurement requires passage of an electric current through the pot wire, the potentionometric transducer is of an active type. A stimulus (displacement) is coupled to the pot wiper, whose movement causes the resistance change (Fig. 5-1.1). In most practical circuits, a resistance measurement is replaced by a measurement of voltage. The voltage across the wiper of a linear pot is proportional to the displacement d

$$V = E\frac{d}{D} ,$$

(5.1)

where D is the full scale displacement, and E is the voltage across the pot (excitation signal). This assumes that there is no loading effect from the interface circuit. If there is an appreciable load, the linear relationship between the wiper position and the output voltage will not hold. Besides, the output signal is proportional to the excitation voltage applied across the sensor. This voltage, if not maintained constant, may be a source of error. It should be noted that resistance of the pot is not a part of the equation, which means that its stability (for instance, over a temperature range) virtually makes no effect on accuracy. For the low power applications, high impedance pots are desirable, however, the loading effect must be always considered. The wiper of the pot is usually electrically isolated from the sensing shaft.

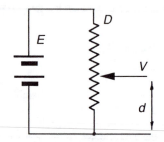

Fig. 5-1.1 Potentiometer as a position sensor

Fig. 5-1.2A shows one problem associated with a wire-wound potentiometer. The wiper may, while moving across the winding, make contact with either one or two wires, thus resulting in uneven voltage steps (Fig. 5-1.2B) or a variable resolution. Therefore, when the coil potentiometer with N turns is used, only the average resolution n should be considered

$$n = 100/N\% \ . \tag{5.2}$$

The force which is required to move the wiper comes from the measured object and the resulting energy is dissipated in the form of heat. Wire-wound potentiometers are fabricated with thin wires having a diameter in the order of 0.01 mm. A good coil potentiometer can provide an average resolution of about 0.1% of FS, while the high-quality resistive film potentiometers may yield an infinitesimal resolution which is limited only by the uniformity of the resistive material and noise floor of the interface circuit. The continuous resolution pots are fabricated with conductive plastic, carbon film, metal film, or a ceramic-metal mix which is known as *cermet*. The wiper of the precision potentiometers are made from precious metal alloys. Displacements sensed by the angular potentiometers range from approximately 10° to over 3000° for the multi-turn pots (with gear mechanisms). While being quite useful in some applications, potentiometers have several drawbacks:

1. Noticeable mechanical load (friction)

2. Need for a physical coupling with the object

3. Low speed

4. Friction and excitation voltage cause heating of the potentiometer

5.1 Low environmental stability.

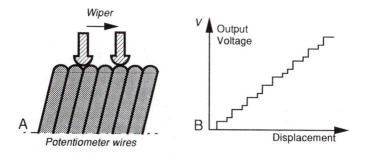

Fig. 5-1.2 Uncertainty caused by wire-wound potentiometer
A: a wiper may contact one or two wires at a time; B: uneven voltage steps

5.2   GRAVITATIONAL SENSORS

Inclination detectors which measure the angle from the direction to the earth's center of gravity are employed in road construction, machine tools, inertial navigation systems and other applications requiring a gravity reference. An old and still quite popular detector of a position is a mercury switch (Fig. 5-2.1A,B). The switch is made of a glass tube having two electrical contacts and a drop of mercury. When the sensor is positioned with respect to the gravity force in such a way as the mercury moves away from the contacts, the switch is open. A change in the switch orientation causes the mercury to move to the contacts and touch both of them, thus closing the switch. One popular application of this design is in a household thermostat, where the mercury switch is mounted on a bimetal coil which serves as an ambient temperature sensor. Winding or unwinding the coil in response to room temperature affects the switch orientation. Opening and closing the switch controls a heating/cooling system. An obvious limitation of this design is its an on-off operation. A mercury switch is a threshold device, which snaps when its rotation angle exceeds a predetermined value. To measure angular displacement with higher resolution, a more complex sensor is required. One elegant design is shown in Fig. 5-2.1C. A small slightly curved glass tube is filled with partly conductive electrolyte. Three electrodes are built into the tube: two small ones at the ends and an extended electrode along the length of the tube. An air bubble resides in the tube and may move along its length as the tube tilts. Electrical resistances between the center electrode and each of the end electrodes depends on the position of the bubble. As the tube shifts away from the balance position, the resistances increase or decrease proportionally. The electrodes are connected into a bridge circuit which is excited with an ac current to avoid damage to the electrolyte and electrodes.

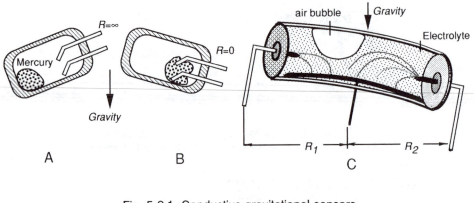

Fig. 5-2.1 Conductive gravitational sensors
A: Mercury switch in the open position
B: Mercury switch in the closed position
C: Electrolytic inclination sensor

A measuring circuit must extract information about the angle of tilt and its polarity. Fig. 5-2.2 shows a simplified circuit which does just that. To protect the electrolyte sensor from damage, a dc component is eliminated by capacitor C_1 from the excitation signal which is produced by a square wave oscillator. The sensor is connected into a Wheatstone bridge which controls the Howland current pump (see the description of the pump in Section 4.3.1). The bridge's differential ac signal is converted into an ac current whose polarity reverses as the sensor tilts in the opposite direction. The current is rectified by a diode full wave rectifier and charges the C_2 capacitor. Thus, a 0.03 µF capacitor receives a unipolar charge. The voltage across the capacitor is amplified by a differential amplifier having a gain of 2. In essence, the C_2 capacitor, a differential amplifier, the analog switch and a comparator form a voltage controlled oscillator. When the voltage across the capacitor becomes high enough, the comparator's output goes high turning on the analog switch, thus discharging the capacitor C_2. The capacitor C_3 provides enough of ac positive feedback around the comparator to allow a complete zero reset of the C_2 capacitor. When the feedback ceases, the comparator's output goes low, opening the analog switch. This allows the capacitor C_2 to charge again by a constant current, and the entire process repeats. The frequency of this oscillator depends on the magnitude of the constant current delivered to the bridge-capacitor configuration. An additional polarity detector compares phases of the pulses from the oscillator and the current pump. The polarity detector produces an output signal which represents the direction of the tilt.

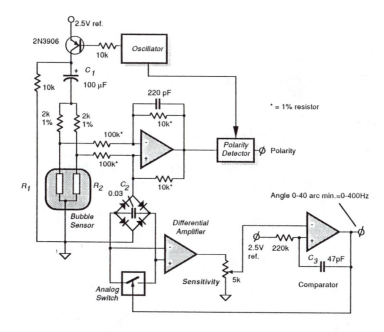

Fig. 5-2.2 A simplified circuit diagram for the electrolytic inclination sensor

A more advanced inclination sensor employs an array of photodetectors [1]. The detector is useful in civil and mechanical engineering for the shape measurements of complex objects with high resolution. Examples include the measurement of ground and road shapes, and the flatness of an iron plate, which can not be done by the conventional methods. The sensor (Fig. 5-2.3A) consists of a light-emitting diode (LED) and a hemispherical spirit level mounted on a pn-junction photodiode array. A shadow of the bubble in the liquid is projected onto the surface of the photodiode array. When the sensor is kept horizontally, the shadow on the sensor is circular as shown in Fig. 5-2.3B, and the area of the shadow on each photodiode of the array is the same. However, when the sensor is inclined, the shadow becomes slightly elliptic as shown in Fig. 5-2.3C, implying that the output currents from the diodes are no longer equal. In a practical sensor, the diameter of the LED is 10 mm and the distance between the LED and the level is 50 mm, the diameters of the hemispherical glass and the bubble are 17 and 9 mm, respectively. The outputs of the diodes are converted into a digital form and calibrated at various tilt angles. The calibration data are compiled into look-up tables which are processed by a computing device. By positioning the sensor at the cross point of the lines drawn longitudinally and latitudinally at an interval on the slanting surface of an object, x and y components of the tilt angle can be obtained and the shape of the object is reconstructed by a computer.

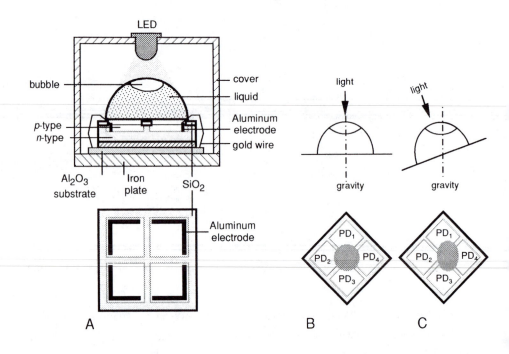

Fig. 5-2.3 Optoelectronic inclination sensor
A: design; B: a shadow at a horizontal position; C: a shadow at the inclined position

5.3 THERMAL SENSORS

It is possible to apply temperature measurement techniques to measure the liquid level in a tank by employing thermal conductivities and thermal capacitances of different phases: gases and liquids. One way to fabricate a fluid level sensor is to use a temperature difference between liquid and gas. A liquid may have a temperature quite different from the environment. For instance, a water boiler may contain water at a temperature near 100°C. A temperature sensor placed outside on the tank wall will register temperature which is a function of several factors: thermal conductivity of the tank walls, pressure inside the tank, water and ambient temperatures. There is a substantial gradient in the temperature of the wall across the water line. (T_2-T_1). The temperature is lower above the water line, even if the vapor has the same temperature, because of the thermal conductivity of the tank walls. This thermal gradient can be detected by an array of temperature detectors placed along the water tank wall (Fig. 5-3.1). The detectors can be RTDs or thermistors whose outputs are multiplexed by a gate circuit (MUX), converted into a digital format (A/D) and analyzed by a microprocessor (μP) to detect the water level. Naturally, a level resolution of the arrangement is equal to the distance between two adjacent temperature detectors. For better accuracy, the sensors should be thermally insulated from the outside environment.

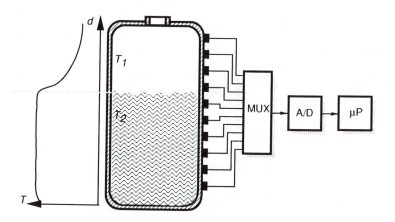

Fig. 5-3.1 An array thermal detector of liquid level

Another approach is based on the active thermal detection where thermal sensor measures heat dissipation through a tank wall (Fig. 5-3.2A). A long sensing strip is positioned over the outside surface of the tank, from bottom to top. The strip contains two resistive components. One possess properties of a thermistor, that is, its resistance is a function of temperature. It serves as a temperature sensor. The other component in a strip is a heating element. The strip is connected into a resistive bridge and an amplifier, which controls electric current through the heating element. This results in a temperature increase of the strip, including the imbedded temperature sensor. Quickly, after applying power, the bridge comes into an equilibrium state which corresponds to a specific constant temperature set by the bridge's fixed resistors. A voltage across the heater depends mainly on three factors: the set temperature, the tank temperature and the liquid level in the tank. Since liquid is a better thermal conductor than a gaseous phase, the higher the liquid level, the higher the thermal loss from the strip through the tank wall and the tank contents. A combined heat flow Q_2 through the wall, and Q_3 through the liquid, is much higher than Q_1 through the wall above the liquid level. Therefore, to maintain the fixed temperature of the strip, the amplifier must deliver a higher voltage for a higher fluid level. Fig. 5-3.2B shows a family of curves establishing a relationship between the liquid level and the voltage across the heating element. The higher temperature of the tank the lower the curve. Hence, for practical purposes, the circuit must incorporate an additional temperature compensating sensor. Naturally, the strip should not necessarily incorporate a thermistor-like sensor. Semiconductors, RTD, and other temperature sensitive devices may be successfully employed. However, a distributed thermistor is more attractive for many applications due to its simplicity and lower cost.

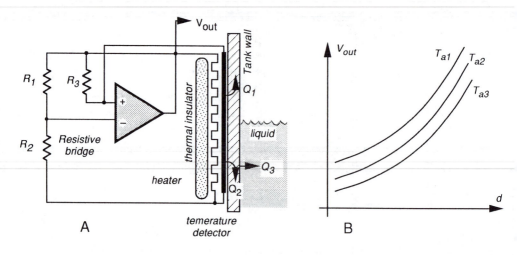

Fig. 5-3.2 Active thermoresistive liquid level sensor
A: Simplified circuit diagram
B: Voltages across heating elements for three different tank temperatures

5.4   CAPACITIVE SENSORS

5.4.1 Gauging capacitive sensors

Equation (3.2.7) defines that the capacitance of a flat capacitor is inversely proportional to the distance between the plates. This dependence can be employed for measuring position, displacement, gauging, or any other similar parameter. The operating principle of a capacitive gauge, proximity, and position sensors is based on either changing the geometry (i.e., a distance between the capacitor plates), or capacitance variations in the presence of conductive or dielectric materials. The ability of capacitive detectors to sense virtually all materials makes them an attractive choice for many applications. When detecting grounded metals, these sensors virtually rely on eddy currents developed in conductors and a typical detecting distance is in the range up to 30 mm. On the other hand, detecting dielectrics can be done at shorter distances depending on the dielectric constants. For instance, glass (κ=4.0) can be detected at 8-10 mm, while polyethylene (κ=2.0) at only 2 mm.

A design of a capacitive sensor is shown in Fig. 5-4.1, where one plate of a capacitor is connected to the central conductor of a coaxial cable, while the other plate is formed by a target. Note that the probe plate is surrounded by a grounded guard to minimize a fringing effect and improve linearity. The capacitive probes operate at frequencies in the 3 MHz range and can detect very fast moving targets, since a typical frequency response of a probe with a built-in electronic interface is in the range of 40 kHz.

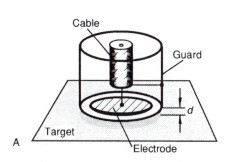
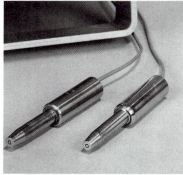

Fig. 5-4.1 A capacitive probe with a guard ring
A - cross-sectional view; B - outside view
(Courtesy of ADE Technologies, Inc., Newton, MA)

5.4.2 Ratiometric displacement sensor

A capacitive displacement sensor operates on a ratiometric principle involving two capacitors, one of a fixed value and another is variable [2]. The capacitance ratio (measured at 2.5 kHz) serves to reject changes in ambient conditions, including humidity, because both capacitors are subjected to the same ambient air and thus their dielectric constants should track (Fig. 5-4.2).

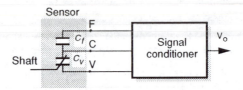

Fig. 5-4.2 Equivalent circuit of a capacitive position sensor

The actual sensor is comprised of a cylindrical body with electrodes C, V, and F whose positions are fixed with respect to the cylinder (Fig. 5-4.3). A tubular dielectric shaft can be moved in and out between the common electrode C and the variable electrode V thus changing the value of a capacitance between them. The moving shaft is attached to the object whose displacement is measured. The principle of operation is quite simple: the value of a variable capacitor C_v (Fig. 5-4.2) which depends on a dielectric shaft displacement, is compared with the value of a fixed capacitor C_f and the ratio of the two values is processed by an electronic circuit. The displacement is ultimately expressed in terms of a voltage and is displayed on a readout device. The output voltage V_o of the triaxial displacement transducer circuit is [2]

$$V_o = KV_2\left(\frac{C_v}{C_f} - 1\right), \qquad (5.3)$$

where V_2 is a known fixed reference voltage and K is the known gain of the circuit.

The full change in the output of the sensor occurring at maximum displacement of 12 mm, is 10 V. It was found that the change in the output due to a relative humidity change from 10% to 85% at 21°C, was 0.035V, or 0.35% FS (full scale), which for many applications is excessively high. In an effort to eliminate this effect, air holes were placed in the device structure to allow the mixing of air between the two capacitors, however this did not noticeably improved the performance. Experimental investigation of a dependence of C_v and C_f of relative humidity [3] has indicated that C_f increases by about 0.3% and C_v by about 0.03% over an entire humidity range. This large change in capacitances (particularly in C_f) is far in excess of the expected change based on the change in the dielectric constant of moist air. It was found that the adsorption of the vapor on the ceramic supports in the region of the common electrode and the fixed electrode has a much greater effect than the adsorption on the electrodes. Moisture redirects electric flux lines from the fixed electrode toward the shield, increasing the flux between the capacitor plates, thus increasing the capacitance.

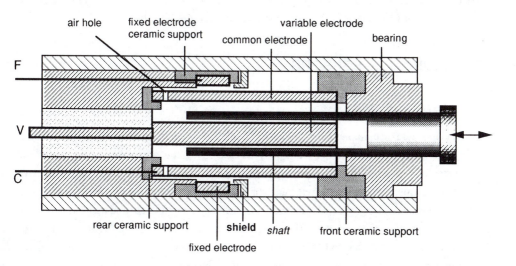

Fig. 5-4.3 Cylindrical capacitive sensor with moving dielectric shaft
Note the shield which prevents redirection of electric field

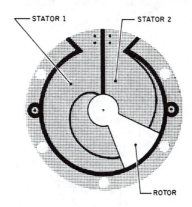

Fig. 5-4.4 A differential angular capacitive sensor
produced by Zi-Tech Instruments Corp.
(Reprinted with permission, Sensors Magazine, ©1992)

To minimize a capacitance dependence of moisture, the capacitors must be shielded from redirecting the flux. This can be done by adding grounded shielding lips as shown in Fig. 5-4.3. This solution is so effective that there was no measurable moisture effect on the output signal. A capacitive sensor also can be used for the angular position measurement. Fig. 5-4.4 shows a differential capacitor where each stator plate forms a capacitance with respect to the rotor plate. As the rotor moves facing two plates of the stator, the capacitance varies linearly.

5.5 INDUCTIVE SENSORS

5.5.1 LVDT and RVDT

Position and displacement may be sensed by methods of electromagnetic induction. A magnetic flux coupling between two coils may be altered by the movement of an object and subsequently converted into voltage. Variable inductance sensors that use a nonmagnetized ferromagnetic medium to alter the reluctance (magnetic resistance) of the flux path are known as variable-reluctance transducers [4]. The basic arrangement of a multiinduction transducer contains two coils — primary and secondary. The primary carries ac excitation (V_{ref}) that induces a steady ac voltage in the secondary coil (Fig. 5-5.1). The induced amplitude depends on flux coupling between the coils. There are two techniques to change the coupling. One is the movement of an object made of ferromagnetic material within the flux path. This changes the reluctance of the path, which, in turn, alters the coupling between the coils. This is the basis for the operation of LVDT (linear variable differential transformer), RVDT (rotary variable differential transformer), and the mutual inductance proximity sensors. The other method is to physically move one coil with respect to another.

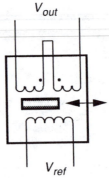

Fig. 5-5.1 Circuit diagram of the inductive sensor

LVDT is a transformer with a mechanically actuated core. The primary is driven by a sine wave (excitation signal) having a stabilized amplitude. Sine wave eliminates error related harmonics in the transformer [5]. An ac signal is induced in the secondary coils. A core made of a ferromagnetic material is inserted coaxially into the cylindrical opening without physically touching the coils. The two secondaries are connected in the opposed phase. When the core is positioned in the magnetic center of the transformer, the secondary output signals cancel and there is no output voltage. Moving the core away from the central position unbalances the induced magnetic flux ratio between the secondaries, developing an output. As the core moves, the reluctance of the flux path changes. Hence, the degree of flux coupling depends on the axial position of the core. At a steady state, the amplitude of the induced voltage is proportional, in the linear operating region, to the core displacement. Consequently, voltage may be used as a measure of a displacement. The LVDT provides the direction as well as magnitude of the displacement. The direction is determined by the phase angle between the primary (reference) voltage and the secondary voltage. Excitation voltage is generated by a stable oscillator. A suitable circuit is shown in Fig. 5-5.2. To exemplify how the sensor works, Fig. 5-5.2 shows the LVDT connected to a synchronous detector which rectifies the sine wave and presents it at the output as a dc signal. The synchronous detector is comprised of an analog multiplexer (MUX) and a zero-crossing detector which converts the sine wave into the square pulses compatible with the control input of the multiplexer. A phase of the zero-crossing detector should be trimmed for the zero output at the central position of the core. The output amplifier can be trimmed to a desirable gain to make the signal compatible with the next stages. The synchronized clock to the multiplexer means that the information presented to the RC-filter at the input of the amplifier is amplitude and phase sensitive. The output voltage represents how far the core is from the center and on which side.

For LVDT to measure transient motions accurately, the frequency of the oscillator must be at least ten times higher than the highest significant frequency of the movement. For slow changing process, stable oscillator may be replaced by coupling to a power line frequency of 60 or 50 Hz.

Advantages of the LVDT and RVDT are the following: 1) the sensor is a noncontact device with no or very little friction resistance with small resistive forces; 2) hysteresis (magnetic and mechanical) are negligible; 3) output impedance is very low; 4) low susceptibility to noise and interferences; 5) its construction is solid and robust, 6) infinitesimal resolution is possible.

One useful application for the LVDT sensor is in the so-called *gauge heads* which are used in tool inspection and gauging equipment. In that case, the inner core of the LVDT is spring loaded to return the measuring head to a preset reference position.

The RVDT operates on the same principle as LVDT, except that a rotary ferromagnetic core is used. The prime use for the RVDT is the measurement of angular displacement. The linear range of measurement is about ±40° with a nonlinearity error of about 1%.

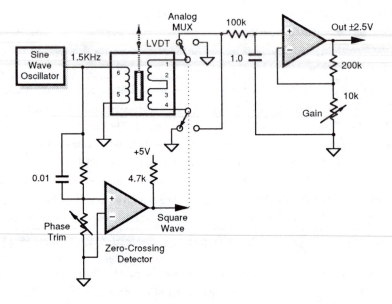

Fig. 5-5.2 A simplified circuit diagram of an interface for an LVDT sensor

5.5.2 Eddy current sensors

To sense the proximity of non-magnetic, but conductive materials the effect of *eddy currents* is used in a dual-coil sensor (Fig. 5-5.3A). One coil is used as a reference, while the other is for the sensing of the magnetic currents induced in the conductive object. Eddy (circular) currents produce the magnetic field which opposes that of the sensing coil, thus resulting in a disbalance with respect to the reference coil. The closer the object to the coil, the larger the change in the magnetic impedance. The depth of the object where eddy currents are produced is defined by

$$\delta = \frac{1}{\sqrt{\pi f \mu \sigma}}, \tag{5.4}$$

where f is the frequency and σ is the target conductivity. Naturally, for the effective operation, the object thickness should be larger than that depth. Hence, eddy detectors should not be used for detecting film metallized or foil objects. Generally, relationship between the coil impedance and distance to the object x is nonlinear and temperature dependent. The operating frequency range of the eddy current sensors range from 50 kHz to 10 MHz.

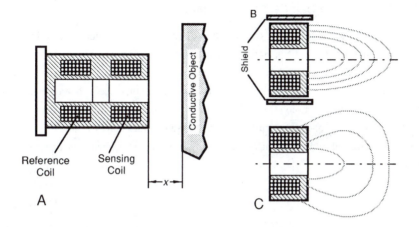

Fig. 5-5.3 A: Electromagnetic proximity sensor
B: Sensor with the shielded front end; C: Unshielded sensor

Figs. 5-5.3 B and C show two configurations of the eddy sensors: with the shield and without one. The shielded sensor has a metal guard around the ferrite core and the coil assembly. It focuses the electromagnetic field to the front of the sensor. This allows the sensor to be imbedded into a metal structure without influencing the detection range. The unshielded sensor can sense at its sides as well as from the front. As a result, the detecting range of an unshielded sensor is usually somewhat greater than that of the shielded sensor of the same diameter. To operate properly, the unshielded sensors require nonmetallic surrounding objects.

In addition to a position detection, eddy sensors can be used to determine material thickness, nonconductive coating thickness, conductivity and plating measurements, and cracks in the material. A crack detection and surface flaws become the most popular applications for the sensors. Depending on the applications, eddy probes may be of many coil configurations: some are very small in diameter (2-3 mm), while others are quite large (25 mm). Some companies even make custom designed probes to meet unique requirements of the customers (Staveley Instruments, Inc., Kennewick, WA). One important advantage of the eddy current sensors is that they do not need magnetic material for the operation, thus they can be quite effective at high temperatures (well exceeding Curie temperature of a magnetic material), and for measuring the distance to or level of conductive liquids, including molten metals. Another advantage of the detectors is that they are not mechanically coupled to the object and thus the loading effect is very low.

5.5.3 Brushless absolute angle sensor

The brushless angle sensing can be achieved using a variable inductance principle to create autotransformer devices[1]. The operating principle of this device in some respects is similar to those of the LVDT and eddy current sensors. If two equivalent inductances are series-connected across an ac source, they will divide the voltage equally. If the inductance of one of them is increased while the other is decreased, the voltages across the halves of the coil will be disbalanced. Fig. 5-5.4 shows this operating principle. The autotransformer coil is wound around the poles of a ferromagnetic stator. The coil has three terminals, where a terminal B is a central point of the transformer.

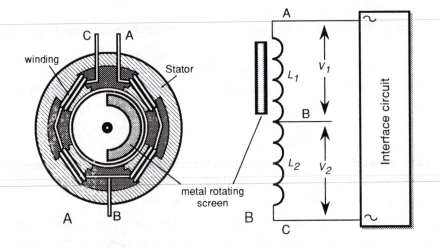

Fig. 5-5.4 Brushless inductive sensor (A) detects rotation angle by comparing voltages V_1 and V_2 across the stator windings; (B) is the equivalent circuit

---

[1] U.S. Patent 4,991,301; British patents 2,223,590 and 2,241,788.

The ferromagnetic stators drives the magnetic flux across an annular air gap. In this gap a metal semicircular screen is rotated by the input shaft. Eddy currents induced in this screen produce opposing flux, and thus reduce the inductance of the corresponding screened poles, by the comparison with those unscreened. The voltage V_2 between the midpoint tap B and terminal C will therefore be more than half the voltage across the supply terminals A and C for the position of the screen shown in the picture. Turning the shaft will progressively transfer the low inductance zone to section A-C, with, with the reduction in voltage V_2 over the range of 180°. The difference between the two voltages is a nominally linear analog curve, which enables measurements of angles up to 180°. By placing additional taps on the transformer, angles of 360° can be measured with the assistance of an appropriate electronic processing of the output voltages. The sensor can be designed using molded ferroplastic magnetic circuits for the economical high-volume production (Fig. 5-5.5). Typical specifications of the sensor are given in Table 5-1.

Table 5-1 Specifications of 11D12 Transyn angular sensor

Characteristic	Value
Dimensions	27 mm dia x 45 mm length
Resolution	5.27 arcmin
Linearity	10.6 arcmin
Temperature range	-10 to +80°C
Temperature coefficient	±20 ppm/°C
Maximum tracking rate	260 rps

Fig. 5-5.5 Construction of a multipole angular sensor
(Courtesy of Transyn, Div. of Radiodetection Ltd., Bristol, U.K.)

5.5.4 Transverse inductive sensor

Another position sensing device is called a *transverse inductive proximity sensor*. It is useful for sensing relatively small displacements of ferromagnetic materials. As the name implies, the sensor measures the distance to an object which alters the magnetic field in the coil. The coil inductance is measured by an external electronic circuit (Fig. 5-5.6). A self-induction principle is the foundation for the operation of such a transducer. When the proximity sensors moves into the vicinity of a ferromagnetic object, it magnetic field changes thus altering the inductance of the coil. The advantage of the sensor is that it is a noncontact device, whose interaction with the object is only through the magnetic field. An obvious limitation is that it is useful only for the ferromagnetic objects at relatively short distances.

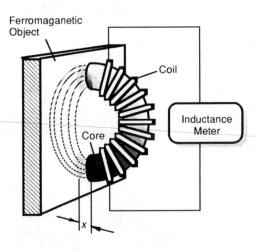

Fig. 5-5.6 A transverse inductive proximity sensor

A modified version of the transverse transducer is shown in Fig. 5-5.7A. To overcome the limitation for measuring only ferrous materials, a ferromagnetic disk is attached to a displacing object while the coil has a stationary position. Alternatively, the coil may be attached to the object and the core is stationary. This proximity sensor is useful for measuring small displacements only, as its linearity is poor in comparison with LVDT. However, it is quite useful as a proximity detector for the indication of close proximity to an object which is made of any solid material. The magnitude of the output signal as function of distance to the disk is shown in Fig. 5-5.7B.

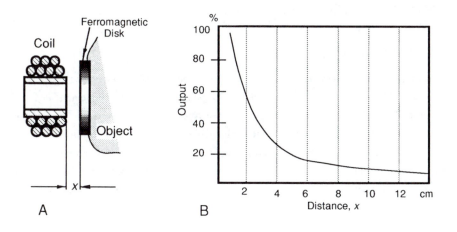

Fig. 5-5.7 Transverse sensor with an auxiliary ferromagnetic disc (A),
and the output signal as function of distance (B)

A sensor for detecting ferrous targets is often housed in a plastic case with a threaded body which allows its easy installation (Fig. 5-5.8). A switching speed is fairly high - in the range of 25kHz, which allows the detection of fast moving object.

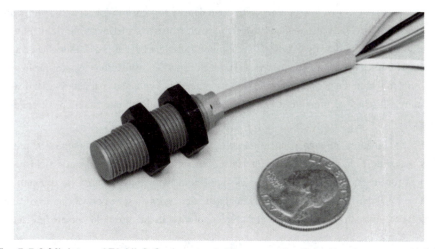

Fig. 5-5.8 Miniature 972 ULC Series proximity sensor for detecting ferrous materials
(Courtesy of Micro Switch, A Honeywell Division)

5.6 MAGNETIC SENSORS

5.6.1 Reed switches

One of the simplest magnetic proximity sensors is a combination of a reed switch and a permanent magnet. A reed switch is a hermetically sealed pair of contacts which are activated when they are properly aligned with the magnetic flux. Their typical applications include security systems, where they are mounted on a door edge for indicating the closing and opening of the door. The reed switches are relatively fast - the operating time is less than 1ms, and have a substantial degree of hysteresis which makes them quite immune to small fluctuations in the magnetic field. Typically they are activated when the magnet approaches the switch by about 5mm, and are released when the magnet moves away by about 10 to 15mm.

5.6.2 Hall effect sensors[1]

There are two types of Hall sensors: linear and threshold (Fig. 5-6.1). A linear sensor usually incorporates an amplifier for the easier interface with the peripheral circuits. In comparison with a basic sensor (Fig. 3-8.1), they operate over a broader voltage range and are more stable in a noisy environment. These sensors are not quite linear (Fig. 5-6.2A) with respect to magnetic field density and, therefore, for the precision measurements require a calibration. The threshold-type sensors in addition to the amplifier, contain a Schmitt trigger detector with a built-in hysteresis. The output signal as a function of a magnetic field density is shown in Fig. 5-6.2B. The signal is two-level and has clearly pronounced hysteresis with respect to the magnetic field. When the applied magnetic flux density exceeds a certain threshold, the trigger provides a clean transient from the OFF to ON position. The hysteresis eliminates spurious oscillations by introducing a dead band zone, in which the action is disabled after the threshold value has passed. The Hall sensors are usually fabricated as monolithic silicon chips and encapsulated into small epoxy or ceramic packages.

For the position and displacement measurements, the Hall effect sensors must be provided with a magnetic field source and an interface electronic circuit. The magnetic field has two important for this application characteristics — a flux density and a polarity (or orientation). It should be noted that for better responsivity, magnetic field lines must be normal (perpendicular) to the flat face of the sensor and must be at a correct polarity. In the threshold sensors fabricated by

---

[1] See Section 3.8 for the operating principle.

Sprague®, the south magnetic pole will cause switching action and the north pole will have no effect.

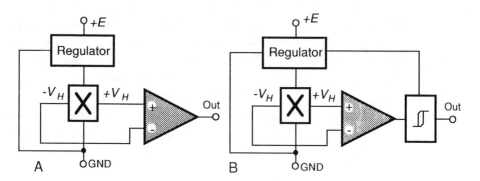

Fig. 5-6.1 Circuit diagrams of a linear (A) and threshold (B) Hall effect sensors

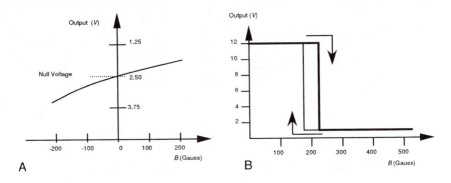

Fig. 5-6.2 Transfer functions of a linear (A) and a threshold (B) Hall effect sensors

Before designing a position detector with a Hall sensor, an overall analysis should be performed in approximately the following manner. First, the field strength of the magnet should be investigated. The strength will be the greatest at the pole face, and will decrease with increasing distance from the magnet. The field may be measured by the gaussmeter or a calibrated Hall sensor. For a threshold type Hall sensor, the longest distance at which the sensor's output goes from ON (high) to OFF (low) is called a *release point*. It can be used to determine a critical distance where the sensor is useful. A magnetic field strength is not linear with distance and depends greatly upon the magnet shape, the magnetic circuit, and the path traveled by the magnet. The Hall conductive strip is situated at some depth within the sensor's housing. This determines the minimum operating distance. A magnet must operate reliably with the total effective air gap in the working environment. It must fit the available space, must be mountable, affordable, and available[1].

The Hall sensors can be used for the interrupter switching with a moving object. In this mode, the activating magnet and the Hall sensor are mounted on a single rugged assembly with a small air gap between them (Fig. 5-6.3). Thus, the sensor is held in the ON position by the activating magnet. If a ferromagnetic plate, or vane, is placed between the magnet and the Hall sensor, the vane forms a magnetic shunt that distorts the magnetic flux away from the sensor. This causes the sensor to flip to the OFF position. The Hall sensor and the magnet could be molded into a common housing, thus eliminating the alignment problem. The ferrous vanes which interrupt the magnetic flux could have linear or rotating motion. An example of such a device is an automobile distributor.

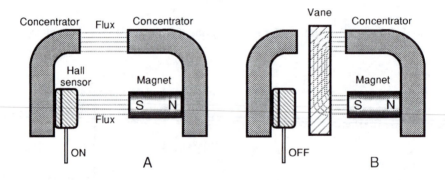

Fig. 5-6.3 The Hall effect sensor in the interrupter switching mode
A: The magnetic flux turns the sensor on; B: the magnetic flux is shunted by a vane
(after [6])

To illustrate how the Hall effect sensor can be used to monitor small displacement, let us describe a level sensor for an automobile fuel tank [7]. A simplified design of the sensor is shown in Fig. 5–6.4. The design is a combination of the following parts: a special plastic, which is chemically inert, tough, but economical, two support springs, a magnet, and a Hall sensor. The float has buoyancy in fuel such that when the fluid level rises from minimum to maximum, the float moves by about 2 mm. A small magnet is positioned on the top of the float near the linear Hall sensor. The float is suspended within the mounting tube by a pair of springs which are rigid radially but flexible longitudinally. The springs allow the float to move only vertically without the usual stiction, hysteresis, and wear associated with the conventional float sensors. Fluid enters the bottom

---

[1] For more information on permanent magnets see section 3.3.

of the float through a series of small holes. Slots near the top of the sensor allow air to escape as the fluid rises, however there is an air bubble maintained near the magnet. This is an added security measure to prevent magnetic particles present in the fluid from contaminating the magnetic field. A 2-mm full scale displacement of a magnet, provides an accurate and repeatable signal at the sensor's output.

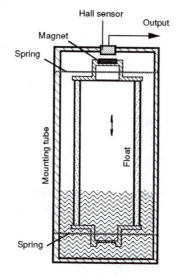

Fig. 5-6.4 Fluid level detector with a Hall effect sensor

5.6.3 Magnetoresistive sensors[1]

These sensors are similar in application to the Hall effect sensors. For functioning, they require an external magnetic field. Hence, whenever the magnetoresistive sensor is used as a proximity, position, or rotation detector it must be combined with a source of a magnetic field. Usually, the field is originated in a permanent magnet which is attached to the sensor. Fig. 5-6.5 shows a simple arrangement for using a sensor-permanent-magnet combination to measure linear displacement. It reveals some of the problems likely to be encountered if proper account is not taken of the effects described below. When the sensor is placed in the magnetic field, it is exposed

---

[1] Information on the KZM10 and KM110 sensors is courtesy of Philips Semiconductors BV. Eindhoven, The Netherlands. For a more detailed description of the sensor operation see section 15.2.4.

to the fields in both the x and y directions. If the magnet is oriented with its axis parallel to the sensor strips (i.e., in the x-direction) as shown in Fig. 5-6.5(a), H_x then provides the auxiliary field, and the variation in H_y can be used as a measure of x displacement. Fig. 5-6.5(b) shows how both H_x and H_y vary with x, and Fig. 5-6.5(c) shows the corresponding output signal. In this example, H_x never exceeds $\pm\hat{\ }H_x$ (the field that can cause flipping of the sensor) and the sensor characteristics remain stable and well-behaved throughout the measuring range. However, if the magnet it too powerful, or the sensor passes too close to the magnet, the output signal will be drastically different.

Suppose the sensor is initially on the transverse axis of the magnet (x=0). H_y will be zero and H_x will be at its maximum value ($>H_x$). So the sensor will be oriented in the +x direction and the output voltage will vary as in Fig. 5-6.6(b). With sensor's movement in +x direction, H_y and V_0 increase, and H_x falls to zero and then increases negatively until H_y exceeds $-H_x$. At this point, the sensor characteristic flips and the output voltage reverses, moving from A to B in Fig. 5-6.6(b). A further increase in x causes the sensor voltage to move along BE. If the sensor is moved in the opposite direction, however, H_x increases until it exceeds $+H_x$ and V_0 moves from B to C. At this point, the sensor characteristic again flips and V_0 moves from C to D. Then, under these conditions, the sensor characteristic will trace the hysteresis loop ABCD, and a similar loop in the -x direction. Fig. 5-6.6(b) is an idealized case, since the reversals are never as abrupt as shown.

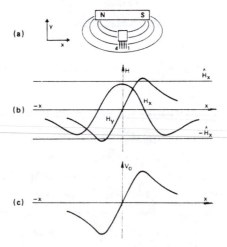

Fig. 5-6.5 Magnetoresistive sensor output in the field of a permanent magnet as a function of its displacement x parallel to the magnetic axis.
The magnet provides both the axillary and transverse fields. Reversal of the sensor relative to the magnet will reverse the characteristic

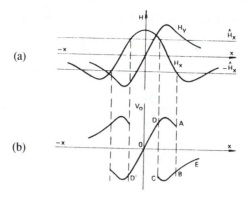

(a)

(b)

Fig. 5-6.6 Sensor output with a too strong magnetic field

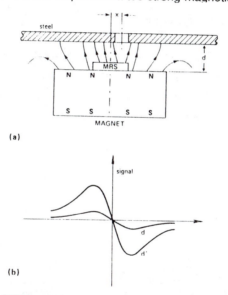

(a)

(b)

Fig. 5-6.7 One point measurement with the KMZ10
(a) the sensor is located between the permanent magnet and the metal plate
(b) Output signals for two distances between the magnet and the plate

Fig. 5-6.7(a) shows how KMZ10B and KM110B magnetoresistive sensors may be used to make position measurements of a metal object. The sensor is located between the plate and a permanent magnet, which is oriented with its magnetic axis normal to the axis of the metal plate. A discontinuity in the plate's structure, such as a hole or a region of nonmagnetic material, will disturb the magnetic field and produce a variation in the output signal from the sensor. Fig. 5-6.7(b) shows the output signal for two values of spacing d. In the point where the hole and the sensor are precisely aligned, the output is zero regardless of the distance d or surrounding temperature.

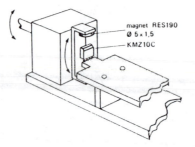

Fig. 5-6.8 Angular measurement with the KMZ10 sensor

Fig. 5-6.8 shows another setup which is useful for measuring angular displacement. The sensor itself is located in the magnetic field produced by two RES190 permanent magnets fixed to a rotable frame. The output of the sensor will then be a measure of the rotation of the frame.

Fig. 5-6.9A depicts the use of a single KM110 sensor for detecting rotation and direction of a toothed wheel. The method of direction detection is based on a separate signal processing for the sensor's two half-bridge outputs.

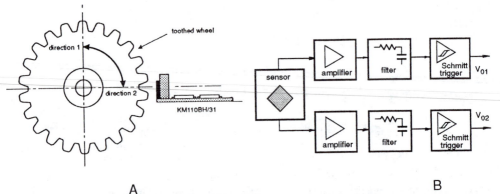

A   B

**Fig. 5-6.9 A: Optimum operating position of a megnetoresistive module.
Note a permanent magnet positioned behind the sensor;
B: Block diagram of the module circuit**

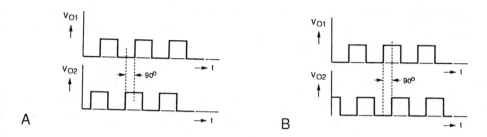

Fig. 5-6.10 Output signal from the amplifiers for direction 1 (A) and 2(B)

The sensor operates like a magnetic Wheatstone bridge measuring nonsymmetrical magnetic conditions such as when the teeth or pins move in front of the sensor. The mounting of the sensor and the magnet is critical, so the angle between the sensor's symmetry axis and that of the toothed wheel must be kept near zero. Further, both axes (sensor's and wheel's) must coincide. The circuit (Fig. 5-6.B) connects both bridge outputs to the corresponding amplifiers, and, subsequently, to the low pass filters and Schmitt-triggers to form the rectangular output signals. A phase difference between both outputs (Fig. 5-6.10A and B) is an indication of a rotation direction. Typical specifications for a module with a magnetoresistive sensor are given in Table 5-2.

Table 5-2 Specifications for KM110BH/31 module (Philips Semiconductors)

CHARACTERISTIC	VALUE
Supply voltage	5V (4 to 10V, ripple up to 40 mV pp.)
Outputs	0 and 5 V (logic)
Temperature range	-40 to +125°C (150°C max. for 500 h)
Measuring distance	0.5 to 2 mm
Frequency range	2 Hz to 20 kHz
Output resistance	100Ω (LOW), 10 kΩ (HIGH)
Min. external load	100 kΩ (without pull-up resistor)
Dimensions	25.4 x 11.8 mm (without terminal pins)

5.7 MAGNETOSTRICTIVE DETECTOR

A transducer which can measure displacement with high resolution across long distances can be built by using magnetostrictive and ultrasonic technologies [8]. The transducer is comprised of two major parts: a long waveguide (up to 7 m long) and a permanent ring magnet (Fig. 5-7.1). The magnet can move freely along the waveguide without touching it. A position of that magnet is the stimulus which is converted by the sensor into an electrical output signal. A waveguide contains a conductor which upon applying an electrical pulse, sets up a magnetic field over its entire length. Another magnetic field produced by the permanent magnet exists only in its vicinity. Thus two magnetic fields may be setup at the point where the permanent magnet is located. A superposition of two fields results in the net magnetic field which can be found from the vector summation. This net field, while being helically formed around the waveguide, causes it to experience a minute torsional strain, or twist at the location of the magnet. This twist is known as Wiedemann effect.

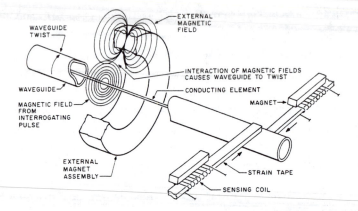

Fig. 5-7.1 A magnetostrictive detector uses ultrasonic waves
to detect position of a permanent magnet
(Reprinted with permission from Sensors Magazine, ©1991)

Therefore, electric pulses injected into the waveguide's coaxial conductor produce mechanical twist pulses which propagate along the waveguide with the speed of sound specific for its material. When the pulse arrives to the excitation head of the sensor, the moment of its arrival is precisely measured. One way to detect that pulse is to use a detector that can convert an ultrasonic twitch into electric output. This can be accomplished by piezoelectric sensors, or as it is shown in Fig. 5-7.1, by the magnetic reluctance sensor. The sensor consists of two tiny coils positioned near two permanent magnets. The coils are physically coupled to the waveguide and can jerk whenever the waveguide experiences the twitch. This sets up short electric pulses across the coils. Time delay of these pulses from the corresponding excitation pulses in the coaxial conductor is the exact measure of the ring magnet position. An appropriate electronic circuit converts time delay into a digital code representative of a position of the permanent magnet on the waveguide. The advantage of this sensor is in its high linearity (on the order of 0.05% of the full scale), good repeatability (on the order of 3 μm), and a long term stability. The sensor can withstand aggressive environments, such as high pressure, high temperature, and strong radiation. Another advantage of this sensor is its low temperature sensitivity which by a careful design can be achieved on the order of 20 ppm/°C.

Applications of this sensor include hydraulic cylinders, injection molding machines (to measure linear displacement for mold clamp position, injection of molding material and ejection of the molded part), mining (for detection of rocks movements as small as 25 μm), rolling mills, presses, forges, elevators, and other devices where fine resolution along large dimensions is a requirement.

5.8   OPTICAL SENSORS

After the mechanical contact and potentiometric sensors, optical sensors are probably the most popular for measuring position and displacement. Their main advantages are simplicity, the absence of the loading effect, and relatively long operating distances. They are insensitive to stray magnetic fields and electrostatic interferences, which makes them quite suitable for many sensitive applications. An optical position sensor usually requires at least three essential components: a light source, a photodetector, and light guidance devices, which may include lenses, mirrors, optical fibers, etc. An example of single and dual mode fiber optic proximity sensors are shown in Figs. 3-16.14 and 3-16.15. Similar arrangements are often implemented without the optical fibers, when light is guided toward a target by focusing lenses, and is diverted back to detectors by the reflectors. Nowadays, this basic technology has been substantially improved. Some more complex and sophisticated products have evolved. The improvements are aimed to better selectivity, noise immunity, and reliability of the optical sensors.

5.8.1 A proximity detector with polarized light

One method of building a better optoelectronic sensor is to use polarized light [9]. Each light photon has specific magnetic and electric field directions perpendicular to each other and to the direction of propagation. The direction of the electric field is the direction of the light *polarization*. Most of the light sources produce light with randomly polarized photons. To make light polarized, it can be directed through a polarizing filter, that is, a special material which transmits light polarized only in one direction and absorbs and reflects photons with wrong polarizations. However, any direction of polarization can be represented as a geometrical sum of two ortogonal polarizations: one is the same as the filter, and the other is nonpassing. Thus, by rotating the polarization of light before the polarizing filter we may *gradually* change the light intensity at the filter's output (Fig. 5-8.1).

When polarized light strikes an object, the reflected light may remain it polarization (specular reflection) or the polarization angle may change. The latter is typical for many nonmetallic objects. Thus, to make a sensor nonsensitive to reflective objects (like metal cans, foil wrappers, and the like), it may include two perpendicularly positioned polarizing filters: one at the light source and the other at the detector (Fig. 5-8.2A and B).

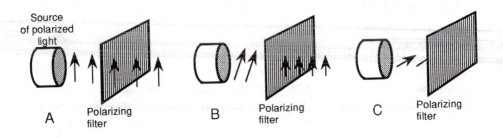

Fig. 5-8.1 Passing polarized light through a polarizing filter
A: direction of polarization is the same as of the filter
B: direction of polarization is rotated with respect to the filter
C: direction of polarization is perpendicular with respect to the filter

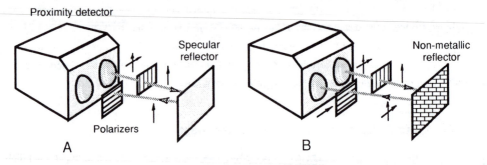

Fig. 5-8.2 Proximity detector with two polarizing filters positioned at 90° angle with respect to one another
A: polarized light returns from the metallic object within the same plane of polarization;
B: nonmetallic object depolarizes light, thus allowing it to pass through the polarizing filter

The first filter is positioned at the emitting lens (light source) to polarize the outgoing light. The second filter is at the receiving lens (detector) to allow passage of only those components of light, which have a 90° rotation with respect to the outgoing polarization. Whenever light is reflected from a specular reflector, its polarization direction doesn't change and the receiving filter will not allow the light to pass to a photodetector. However, when light is reflected in a non-specular manner, its components will contain a sufficient amount of polarization to go through the receiving filter and activate the detector. Therefore, the use of polarizers reduces false-positive detections of non-metallic objects.

5.8.2 Fiber-optic sensors

Fiber-optic sensors can be used quite effectively as proximity and level detectors. One example of the displacement sensor is shown in Fig. 3-16.6B, where the intensity of the reflected light is modulated by distance d to the reflective surface.

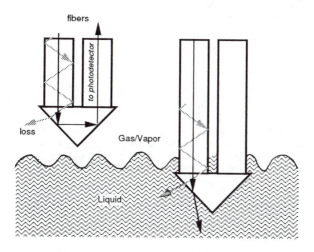

Fig. 5-8.3 An optical liquid level detector utilizing a change in the refractive index

A liquid level detector with two fibers and a prism is shown in Fig. 5-8.3. It utilizes the difference between refractive indices of air (or gaseous phase of a material) and the measured liquid. When the sensor is above the liquid level, a transmitting fiber (on the left) sends most of its light to the receiving fiber (on the right) due to a total internal reflection in the prism. However, some light rays approaching the prism reflective surface at angles less than the angle of total internal reflection are lost to the surrounding. When the prism reaches the liquid level, the angle of total internal reflection changes because the refractive index of a liquid is higher than that of air. This results in much greater loss in the light intensity which can be detected at the other end of the receiving fiber. The light intensity is converted into an electrical signal by any appropriate photodetector. Another version of the sensor is shown in Fig. 5-8.4 which shows a sensor fabricated by Gems Sensors (Plainville, CT). The fiber is U-shaped and upon being immersed into liquid, modulates the intensity of passing light. The detector has two sensitive regions near the bends where the radius of curvature is the smallest. An entire assembly is packaged into a 5-mm diameter probe and has a repeatability error of about 0.5 mm. Note that the shape of the sensing element draws liquid droplets away from the sensing regions when the probe is elevated above the liquid level.

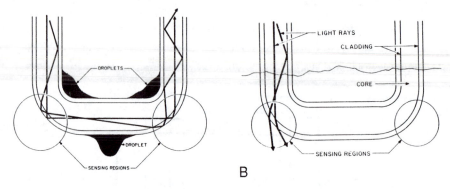

Fig. 5-8.4 A U-shaped fiber optic liquid level sensor
A: When the sensor is above the liquid level, the light at the output is strongest;
B: when the sensitive regions touch liquid, the light propagated through the fiber drops
(Reprinted with permission from Sensors Magazine, ©1989)

5.8.3   Grating sensors

An optical displacement transducer can be fabricated with two overlapping gratings which serve as a light intensity modulator (Fig. 5-8.5A). The incoming pilot beam strikes the first, stationary grating which allows only about 50% of light to pass toward the second, moving grating. When the opaque sectors of the moving grating are precisely aligned with the transmitting sectors of the stationary grating, the light will be completely dimmed out. Therefore, the transmitting light beam intensity can be modulated from 0% to 50% of the pilot beam (Fig. 5-8.5B). The transmitted beam is focused on a sensitive surface of a photodetector which converts light into electric current.

The full scale displacement is equal to the size of an opaque (or clear) sector. There is a trade-off between the dynamic range of the modulator and its sensitivity. That is, for the large pitch of the grating (large sizes of the transparent and opaque sectors) the sensitivity is low, however, the full scale displacement is large. For the higher sensitivity, the grating pitch can be made very small, so that the minute movements of the grating will result in a large output signal. This type of a modulator was used in a sensitive hydrophone [10] to sense displacements of a diaphragm. The grating pitch was 10μm which means that the full scale displacement was 5μm. The light source was a 2-mW He-Ne laser whose light was coupled to the grating through an optical fiber. The tests of the hydrophone have demonstrated that the device is sensitive with a dynamic range of 125dB of pressure as referenced to 1μPa, with a frequency response up to 1kHz.

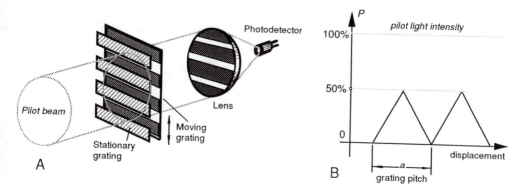

Fig. 5-8.5 Optical displacement sensor with grating light modulator
A: schematic; B: transfer function

A grating principle of light modulation is employed in very popular rotating or linear encoders, where a moving mask (usually fabricated in the form of a disk) has transparent and opaque sections (Fig. 5-8.6).

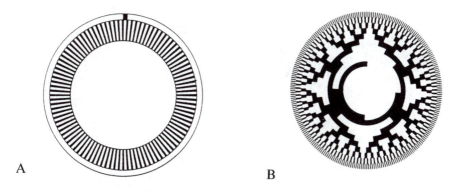

Fig. 5-8.6 Incremental (A) and absolute (B) optical encoding disks

The encoding disk functions as an interrupter of light beams within an optocoupler. That is, when the opaque section of the disk breaks the light beam, the detector is turned off (indicating digital ZERO), and when the light passes through a transparent section, the detector is on (indicating digital ONE). The optical encoders typically employ infrared emitters and detectors operating in the spectral range from 820 to 940 nm. The disks are made from laminated plastic and the opaque lines are produced by a photographic process. These disks are light, have low inertia, low cost, and excellent resistance to shock and vibration. However, they have a limited operating temperature range. Disks for a broader temperature range are fabricated of etched metal.

There are two types of encoding disks: the incremental, which produces a transient whenever it is rotated for a pitch angle, and the absolute, whose angular position is encoded in a combination of opaque and transparent areas along the radius. The encoding can be based on any convenient digital code. The most common are the gray code, the binary, and the BCD (binary coded decimals).

The incremental encoding systems are more commonly used than the absolute systems, because of their lower cost and complexity, especially in applications, where count is desirable instead of a position. When employing the incremental encoding disks, the basic sensing of movement can be made with a single optical channel (an emitter-detector pair), while the speed and incremental position, and direction sensing must use two. The most commonly used approach is a quadrature sensing, where the relative position of the output signals from two optical channels are compared. The comparison provides the direction information, while either of the individual channels gives the transition signal which is used to derive either count or speed information (Fig. 5-8.7).

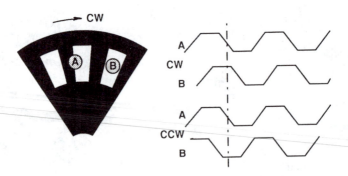

Fig. 5-8.7 Direction sensing with two optocouplers
When the wheel rotates clockwise (CW), channel A signal leads B by 90°;
When the wheel rotates counter-clockwise (CCW), channel B signal leads A by 90°

5.8.4 Linear optical sensors (PSD)

For precision position measurements over short and long ranges, optical systems operating in the near infrared can be quite effective. An example is a *position sensitive detector* (PSD) produced for the precision position sensing and autofocusing in photographic and video cameras. The position measuring module is of an active type: it incorporates a light emitting diode (LED) and a photodetective PSD. The position of an object is determined by applying the principle of a triangular measurement. Fig. 5-8.8 shows that the near infrared LED through a collimator lens produces a narrow-angle beam (<2°). The beam is a 0.7 ms wide pulse. On striking the object, the beam is reflected back to the detector. The received low intensity light is focused on the sensitive surface of the PSD. The PSD then generates the output signal (currents I_B and I_A) which is proportional to distance x of the light spot on its surface, from the central position. The intensity of a received beam greatly depends on the reflective properties of an object. Diffusive reflectivity in the near infrared spectral range is close to that in the visible range, hence, the intensity of the light incident on PSD has a great deal of variations. Nevertheless, the accuracy of measurement depends very little on the intensity of the received light.

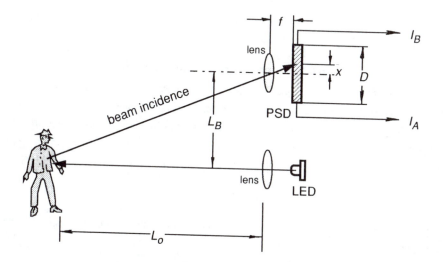

Fig. 5-8.8 The PSD sensor measures distance by applying a triangular principle

A PSD operates on the principle of photoeffect. It makes use of a surface resistance of a silicon photodiode. Unlike MOS and CCD sensors integrating multielement photodiode arrays, the

PSD has a nondiscrete sensitive area. It provides one-dimensional, or two-dimensional [11] position signals on a light spot traveling over its sensitive surface. A sensor is fabricated of a piece of high resistance silicon with two layers (p and n^+ types) built on its opposite sides (Fig. 5-8.9). A one-dimensional sensor has two electrodes (A and B) formed on the upper layer to provide electrical contacts to the p-type resistance. There is a common electrode (C) at the center of the bottom layer. Photoelectric effect occurs in the upper pn-junction. The distance between two upper electrodes is D, and the corresponding resistance between these two electrodes is R_D.

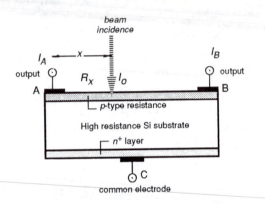

Fig. 5-8.9 Design of a one-dimensional PSD

Let us assume that the beam incidence strikes the surface at distance x from the A electrode. Then, the corresponding resistance between that electrode and the point of incidence is, respectively, R_x. The photoelectric current I_o produced by the beam is proportional to its intensity. That current will flow to both outputs (A and B) of the sensors in the corresponding proportions to the resistances and, therefore, to the distances between the point of incidence and the electrodes

$$I_A = I_o \frac{R_D - R_x}{R_D} \text{ and } I_B = I_o \frac{R_x}{R_D} \ . \tag{5.5}$$

If the resistances-versus-distances are linear, they can be replaced with the respective distances on the surface

$$I_A = I_o \frac{D - x}{D} \text{ and } I_B = I_o \frac{x}{D} \ . \tag{5.6}$$

To eliminate the dependence of the photoelectric current (and of the light intensity), we can use a ratiometric technique, that is, we take the ratio of the currents

$$P = \frac{I_A}{I_B} = \frac{D}{x} - 1 \quad, \tag{5.7}$$

which we can rewrite for value of x:

$$x = \frac{D}{P + 1} \quad . \tag{5.8}$$

Fig. 5-8.8 shows geometrical relationships between various distances in the measurement system. Solving two triangles for L_o yields

$$L_o = f \frac{L_B}{x} \quad, \tag{5.9}$$

where f is the focal distance of the receiving lens. Substituting Eq. 5.8 we obtain the distance in terms of the current ratio

$$L_o = f \frac{L_B}{D} (P+1) = k (P + 1) \quad, \tag{5.10}$$

where k is called the module geometrical constant. Therefore, the distance from the module to the object linearly affects the ratio of the PSD output currents.

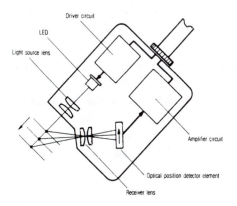

Fig. 5-8.10 Optical position displacement sensor
(From Keyence Corp. of America, Fair Lawn, N.J.)

A similar operating principle is implemented in an industrial optical displacement sensor (Fig. 5-8.10) where PSD is used for measurement small displacements at operating distances of several centimeters. Typical specifications of the sensing module with a PSD are given in Table 5-3. Such optical sensors are highly efficient for the on-line measurements of height of a device (PC-board inspection, liquid and solids level control, laser torch height control, etc.), for measurement of eccentricity of a rotating object, for thickness and precision displacement measurements, for detection of presence or absence of an object (medicine bottle caps), etc. A great advantage of an optical displacement sensor with a PSD is that its accuracy may be much greater than the accuracy of the PSD itself [12].

Table 5-3 Typical specifications of an optical displacement sensor utilizing a PSD assembly

Characteristic	Value
Measurement distance	40 mm
Measurement range	±5 mm
Spot diameter	3 mm max.
Resolution	10 µm
Accuracy	± 20 µm
Temperature fluctuation	0.05%/°C
Ambient operating illumination	1,000 lx max.
Ambient operating temperature	0 to 50°C

The PSD elements are produced of two basic types: one- and two-dimensional. Equivalent circuits of both are shown in Fig. 5-8.11. Since the equivalent circuit has a distributed capacitance and resistance, the PSD time constant varies depending on the position of the light spot. In response to an input step function, a small area PSD has rise time in the range of 1-2 µs. Its spectral response is approximately from 320 to 1100 nm, that is, the PSD covers UV, visible and near infrared spectral ranges. Small area one-dimensional PSDs have sensitive surfaces ranging from 1x2 to 1x12 mm, while the large area two-dimensional sensors have square areas with a side ranging from 4 to 27 mm (Fig. 5-8.12). Technical specifications for a miniature epoxy-molded PSD are given in Table 5-4.

Table 5-4 Characteristics of one-dimensional S3274 PSD
(Hamamatsu Photonics K.K.)

Characteristic	Value
Active area (mm)	1 x 3.5
Sensor dimensions (mm)	4.1 x 5.0
Peak spectral response (nm)	920
Refractive index of package	1.53
Center position accuracy (µm)	±200
Position detection error (µm)	±35 max.
Interelectrode resistance (kΩ)	140 typ.
Dark current (nA)	0.05 typ.
Terminal capacitance (pF)	15 typ.
Rise time (µs)	10 typ.

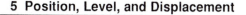

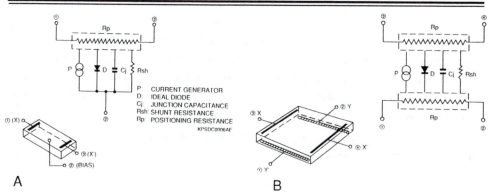

P: CURRENT GENERATOR
D: IDEAL DIODE
Cj: JUNCTION CAPACITANCE
Rsh: SHUNT RESISTANCE
Rp: POSITIONING RESISTANCE
KPSDC0006AE

A   B

Fig. 5-8.11 Equivalent circuits for the (A) one- and (B) two-dimensional position sensitive detectors
(Courtesy of Hamamatsu Photonics K.K., Japan)

Fig. 5-8.12 One- and two-dimensional position sensitive detectors (PSD)
(Courtesy of Hamamatsu Photonics, K.K., Japan)

5.9   ULTRASONIC SENSORS

For noncontact distance measurements, an active sensor which transmits some kind of a pilot signal and receives a reflected signal can be designed. The transmitted energy may be in a form of any radiation, for instance, optical (like in a PSD which is described above), electromagnetic, acoustic, etc. Transmission and reception of the ultrasonic energy is a basis for very popular ultrasonic range meters, and velocity detectors. Ultrasonic waves are mechanical, that is, acoustic waves covering frequency range well beyond the capabilities of human ears, i.e., over 20kHz. However, these frequencies may be quite perceptive by smaller animals, like dogs, cats, rodents, insects. Indeed, the ultrasonic detectors are the biological ranging devices for bats and dolphins.

When the waves are incident on an object, part of their energy is reflected. In many practical cases, the ultrasonic energy is reflected in a diffuse manner. That is, regardless of the direction where the energy comes from, it is reflected almost uniformly within a wide solid angle, which may approach 180°. If an object moves, the frequency of the reflected waves will differ from the transmitted waves. This is called the Doppler effect[1].

A distance L_o to the object can be calculated through the speed v of the ultrasonic waves in the media, and the angle, Θ (Fig. 5-9.1A)

$$L_o = \frac{vt Cos\Theta}{2} , \qquad (5.11)$$

where t is the time for the ultrasonic waves to travel to the object and back to the receiver. If a transmitter and a receiver are positioned close to each other as compared with the distance to the object, then $Cos\Theta \approx 1$. Ultrasonic waves have an obvious advantage over the microwaves: they propagate with the speed of sound, which is much slower than the speed of light at what microwaves propagate. Thus, time t is much longer and its measurement can be accomplished easier and cheaper.

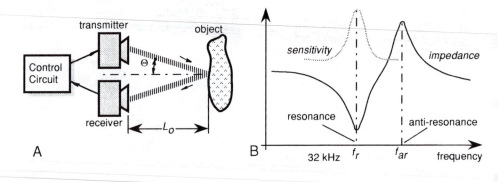

Fig. 5-9.1 Ultrasonic distance measurement
A - basic arrangement; B - impedance characteristic of a piezoelectric transducer

To generate any mechanical waves, including ultrasonic, the movement of a surface is required. This movement creates compression and expansion of medium which can be gas (air), liquids or solids[2]. The most common type of the excitation device which can generate surface movement in the ultrasonic range, is a piezoelectric transducer operating in the so-called *motor*

---

[1] See Section 6.2 for the description of the Doppler effect for the microwaves. The effect is fully applicable to the propagation of any energy having wave nature, including ultrasonic.
[2] See Section 3.12 for a description of sound waves.

mode. The name implies that the piezoelectric device directly converts electrical energy into mechanical energy.

Fig. 5-9.2A shows that the input voltage applied to the ceramic element causes it to flex and transmit ultrasonic waves. Because piezoelectricity is a reversible phenomenon, the ceramic generates voltage when incoming ultrasonic waves flex it. A typical operating frequency of the transmitting piezoelectric element is near 32 kHz. For better efficiency, the frequency of the driving oscillator should be adjusted to the resonant frequency f_r of the piezoelectric ceramic (Fig. 5-9.1B) where the sensitivity and efficiency of the element is the best. When the measurement circuit operates in a pulsed mode, the same piezoelectric element is used for both transmission and receiving. When the system requires continues transmission of ultrasonic waves, separate piezoelectric elements are employed for the transmitter and receiver. A typical design of an air-operating sensor is shown in Fig. 5-9.2B.

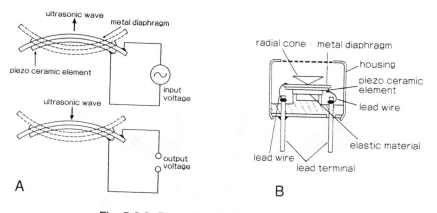

Fig. 5-9.2 Piezoelectric ultrasonic transducer
A - input voltage flexes the element and transmits ultrasonic waves, while incoming waves produce output voltage;
B - Open aperture type of ultrasonic transducer for operation in air
(Courtesy of Nippon Ceramic, Japan)

5.10 GYROSCOPES

Next to a magnetic compass, a gyroscope probably is the most common navigation sensor. In many cases, where a geomagnetic field is either absent (in space), or is altered by the presence of some disturbances, a gyroscope is an indispensable sensor for defining the position of a vehicle. A gyroscope, or a *giro* for short, is a "keeper of direction", like a pendulum in a clock is a "keeper of time". A gyro operation is based on the fundamental principle of the conservation of angular momentum: *in any system of particles, the total angular momentum of the system relative to any point fixed in space remains constant, provided no external forces act on the system.*

5.10.1   Mechanical gyroscope

A mechanical gyro is comprised of a massive disk free to rotate about a spin axis (Fig. 5-10.1) which itself is confined within a framework that is free to rotate about one or two axes. Hence, depending on the number of rotating axes, gyros can be either of a single, or two-degree-of-freedom types. The two qualities of a gyro account for it usefulness are: 1) the spin axis of a free gyroscope will remain fixed with respect to space, provided there are no external forces to act upon it, and 2) a gyro can be made to deliver a torque (or output signal) which is proportional to the angular velocity about an axis perpendicular to the spin axis.

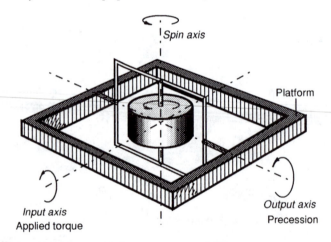

Fig. 5-10.1 Mechanical gyroscope with a single degree-of-freedom

When the wheel (rotor) freely rotates, it tends to preserve its axial position. If the gyro platform rotates around the input axis, the gyro will develop a torque around a perpendicular (output) axis, thus turning its spin axis around the output axis. This phenomenon is called *precession* of a gyro. It can be explained by Newton's law of motion for rotation: *the time rate of change of angular momentum about any given axis is equal to the torque applied about the given axis.* That is, when a torque T is applied about the input axis, and the speed ω of the wheel is held constant, the angular momentum of the rotor may be changed only by rotating the projection of the spin axis with respect to the input axis, i.e., the rate of rotation of the spin axis about the output axis is proportional to the applied torque

$$T = I\omega\Omega \ , \qquad\qquad\qquad\qquad (5.12)$$

where Ω is the angular velocity about the output axis and I is the inertia of a gyro wheel about the spin axis. To determine the direction of precession, the following rule can be used: *precession is always in such a direction as to align the direction of rotation of the wheel with the direction of rotation of the applied torque.*

The accuracy of mechanical gyros greatly depend on the effects which may cause additional unwanted torques and cause drifts. The sources of these are friction, imbalanced rotor, magnetic effects, etc. One method which is widely used to minimize rotor friction is to eliminate the suspension entirely by floating the rotor and the driving motor in a viscous, high-density liquid, such as one of the fluorocarbons. This method requires close temperature control of the liquid and also may suffer from aging effects. The other method of friction reduction is to use the so-called gas bearings, where the shaft of the rotor is supported by high pressure helium, hydrogen or air. An even better solution is to support the rotor in vacuum by an electric field (electrostatic gyros). A magnetic gyro consists of a rotor supported by a magnetic field. In that case, the system is cryogenically cooled to temperatures where the rotor becomes superconductive. Then, an external magnetic field produces enough counter-field inside the rotor that the rotor floats in a vacuum. These magnetic gyroscopes also are called cryogenic.

5.10.2 Monolithic silicon gyroscope

During recent years there have been attempts to develop low cost inertial instruments (gyros and accelerometers) suitable for mass production. A promising concept of such a device is a micromachined vibrating rate gyro [13]. The device is a two gimbal structure supported by torsional flexures (Fig. 5-10.2). It is undercut and free to move in the active area. In operation, the outer, or "motor" is driven at a constant amplitude by electrostatic torquing using electrodes placed in close proximity. This oscillatory motion is transferred to the inner gimbal along the stiff axis of the inner flexures, setting up an oscillating momentum vector with the inertial element. In the presence of an angular rotational rate normal to the plane of the device, the Coriolis force will cause the inner gimbal to oscillate about its weak axis with a frequency equal to the drive frequency and with an amplitude proportional to the inertial input rate. Maximum resolution is obtained when the outer gimbal is driven at a resonant frequency of the inner gimbal. The readout of the output motion is accomplished by setting the differential change in capacitance between the inner gimbal and a pair of electrodes. When operated open loop, the angular displacement of the inner gimbal about the output axis is proportional to the input rate. That is, the output angle Θ is proportional to an inertia ration term, the drive angle, ϕ_0, the mechanical Q, and the input rate Ω. It is inversely proportional to the drive frequency ω_n

$$\Theta = \left[\frac{I_x + I_y - I_z}{I_x}\right]\frac{\phi_0\Omega Q}{\omega_n} \quad . \tag{5.13}$$

In a practical application, the device is operated closed loop and the inner gimbal is rebalanced to null in phase and in quadrature. A detailed description of the gyroscope may be found elsewhere [14].

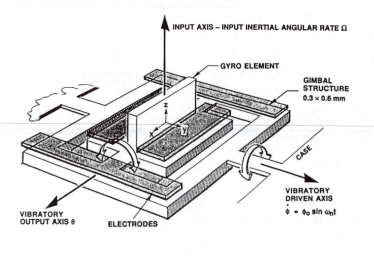

Fig. 5-10.2 Vibratory rate gyro concept
(from [14])

5.10.3. Optical gyroscopes

Modern development of sensors for guidance and control applications are based on employing the so-called Sagnac effect, which is illustrated in Fig. 5-10.3 [15]. Two beams of light generated by a laser propagate in opposite directions within an optical ring having refractive index n and radius R. One beam goes in clockwise (CW) direction, while the other in a counterclockwise (CCW) direction. The amount of time which takes light to travel within the ring takes $\Delta t = 2\pi R/nc$, where c is the speed of light. Now, let us assume that the ring rotates with angular rate Ω in the clockwise direction. In that case light will travel different paths at two directions. The CW beam will travel $l_{cw} = 2\pi R + \Omega R \Delta t$, while the CCW beam will travel $l_{ccw} = 2\pi R - \Omega R \Delta t$. Hence, the difference between the paths is

$$\Delta l = \frac{4\pi\Omega R^2}{nc} \quad . \tag{5.14}$$

Therefore, to accurately measure Ω, a technique must be developed to determine Δl. There are three basic methods are known for the path detection: 1) optical resonators, 2) open-loop interferometers, and 3) closed-loop interferometers.

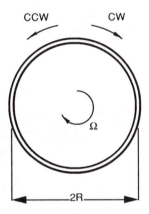

Fig. 5-10.3 Sagnac effect

For the ring laser gyro, measurements of Δl are made by taking advantages of the lasing characteristics of an optical cavity (that is, of its ability to produce coherent light). For lasing to occur in a closed optical cavity, there must be an integral number of wavelengths about the complete ring. The light beams which do not satisfy this condition, interfere with themselves as they make subsequent travel about the optical path. In order to compensate for a change in the perimeter due to rotation, the wavelength λ and frequency v of the light must change

$$-\frac{dv}{v} = \frac{d\lambda}{\lambda} = \frac{dl}{l} \quad . \tag{5.15}$$

The above is a fundamental equation relating frequency, wavelength, and perimeter change in the ring laser. If the ring laser rotates at a rate Ω, then Eq. (5.14) indicates that light waves stretch in one direction and compress in the other direction to meet the criteria for the lasing of an integral number of wavelengths about the ring. This, in turn, results in a net frequency difference between the light beams. If the two beams are bit together (mixed), the resulting signal has frequency

$$F = \frac{4A\Omega}{\lambda nl} \quad ,$$ (5.16)

where A is the area enclosed by the ring.

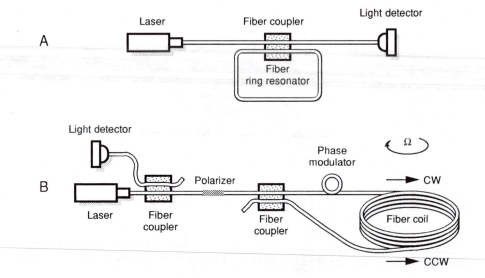

**Fig. 5-10.4 A: Fiber optic ring resonator
B: Fiber optic analog coil gyro (adapted from [15])**

In practice, optic gyros are designed with either a fiber ring resonator, or the fiber coil where the ring has many turns of the optical fiber [16]. The optic ring resonator is shown in Fig. 5-10.4A It consists of a fiber loop formed by a fiber beamsplitter that has a very low cross-coupling ratio. When the incoming beam is at the resonant frequency of the fiber ring, the light couples into the fiber cavity and the intensity in the exiting light drops. The coil fiber gyro (Fig. 5-10.4B) contains a light source and the detector coupled to the fiber. The light polarizer is positioned between the detector and the second coupler to insure that both counter-propagating beams traverse the same path in the fiber optic coil [17]. The two beams mix and impinge onto the detector, which monitors the cosinusoidal intensity changes caused by rotationally induced phase changes between the beams. This type of optical gyro provides a relatively low cost, small size rotation sensitive sensor with a dynamic range up to 10,000. Applications include yaw and pitch measurements, attitude stabilization and gyrocompassing. A major advantage of optical gyros is their ability to operate under hostile environments that would be difficult, if not impossible, for the mechanical gyros.

REFERENCES

1. Kato, H., Kojima, M., Gattoh, M., Okumura, Y. and Morinaga, S. Photoelectric inclination sensor and its application to the measurement of the shapes of 3-D objects. *IEEE Trans. Instrum. Meas.* vol. 40, no. 6, pp: 1021-26, Dec. 1991

2. Bertone, G. A. and Meiksin, Z. H. Elimination of the anomalous humidity effect in precision capacitance based transducers. *IEEE Trans. Instrum. Meas.*, vol. 40, no. 6, pp. 897-901, Dec. 1991

3. Bertone, *et al.* Investigation of a capacitive based displacement transducer. *IEEE Trans. Instrum. Meas.*, vol. 39, no. 2, pp. 424-28, April 1990

4. De Silva, C. W. Control sensors and actuators, Prentice Hall, Englewood Cliffs, NJ, 1989

5. Linear Application Handbook, Linear Technology, AN3-9, 1990

6. Sprague, CN-207 Hall Effect IC Applications, 1986

7. Donahoe, T. Fluid level sensor for automobiles or can a sensor compete with a dipstick? *Sensors Expo Proceedings*, ©Helmers Publishing, Inc., 1991.

8. Magnetostrictive, ultrasonic transducer measures displacement and velocity. *Sensors*, June, pp: 20-22, 1991.

9. Juds, S. Thermoelectricity: an introduction to the principles. John Wiley & Sons, New York

10. Spillman, W.B., Jr. Multimode fiber-optic hydrophone based on a schlieren technique. *Appl. Opt.* Vol. 20, p: 465, 1981

11. Morikawa, Y. and Kawamura, K. A small distortion two-dimensional position sensitive detector (PSD) with on-chip MOSFET switches. In: *Transducers'91. International conference on solid-state sensors and actuators. Digest of technical papers.*, pp: 723-726, ©IEEE, 1991

12. van Drecht, J. and Meijer G.C.M. Concepts for the design of smart sensors and smart signal processors and their applications to PSD displacement transducers. In: *Transducers'91. International conference on solid-state sensors and actuators. Digest of technical papers.*, pp: 475-478, ©IEEE, 1991

13. Greiff, P., Boxenhorn, B., King, T., and Niles, L. Silicon monolithic micromechanical gyroscope. In: *Transducers'91. International conference on solid-state sensors and actuators. Digest of technical papers*, pp: 966-968, ©IEEE, 1991

14. Boxenhorn, B.B., Dew. B., and Greiff, P. The micromechanical inertial guidance system and its applications. In: *14th Biennial guidance test symposium*, 6588th Test Group, Holloman AFB, New Mexico, Oct. 3-5, 1989

15. Udd, E. Fiber optic sensors based on the Sagnac interferometer and passive ring resonator. In: Fiber optic sensors. Eric Udd, ed. © John Wiley & Sons, Inc., pp: 233-269, 1991.

16. Ezekiel, S. and Arditty, H, J., eds. Fiber-optic rotation sensors. Springer series in optical sciences., vol. 32, Springer-Verlag, NY, 1982

17. Fredericks, R. J., Ulrich, R. Phase error bounds of fiber gyro with imperfect polarizer/depolarizer. *Electron. Lett.*, vol. 29, p: 330, 1984

6

Occupancy and Motion Detectors

The occupancy sensors detect the presence of people in a monitored area. Motion detectors respond only to moving objects. A distinction between the two is that the occupancy sensors produce signals whenever an object is stationary or not, while the motion detectors are selectively sensitive to moving objects. The applications of these sensors include security, surveillance, energy management (electric lights control), personal safety, friendly home appliances, interactive toys, novelty products, etc. Depending on the applications, the presence of humans may be detected through any means that is associated with some kind of a human body's property or body's actions [1]. For instance, a detector may be sensitive to body weight, heat, sounds, dielectric constant, etc. The following types of detectors are presently used for the occupancy and motion sensing of people:

1. *Capacitive* - detectors of human body capacitance;

2. *Acoustic* - detectors of sound produced by people;

3. *Photoelectric* - interruption of light beams by moving objects;

4. *Optoelectric* - detection of variations in illumination or optical contrast in the protected area;

5. *Pressure mat switches* - pressure sensitive long strips used on floors beneath the carpets to detect weight of an intruder;

6. *Stress detectors* - strain gauges imbedded into floor beams, staircases, and other structural components;

7. *Switch sensors* - electrical contacts connected to doors and windows;

8. *Magnetic switches* - a noncontact version of switch sensors;

9. *Vibration detectors* - react to the vibration of walls or other building structures;

10. *Glass breakage detectors* - sensors reacting to specific vibrations produced by shattered glass;

11. *Infrared motion detectors* - devices sensitive to heat waves emanated from warm or cold moving objects;

12. *Microwave detectors* - active sensors responsive to microwave electromagnetic signals reflected from objects;

13. *Ultrasonic detectors* - are similar to microwaves except that instead of electromagnetic radiation, ultrasonic waves are used;

14. *Video motion detectors* - is a video equipment which compares a stationary image stored in memory with the current image from the protected area;

15. *Laser system detectors* - similar to photoelectric detectors, except that they use narrow light beams and combinations of reflectors;

16. *Triboelectric detectors* - sensors capable of detecting static electric charges carried by moving objects.

One of the major aggravations in detecting the occupancy or intrusion is a false positive detection. The term "false positive" means that the system indicates an intrusion when there is none. In some noncritical applications where false positive detections occur once in a while, for instance, in a toy or a motion switch controlling electric lights in a room, this may be not a serious problem: the lights will be erroneously turned on for a short time, which unlikely do any harm[1]. In other systems, especially used for the security purposes, the false positive detections, while generally not as dangerous as false negative ones (missing an intrusion), may become a serious problem. While selecting a sensor for critical applications, considerations should be given to its reliability, selectivity, and noise immunity. It is often a good practice to form a multiple sensor arrangement with symmetrical interface circuits. It may dramatically improve a reliability of a system, especially in the presence of external transmitted noise. Another efficient way to reduce erroneous detections is to use sensors operating on different physical principles [2], for instance, combining capacitive and infrared detectors is an efficient combination as they are receptive to different kinds of transmitted noise.

6.1 ULTRASONIC SENSORS

These detectors are based on the transmission to the object and receiving reflected acoustic waves. A description of the ultrasonic detectors can be found in Section 5.9. For the motion detectors, they may require a somewhat longer operating range and a wider angle of coverage.

---

[1] Probably, just steering up some agitation about the presence of a ghost.

6.2 MICROWAVE MOTION DETECTORS

The microwave detectors offer an attractive alternative to other detectors, when it is required to cover large areas and to operate over an extended temperature range under influence of strong interferences, such as wind, acoustic noise, fog, dust, moisture, etc. The operating principle of the microwave detector is based on radiation of electromagnetic radio-frequency (RF) waves toward a protected area. The most common frequencies are 10.525GHz (X band) and 24.125GHz (K band)[1]. These wavelengths are long enough (λ=3cm at X band) to pass freely through most contaminants, such as airborne dust, and short enough for being reflected by larger objects. The microwave part of the detector consists of a Gunn oscillator, an antenna, and a mixer diode. The Gunn oscillator is a diode mounted in a small precision cavity which, upon application of power, oscillates at microwave frequencies. The oscillator produces electromagnetic waves (frequency f_o), part of which is directed through an iris into a waveguide and focusing antenna which directs the radiation toward the object. Focusing characteristics of the antenna are determined by the application. As a general rule, the narrower is the directional diagram of the antenna, the more sensitive it is (the antenna has a higher gain). Another general rule is that a narrow beam antenna is much larger, while a wide angle antenna can be quite small. A typical radiated power of the transmitter is 10-20 mW. A Gunn oscillator is sensitive to a stability of applied dc voltage and, therefore, must be powered by a good quality voltage regulator. The oscillator may run continuously, or it can be pulsed, which reduces a power consumption from the power supply.

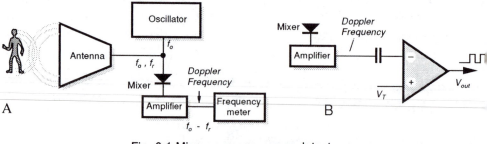

Fig. 6-1 Microwave occupancy detector
A - a circuit for measuring doppler frequency
B - a circuit with a threshold detector

A smaller part of the microwave oscillations is coupled to the Schottkey mixing diode and serves as a reference signal (Fig. 6-1A). In many cases, the transmitter and the receiver are contained in one module called a transceiver. A target reflects some waves back toward the antenna

---

[1] Power of radiation must be sufficiently low not to present any health hazards

which directs the received radiation toward the mixing diode whose current contains a harmonic with a phase differential between the transmitted and reflected waves. The phase difference is in a direct relationship with a distance to the target. The phase-sensitive detector is useful mostly for detecting a distance to an object. For the occupancy and motion detector, a Doppler effect is a basis for the operation of the microwave and ultrasonic detectors. It should be noted that the Doppler-effect device is a true motion detector because it is responsive only to moving targets.

An antenna transmits the frequency f_o which is defined by the wavelength λ_0 as

$$f_o = \frac{c_o}{\lambda_0} \, , \tag{6.2.1}$$

where c_o is the speed of light. When the target moves toward or away from the transmitting antenna, the frequency of the reflected radiation will change. Thus, if the target is moving away with velocity v the reflected frequency will decrease and it will increase for the approaching targets. This is called the *Doppler effect*, after the Austrian scientist Christian Johann Doppler (1803-1853)[1]. While the effect first was discovered for sound, it is applicable to an electromagnetic radiation as well. However, in contrast to sound waves that may propagate with velocities dependent on movement of the source of the sound, electromagnetic waves propagate with speed of light which is an absolute constant. The frequency of reflected electromagnetic waves can be predicted by the theory of relativity as

$$f_r = f_o \frac{\sqrt{1 - (v/c_o)^2}}{1 + v/c_o} \, . \tag{6.2.2}$$

For practical purposes, however, quantity $(v/c_o)^2$ is very small as compared with unity, hence, it can be ignored. Then, the equation for the frequency of the reflected waves becomes identical to that for the acoustic waves:

$$f_r = f_o \frac{1}{1 + v/c_o} \, . \tag{6.2.3}$$

Due to a Doppler effect, the reflected waves have a different frequency f_r. A mixing diode combines the radiated (reference) and reflected frequencies and, being a nonlinear device, produces a signal which contains multiple harmonics of both frequencies. The electric current through the diode may be represented by a polynomial

---

[1] Since 150 years ago acoustical instruments for precision measurements were not available yet, to prove his theory, Doppler placed trumpeters on a railroad flat car and musicians with a sense of absolute pitch near the tracks. A locomotive engine pulled the flat car back and forth at different speeds for two days. The musicians on the ground "recorded" the trumpet notes as the train approached and is receded. The equations held up.

$$i = i_o + \sum_{k=1}^{n} a_k (U_1\cos2\pi f_o t + U_2\cos2\pi f_r t)^k \quad , \qquad (6.2.4)$$

where i_o is a d.c. component, a_k are a harmonic coefficients which depend on a diode operating point, U_1 and U_2 are amplitudes of the reference and received signals, respectively, and t is time. A current through a diode contains an infinite number of harmonics, among which there is a harmonic of a differential frequency: $a_2 U_1 U_2 \cos2\pi(f_o-f_r)t$, which is called a doppler frequency Δf.

The doppler frequency in the mixer can be found from Eq. (6.2.3)

$$\Delta f = f_o-f_r = f_o \frac{1}{c_o/v + 1} \quad , \qquad (6.2.5)$$

and since $c_o/v \gg 1$, the following holds after substituting (6.2.1):

$$\Delta f \approx \frac{v}{\lambda_o} \quad . \qquad (6.2.6)$$

Therefore, the signal frequency at the output of the mixer is linearly proportional to the velocity of a moving target. For instance, if a person walks toward the detectors with the velocity of 0.6 m/s, a doppler frequency for the X-band detector is $\Delta f = 0.6/0.03 = 20$ Hz.

Eq. (6.2.6) holds true only for movements in the normal direction. When the target moves at angles Θ with respect to the detector, the doppler frequency is

$$\Delta f \approx \frac{v}{\lambda_o} \cos\Theta \quad . \qquad (6.2.7)$$

This implies that doppler detectors theoretically become insensitive when a target moves at angles approaching 90°. In the velocity meters, to determine the velocity of a target, what is required is to measure the doppler frequency and the phase, to determine the direction of the movement (Fig. 6-1A). This method is used in police radars. For the supermarket door openers and the security alarms, instead of measuring the frequency, a threshold comparator is used to indicate the presence of a moving target (Fig. 6-1B). It should be noted that even if formula (6.2.7) predicts that the doppler frequency is near zero for targets moving at angles $\Theta=90°$, the entering of a target into the protected area at any angle, results in an abrupt change in the received signal amplitude, and the output voltage from the mixer changes accordingly. Usually, this is sufficient to trigger a response of a threshold detector.

The signal from the mixer is in the ranges from microvolts to millivolts, so the amplification is needed for signal processing. Since the doppler frequency is in the audio range, the amplifier is relatively simple, however, it generally must be accompanied by the so-called notch filters which reject a power line frequency and the main harmonic from full wave rectifiers and fluorescent light fixtures: 60 and 120 Hz (50 and 100 Hz). For the normal operation, received power must be sufficiently high. It depends on several factors, including the antenna aperture area A target area a and distance to the target r

$$P_r = \rho \frac{P_o A^2 a}{4\pi\lambda^2 r^4} \quad . \tag{6.2.8}$$

where P_o is the transmitted power. For the effective operation, target's cross-sectional area a must be relatively large, because for $\lambda^2 \leq a$, a received signal is drastically reduced. Further, reflectivity ρ of a target in the operating wavelength is also very important for the magnitude of the received signal. Generally, conductive materials and objects with high dielectric constants are good reflectors of electromagnetic radiation, while many dielectrics absorb energy and reflect very little. Plastics and ceramics are quite transmissive and can be used as windows in the microwave detectors. The best target for a microwave detector is a smooth, flat conductive plate positioned normally toward the detector. A flat conductive surface makes a very good reflector, however, it may render the detector inoperable at angles other than 0°. Thus, an angle $\Theta=45°$ can completely divert a reflective signal from the receiving antenna. This method of diversion was used quite effectively in the design of the Stealth bomber which is virtually invisible on radar screens.

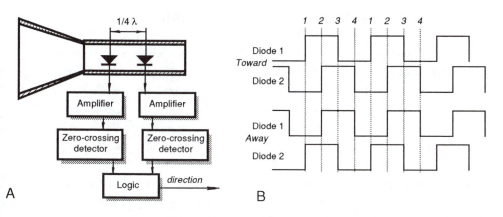

Fig. 6-2 Block diagram (A) and timing diagrams (B)
of a microwave doppler motion detector with a directional sensitivity

To detect whether a target moves toward or away from the antenna, a Doppler concept can be extended by adding another mixing diode to the transceiver module. The second diode is located in the waveguide in such a manner that the doppler signals from both diodes differ in phase by 1/4 of wavelength or by 90° (Fig. 6-2A). These outputs are amplified separately and converted into square pulses which can be analyzed by a logic circuit. The circuit is a digital phase discriminator that determines the direction of motion (Fig. 6-2B). Door openers and traffic control are two major applications for this type of module. Both applications need the ability to acquire a great deal of information about the target for discrimination before enabling a response. In door openers, limiting the field of view and transmitted power may substantially reduce the number of false positive detections. While for door openers a direction discrimination is optional, for traffic control it is a necessity to reject signals from the vehicles moving away. If the module is used for intrusion detection, vibration of building structures may cause a large number of false positive detections. A direction discriminator will respond to vibration with an alternate signal, while to an intruder with a steady logic signal. Hence, the direction discriminator is an efficient way to improve the reliability of the detection.

Whenever a microwave detector is used in the U.S.A., it must comply with the strict requirements (for instance: MSM20100) imposed by the Federal Communication Commission. Similar regulations are enforced in many other countries. Also, the emission of the transmitter must be below 10mW/cm^2 as averaged over any 0.1 hour period, as specified by OSHA 1910.97 for the frequency range from 100 MHz to 100 GHz.

6.3 CAPACITIVE OCCUPANCY DETECTORS

Being a conductive medium with a high dielectric constant, a human body develops a coupling capacitance to its surroundings[1]. This capacitance greatly depends on such factors as the body size, clothing, materials, and type of the surrounding objects, weather, etc. However wide the coupling range is, the capacitance may vary from a few picofarads to several nanofarads. When a person moves, the coupling capacitance changes, thus making possible to discriminate static objects from the moving ones. In effect, all objects form some degree of a capacitive coupling with respect to one another. If a human (or for that purpose, an animal) moves into the vicinity of the objects whose coupling capacitance with each other has been previously established, a new capacitive value arises between the objects as a result of the presence of an intruding body. Fig. 6-3 shows that the capacitance between a test plate, and earth[2] is equal to C_1. When a person moves into the vicinity of the plate, it forms two additional capacitors: one between the plate and its own body C_a and the other between the body and the earth C_b. Then, the resulting capacitance C between the plate and the earth becomes larger by ΔC

---

[1] At 40 MHz, the dielectric constant of muscle, skin, and blood is about 97. For fat and bone it is near 15 [3].
[2] Here, by "earth" we mean any large object, such as earth, lake, metal fence, car, ship, airplane, etc.

$$C = C_1 + \Delta C = C_1 + \frac{C_a C_b}{C_a + C_b} \;. \tag{6.3.1}$$

With the appropriate apparatus, this phenomenon can be used for the occupancy detection. What is required, is to measure a capacitance between a test plate (the probe) and a reference plate (the earth).

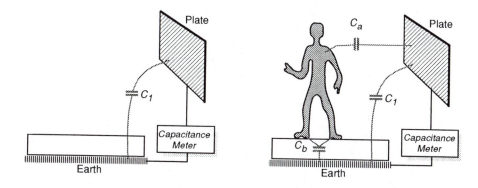

Fig. 6-3 An intruder brings in an additional capacitance to a detection circuit

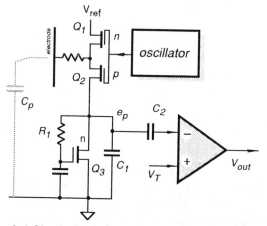

Fig. 6-4 Circuit diagram for a capacitive intrusion detector

Fig. 6-4 illustrates a circuit diagram for detecting variations in the probe capacitance C_p [4]. That capacitance is charged from a reference voltage source V_{ref} through a gate formed by transistor Q_1 when the output voltage of a control oscillator goes low. When it goes high, transistor Q_1 closes while Q_2 opens. The probe capacitance C_p discharges through a constant current sink that is constructed with a transistor Q_3. A capacitor C_1 filters the voltage spikes across the transistor. The average voltage, e_p, represents a value of the capacitor C_p. When an intruder approaches the probe, the latter's capacitance increases, which results in a voltage rise across C_1. The voltage deflection passes through the capacitor C_2 to the input of a comparator with a fixed threshold V_T. The comparator produces the output signal V_{out} when the input voltage exceeds the threshold value.

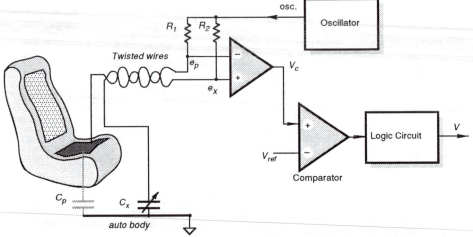

Fig. 6-5 An automotive capacitive intrusion detector

Fig. 6-5 illustrates a capacitive security system for an automobile [5]. A sensing probe is imbedded into a car seat. It can be fabricated as a metal plate, metal net, a conductive fabric, etc. The probe forms one plate of a capacitor C_p. The other plate of the capacitor is formed either by a body of an automobile, or by a separate plate positioned under a floor mat. A reference capacitor C_x is composed of a simple fixed or trimming capacitor which should be placed close to the seat probe. The probe plate and the reference capacitor are, respectively, connected to two inputs of a charge detector (resistors R_1 and R_2). The conductors preferably should be twisted to reduce the introduction of spurious signals as much as possible. For instance, strips of a twinflex cabling were found quite adequate. A differential charge detector is controlled by an oscillator which produces square pulses (Fig. 6-6). Under the no-seat-occupied conditions, the reference capacitor is adjusted to be approximately equal to C_p. Resistors and the corresponding capacitors define time constants of the networks. Both RC circuits have equal time constants τ_1. Voltages across the resistors are fed into the inputs of a differential amplifier, whose output voltage V_c is near zero. Small spikes at the output is the result of some imbalance. When a person is positioned on the seat, his (her) body forms an additional capacitance in parallel with C_p, thus increasing a time constant of the R_1C_p-network from τ_1 to τ_2. This is indicated by the increased spike amplitudes at the output of a differential amplifier. The comparator compares V_c with a predetermined threshold voltage V_{ref}. When the spikes exceed the threshold, the comparator sends an indication signal to the logic circuit that generates signal V manifesting the car occupancy. It should be noted that a capacitive detector is an active sensor, because it essentially required an oscillating test signal to measure the capacitance value.

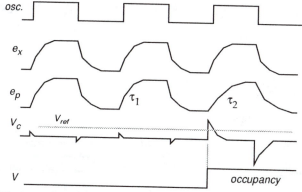

Fig. 6-6 Timing diagrams for a capacitive intrusion detector

When a capacitive occupancy (proximity) sensor is used near or on a metal device, its sensitivity may be severely reduced due to a capacitive coupling between the electrode and the device's metallic parts. An effective way to reduce that stray capacitance is to use driven shields. Fig. 6-7A shows a robot with a metal arm. The arm moves near people and other potentially conductive objects with which it could collide if the robot's control computer is not provided with an advance information on the proximity to the obstacles. An object, while approaching the arm, forms a capacitive coupling with it, which is equal to C_{so}. An arm is covered with an electrically isolated conductive sheath which is called an *electrode*. As Fig. 6-3 shows, a coupling capacitance can be used to detect the proximity. However, the nearby massive metal arm (Fig. 6-7B) forms a much stronger capacitive coupling with the electrode which drugs the electric field from the object. An elegant solution[1] is to shield the electrode from the arm by an intermediate shield as shown in Fig. 6-7C. The sensor's assembly is a multilayer cover for the robot's arm, where the bottom layer is an insulator, then there is a large electrically conductive shield, then another layer of insulation, and on the top is a narrower sheet of the electrode. To reduce a capacitive coupling between the electrode and the arm, the shield must be at the same potential as the electrode, that is, its voltage must be driven by the electrode voltage (thus the name is *driven shield*). Hence, there would be no electric field

[1] This device was developed for NASA's Jet Propulsion Labioratory by M.S. Katow at Palnning Research Corp.

between them, however, there will be a strong electric field between the shield and the arm. The electric field is squeezed out from beneath the electrode and distributed toward the object. Fig. 6-8 shows a simplified circuit diagram of a square-wave oscillator whose frequency depends on the net input capacitance, comprised of C_{sg} (sensor-to-ground), C_{so} (sensor-to-object), and C_{og} (object-to-ground). The electrode is connected to the shield through a voltage follower. A frequency-modulated signal is fed into the robot's computer for controlling the arm movement. This arrangement allows us to detect the proximity to conductive objects over the range of 30 cm.

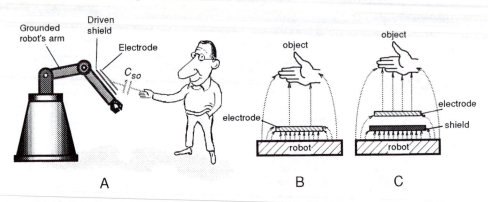

Fig. 6-7 A capacitive proximity sensor
A driven shield is positioned on the metal arm of a grounded robot (A).
Without the shield, the electric field is mostly distributed between the electrode and the robot (B), while a driven shield directs electric field from the electrode toward the object (C)

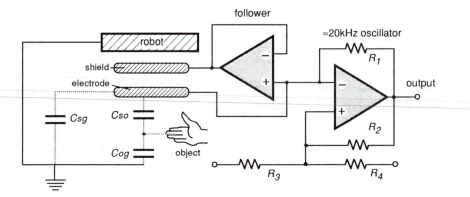

Fig. 6-8 Simplified circuit diagram of a frequency modulator
controlled by the input capacitances.

6.4   TRIBOELECTRIC DETECTORS

Any object can accumulate, on its surface, static electricity. These naturally occurring charges arise from the triboelectric effect that is a process of charge separation due to object movements, friction of clothing fibers, air turbulence, atmosphere electricity, etc. (see Section 3.1). Usually, air contains either positive or negative ions that can be attracted to the human body, thus changing its charge. Under the idealized static conditions, an object is not charged - its bulk charge is equal to zero. In reality, any object which at least temporarily is isolated from the ground can exhibit some degree of its bulk charge imbalance. In other words, it becomes a carrier of electric charges.

An electronic circuit is made of conductors and dielectrics. If a circuit is not shielded, all its components exhibit a certain capacitive coupling to the surrounding objects. In practice, the coupling capacitance may be very small: on the order of 1pF or less. An *electrode* can be added to the circuit's input to increase its coupling to the environment, very much like in the capacitive detectors which were covered in a previous Section. The electrode can be fabricated in a form of a conductive surface which is well isolated from the ground.

An electric field is established between the surrounding objects and the electrode whenever at least one of them carries electric charges. In other words, all distributed capacitors formed between the electrode and the environmental objects are charged by the static or slow changing electric fields. The charge magnitude depends on the atmospheric conditions and the nature of the objects. For instance, a person in dry manmade clothes carries a million times stronger charge than a wet swimmer who got out of a swimming pool. Under the no-occupancy conditions, the electric field in the electrode vicinity is either constant or changes relatively slow.

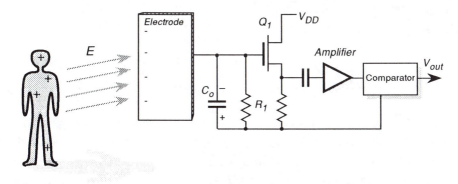

Fig. 6-9 A monopolar triboelectric motion detector

If a charge carrier (a human or an animal) changes its position: moves away or a new charge carrying an object enters into the vicinity of the electrode, the static electric field is disturbed. This results in a redistribution of charges between the coupling capacitors, including those which are formed between the input electrode and the surroundings. An electronic circuit can be adapted to sense these variable charges at its input. In other words, it can be made capable of converting the induced variable charges into electric signals that may be amplified and further processed. Thus, static electricity, which is a naturally occurring phenomenon, can be utilized to generate alternating signals in the electronic circuit to indicate the movement of objects.

Fig. 6-9 shows a monopolar triboelectric motion detector. It is comprised of a conductive electrode connected to an analog impedance converter made with a MOS transistor Q_1 a bias resistor R_1 an input capacitance C_o a gain stage, and a window comparator [6]. While the rest of the electronic circuit may be shielded, the electrode is exposed to the environment and forms a coupling capacitor C_p with the surrounding objects.

In Fig. 6-9, static electricity is exemplified by positive charges distributed along the person's body. Being a charge carrier, the person generates an electric field, having intensity E. The field induces a charge of the opposite sign in the electrode. Under the static conditions, when the person doesn't move, the field intensity is constant and the input capacitance C_o is discharged through a bias resistor R_1. That resistor must be selected of a very high value: on the order of 10^{10} Ω or higher, to make the circuit sensitive to relatively slow motions.

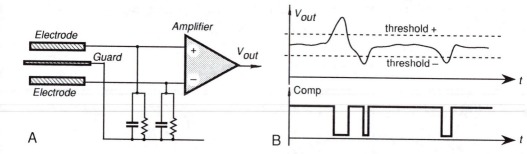

Fig. 6-10 Differential triboelectric motion detector (A) and timing diagrams (B)

When the person moves, intensity E of the electric field changes. This induces the electric charge in the input capacitor C_o and results in the appearance of a variable electric voltage across the bias resistor. The alternating voltage is fed through the coupling capacitor into the gain stage whose output signal is applied to a window comparator. The comparator compares the signal with two thresholds, as it is illustrated in a timing diagram of Fig. 6-10B. A positive threshold is normally higher than the static signal, while the other threshold is lower. During human movement, a signal at the comparator's input deflects either up or down, crossing one of the thresholds. The output signals from the comparator are square pulses which can be utilized and further processed by the conventional data processing devices. It should be noted that contrary to a capacitive motion detector which is an active sensor, a triboelectric detector is passive. That is, it does not generate or emit any signal.

There are several possible sources of interference which may cause spurious detections by the triboelectric detectors. That is, the detector may be subjected to transmitted noise resulting in false positive detection. Among the noise sources are 60 (or 50) Hz power line signals, electromagnetic fields generated by radio stations, power electric equipment, lightnings, etc. Most of these interferences generate electric fields which are distributed around the detector quite uniformly and can be compensated for by a symmetrical input circuit. Fig. 6-10A shows a differential input amplifier with a high common mode rejection ratio. The input stage must have a very high input impedance. A JFET or CMOS circuit preferably should be used. Both the positive and negative inputs are terminated to ground by the identical networks, consisting of resistors, and capacitors. The inputs of the amplifier are connected to two electrodes which are separated from one another by a grounded shield. Each electrode is aimed to the desired direction. If the electrodes are symmetrical, they exhibit maximum sensitivity at normal direction and the lowest sensitivity in the direction where both electrodes are equally exposed to a moving object.

Fig. 6-11 Electrode for a triboelectric motion detector can be camouflaged as a metallized pattern on a surface of a vase

The electrode shape is an important factor in the formation of the sensitivity pattern. It practice, it may be desirable to conceal a motion detector and electrodes, or camouflage them for the security, aesthetics or other reasons. Since the electric field can propagate through many materials having relatively low dielectric constants (like wood, plastics, concrete, etc.), the triboelectric detector virtually may "see" through optically opaque objects. Therefore, the electrodes and the detector could be hidden inside a wall, in the window or door frames. They also could be located inside book covers, nonmetal file cabinets, desks, etc. The electrodes could be shaped in various forms, like wires, tapes, spheres, panels, etc. They also could have shapes of various decorative things, like paper weights, vases, desk lamps, picture frames, toys, etc. Thin conductive transparent coatings on surfaces of glass, ceramic or plastic objects also can function as electrodes. As an example, Fig. 6-11 shows a vase whose surface is metallized forming an electrode which is connected to a detection circuit hidden in the base. Practically, any conductive media may be used as an electrode. For instance, water in a fish tank can function as an electrode if connected to a triboelectric motion detector through an immersed conductor. If a symmetrical circuit is used, like the one shown in Fig. 6-10A, the areas of two electrodes should be identical to assure a good interference reduction.

6.5   OPTOELECTRONIC MOTION DETECTORS

Optoelectronic motion detectors rely on electromagnetic radiation in the optical range, specifically having wavelengths from 0.4 to 20 micrometers. This covers visible, near and far infrared spectral ranges. The detectors are primarily used for the indication of movement of people and animals. They operate over distance ranges up to several hundred meters and, depending on the particular need, may have either a narrow or wide field of view. An object whose movement is to be detected emanates from its surface electromagnetic radiation into the surrounding space. Such radiation may be originated either by an external light source and reflected by the object or it may be produced by the object itself.

First of all, we must consider the limitations of the optoelectronic detectors as opposed to such devices as microwave or ultrasonic devices. Presently, optoelectronic detectors are used almost exclusively to detect the presence or absence of movement qualitatively rather than quantitatively. In other words, the optoelectronic detectors are very useful to indicate whether an object moves or not, while they can not distinguish one moving object from another and they can't be utilized to accurately measure the distance to a moving object or its velocity. The major application areas for the optoelectronic motion detectors are in security systems (to detect intruders), in energy management (to turn lights on and off in a room) and in the so-called "smart" homes where they can control various appliances, such as air conditioners, cooling fans, stereo players, etc. They also may be used in robots, toys, and novelty products. The most important advantage of an optoelectronic motion detector is simplicity and low cost.

6.5.1 Sensor structures

A general structure of an optoelectronic motion detector is shown in Fig. 6-12A. Regardless what kind of sensing element is employed, the following components are essential: a focusing device (a lens or a focusing mirror), a light detecting element, and a threshold comparator. An optoelectronic motion detector resembles a photographic camera. Its focusing components create an image of its field of view on a focal plane. There is no mechanical shutter like in a camera, however, in place of the film, a light sensitive element is used. The element converts focused light into an electric signal. Let us assume that the motion detector is mounted in a room. A focusing lens creates an image of the room on a focal plane where the light sensitive element is positioned. If the room is unoccupied, the image is static and the output signal from the element is steady stable. When an "intruder" enters the room and keeps moving, his image on the focal plane also moves. In a certain moment, the intruder's body is displaced for an angle α and the image overlaps with the element. Assuming that the intruder's body creates an image whose electromagnetic flux is different from that of the static surroundings, the light sensitive element responds with deflecting voltage V. In other words, a moving image must have a certain degree of an optical contrast with its surroundings.

Fig. 6-12B shows that the output signal is compared with two thresholds in the window comparator. The purpose of the comparator is to convert the analog signal V into two logic levels: $\emptyset$ - no motion detected and 1 - motion is detected. In most cases, signal V from the element first must be amplified and conditioned before it becomes suitable for the threshold comparison. The window detector contains both the positive and negative thresholds, while signal V is positioned inbetween. Whenever image of a moving object overlaps with the light sensitive element, voltage V deflects from its steady-state position and crosses one of two thresholds. The comparator generates a positive voltage (1), thus indicating a detection of movement in a field of view. The operation of this circuit is identical to the threshold circuits described earlier for other types of occupancy detectors.

It may be noted from Fig. 6-12 that the detector has quite a narrow field of view: if the intruder keeps moving, his image will overlap with the sensor only once, after that the window comparator output will produce steady $\emptyset$. This is a result of a small area of the sensing element. In some instances, a narrow field of view is required, however, in the majority of cases, a much wider field of view is desirable. This can be achieved by several methods.

Multiple sensors

An array of sensors may be placed in the focal plane of a focusing mirror or lens. All sensors in the array either must be multiplexed or otherwise interconnected to produce a combined detection signal.

Complex sensor shape

If the sensor's surface area is sufficiently large to cover an entire angle of view, it may be broken into smaller elements, thus creating an equivalent of a multiple sensor. To break the surface area into several parts, one may shape the sensing element in an odd pattern like those shown in Fig. 6-13A and B. Each part of the element acts as a separate sensor. All such sensors are virtually connected either in series in a serpentine pattern (Fig. 6-13A) or in a parallel gridlike shape (Fig. 6-13B).

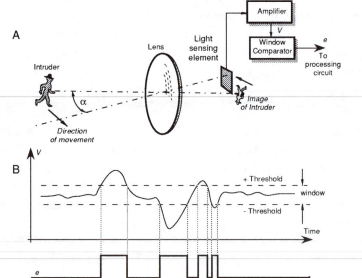

Fig. 6-12 General arrangement of an optoelectronic motion detector
A lens forms an image of a moving object (intruder). When the image crosses the optical axis of the sensor, it superimposes with the sensitive element (A). The sensor responds with the signal which is amplified and compared with two thresholds in the window comparator (B)

The parallel or serially connected sensors generate a combined output signal, for instance, voltage v. An image of the object moves along the sensor's surface crossing alternatively sensitive and nonsensitive areas. This results in an alternate signal v at the sensor terminals. Each sensitive and nonsensitive area must be sufficiently large to overlap with most of the object's image.

Image distortion

Instead of making the sensor in a complex shape, an image of an entire filed of view may be broken into several parts. This can be done by placing a distortion mask in front of the sensor having a sufficiently large area as it is depicted in Fig. 6-13C. The mask is opaque and allows to form an image on the sensor's surface only within its clearings. The mask operation is analogous to the complex sensor's shape as described above.

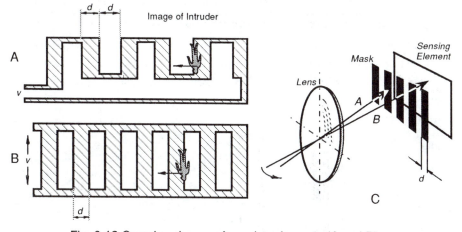

Fig. 6-13 Complex shapes of sensing elements (A and B); an image distortion mask (C)

Facet focusing element

A focusing mirror or a lens may be divided into an array of smaller mirrors or lenses called facets. Each facet creates its own image resulting in multiple images as shown in Fig. 6-14A. When the object moves, the images also move across the sensor, resulting in an alternate signal. By combining multiple facets it is possible to create any desirable detecting pattern in the field of view. Positioning of the facet lens, focal distances, number and a pitch of the facets (a distance between the optical axes of two adjacent facets) may by calculated in every case by applying rules of geometrical optics. The following practical formulas may be applied to find the focal length of a facet

$$f = \frac{Ld}{\Delta} ,$$

(6.5.1)

and the facet pitch is

$$p = 2nd \quad , \tag{6.5.2}$$

where L is the distance to the object, d is the width of the sensing element, n is the number of sensing elements (evenly spaced), and Δ is the object's minimum displacement which must result in a detection.

> For instance, if the sensor has two sensing elements of d=1 mm each which are positioned at 1 mm apart, and the object's minimum displacement Δ=25 cm at a distance L=10 m, the facet focal length is calculated as f=1000 cm ·0.1 cm./25 cm = 4 cm, and the facets should be positioned with a pitch of p=8 mm from one another

By combining facets, one may design a lens which covers a large field of view (Fig. 6-14B) where each facet creates a sensitive zone. Each zone projects an image of an object into the same sensing element. When the object moves, it crosses zone boundaries, thus modulating the sensor's output.

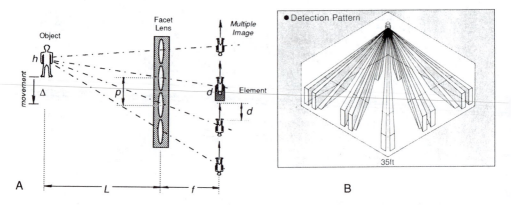
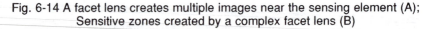

Fig. 6-14 A facet lens creates multiple images near the sensing element (A); Sensitive zones created by a complex facet lens (B)

The above methods of motion detection are generalized over a broad spectral range, specifically, for visible and far infrared (See Fig. 3-14.3). In turn, the far infrared motion detectors may be of a passive or active nature. Below we examine all three types of the detectors.

6.5.2 Visible and near IR motion detectors

Most of the objects radiate electromagnetic waves only in a far infrared spectral range. Hence, visible and near infrared light motion detectors rely on the light which is reflected by the object's body toward the focusing lens or mirror. As a result, the detector essentially requires an external light source for its operation. Such illumination may be sunlight or an invisible infrared light from an additional near infrared light source (a projector). The use of a visible light for detecting moving objects goes back to 1932 when in the preradar era, inventors were looking for ways to detect flying airplanes. In one invention, an airplane detector was built in a form of a photographic camera where the focusing lens was aimed at the sky. A moving plane's image was focused on a selenium photodetector which reacted to a changing contrast in the sky image. Naturally, such a detector could operate only at day time to detect planes flying below clouds. Obviously, those detectors were not too practical. Another version of a visible light motion detector was patented for less critical applications: controlling lights in a room [7] and to make interactive toys [8].

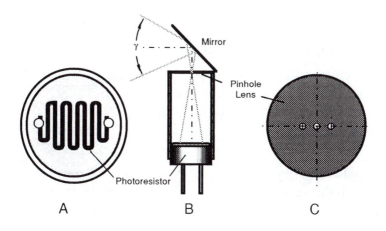

Fig. 6-15 A simple optical motion detector for a light switch and toys
A - a sensitive surface of a photoresistor forms a complex sensing element; B - a flat mirror and a pinhole lens form an image on a surface of the photoresistor; C - a pinhole lens

To turn the lights off in a nonoccupied room, the motion detector ("A motion switch" manufactured by Intermatic, Inc., IL) was combined with a timer and a power solid-state relay. The detector is activated when the room is illuminated. Visible light carries a relatively high energy and may be detected by quantum photovoltaic or photoconductive detectors whose detectivity is high. Thus, the optical system may be substantially simplified. In the "motion switch", the focusing part was built in a form of a pin-hole lens (Fig. 6-15B). That lens is just a tiny hole in an opaque foil. To avoid a light wave diffraction, the hole diameter must be substantially larger than the longest detectable wavelength. Practically, the Motion Switch has a three-facet pin-hole lens where each hole has an aperture of 0.2 mm (Fig. 6-15C). Such a lens has a theoretically infinitely deep focusing range, hence, the photodetector can be positioned at any distance from it. That distance was calculated for a maximum of the object's displacement and a photoresistor used in the design. The photoresistor was selected with a serpentine pattern of a sensing element (Fig. 6-15A) and was connected into a circuit which responds only to its variable component. When the room is illuminated, the sensor acts as a miniature photographic camera: an image of its field of view is formed on a surface of the photoresistor. Moving people in the room cause the image to change in such a way as the contrast changes across the serpentine pattern of the photoresistor. In turn, its resistive value changes which results in the modulation of electric current. This signal is further amplified and is compared with a predetermined threshold. Upon crossing that threshold, the comparator generates electric pulses which reset a 15 minute timer. If no motion is detected within 15 minutes from the last movement, the timer turns lights in the room off. Then, it may be turned on again only manually, because the motion detector doesn't function in darkness.

6.5.3 Far infrared motion detectors

Another version of a motion detector operates in the optical range of thermal radiation, the other name for which is far infrared. Such detectors are responsive to radiative heat exchange between the sensing element and the moving object. Here, we will discuss a detection of moving people, however the technique which is described below may be modified for other warm or cold objects.

The principle of thermal motion detection is based on the physical theory of emission of electromagnetic radiation from any object whose temperature is above absolute zero. The fundamentals of this theory are described in Section 3.14.3. We recommend that the reader first familiarize oneself with that section before going further.

For motion detection, it is essential that a surface temperature of an object be different from that of the surrounding objects, so a thermal contrast would exist. All objects emanate thermal radiation from their surfaces and the intensity of that radiation is governed by the Stefan-Boltzmann law [Eq. (3.14.14)]. If the object is warmer than the surroundings, its thermal radiation is shifted toward shorter wavelengths and its intensity becomes stronger. Many objects whose movement is to be detected are nonmetals, hence they radiate thermal energy quite uniformly within a hemisphere (Fig. 3-14.7A). Moreover, the dielectric objects have generally a high emissivity. Human skin is one of the best emitters whose emissivity is over 90% (See Table 3-15), while most of the fabrics have also high emissivities between 0.74 and 0.95. Below we describe two types of far infrared motion detectors. The first utilizes a passive infrared (PIR) sensor while the second has active far infrared (AFIR) elements.

PIR motion detectors

These detectors became extremely popular for the security and energy management system. The PIR sensing element must be responsive to far infrared radiation within a spectral range from 4 to 20 μm where most of the thermal power emanated by humans is concentrated. There are three types of sensing elements which are potentially useful for that detector: thermistors, thermopiles, and pyroelectrics, however the pyroelectric elements are used almost exclusively for the motion detection thanks to their simplicity, low cost, high responsivity, and a broad dynamic range. A pyroelectric effect is described in Section 3.7 and some detectors are covered in Section 13.7.2. Here, we are going to see how that effect may be used in a practical sensor design.

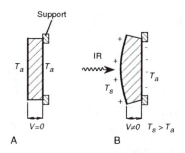

Fig. 6-16 A simplified model of a pyroelectric effect
as a secondary effect of piezoelectricity
Initially, the element has a uniform temperature (A); upon exposure to thermal radiation, its front side expands (B), causing a stress induced charge

A pyroelectric material generates an electric charge in response to thermal energy flow through its body. In a very simplified way it may be described as a secondary effect of a thermal expansion (Fig. 6-16). Since all pyroelectrics are also piezoelectrics, the absorbed heat causes the front side of the sensing element to expand. The resulting stress leads to the development of a piezoelectric charge on the element electrodes. This charge is manifested as voltage across the electrodes deposited on the opposite sides of the material. Unfortunately, the piezoelectric properties of the element have also a negative effect. If the sensor is subjected to a minute mechanical stress due to any external force, it also generates a charge which is indistinguishable from that caused by the infrared heat waves.

To separate thermally induced charges from the piezoelectrically induced charges, a pyroelectric sensor is usually fabricated in a symmetrical form (Fig. 6-17A). In other words, two identical elements are positioned inside the sensor's housing. They are connected to the electronic circuit in such a manner as to produce the out-of-phase signals. One way to fabricate a symmetrical sensor is to deposit two pairs of electrodes on both sides of a pyroelectric element. Each pair forms a capacitor which may be charged either by heat or by a mechanical stress. The electrodes on the upper side of the sensor are connected together forming one continuous electrode, while the two bottom electrodes are separated, thus creating opposite-serially connected capacitors. Depending on the side where the electrodes are positioned, the output signal will have either a positive or negative polarity for the thermal influx. If the sensor requires more than two pairs of electrodes, it still should have an even number of pairs where positions of the pairs alternate for better geometrical symmetry. Sometimes, such an alternating connection is called an interdigitized electrode.

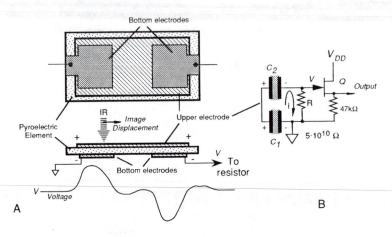

Fig. 6-17 A dual pyroelectric sensor

A: A sensing element with a front (upper) electrode and two bottom electrodes deposited on a common crystallyne substrate. A moving thermal image travels from left part of the sensor to the right generating an altrenate voltage across bias resistor, R (B)

A symmetrical sensing element should be mounted in such a way as to assure that both parts of the sensor generate the same signal if subjected to the same external factors. At any moment, the optical component must focus a thermal image of an object on the surface of one part of the sensor only, which is occupied by a single pair of electrodes. The element generates a charge only across the electrode pair which is subjected to a heat flux. When the thermal image moves from one electrode to another, the current i flowing from the sensing element to the bias resistor R(Fig. 6-17B) changes form zero, to positive, then to zero, to negative, and again to zero (Fig. 6-17A lower portion). A JFET transistor Q is used as an impedance converter. Resistor R value must be very high. For example, a typical alternate current generated by the element in response to a moving person is on the order of 1pA (10^{-12} A). If a desirable output voltage for a specific distance is $v=50$mV, according to Ohm's law the resistor value is R=v/i=50GΩ ($5 \cdot 10^{10}$Ω). Such a resistor can not be directly connected to a regular electronic circuit, hence transistor Q serves as a voltage follower (the gain is close to unity). Its typical output impedance is on the order of several kilohm.

Table 3-5 lists several crystalline materials which possess a pyroelectric effect and can be used to fabricate a sensing element. Among them is a polymer film PVDF which while being not as sensitive as most of the solid-state crystals, has advantages of being flexible and inexpensive. Besides, it can be produced in any size, and may be bent or folded in any desirable fashion.

Besides the sensing element, an infrared motion detector needs a focusing device. Some detectors employ parabolic mirrors while the Fresnel plastic lenses (Section 3.16.4) become more and more popular because they are inexpensive, may be curved to any desirable shape and, in addition to focusing, protect the interior of the detector from outside moisture and pollutants.

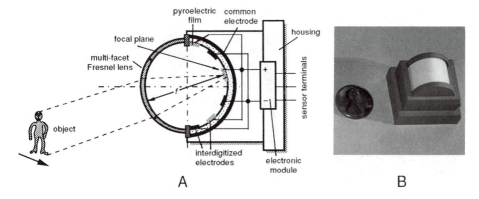

Fig. 6-18 Far infrared motion detector with a curved Fresnel lens and a pyroelectric film
A - internal structure of the sensor
B - external appearance of the sensor

To illustrate how a plastic Fresnel lens and a PVDF film can work together, let's look at the motion detector depicted in Fig. 6-18A. It uses a polyethylene multifaceted curved lens and a curved PVDF film sensor [9]. The sensor design combines two methods described above: a facet lens and a complex electrode shape. The lens and the film are curved with the same radii of curvature equal to one half of the focal distance f thus assuring that the film is always positioned in the focal plane of the corresponding facet of the lens. The film has a pair of large interdigitized electrodes which are connected to the positive and negative input of a differential amplifier located in the electronic module. The amplifier rejects common-mode interference and amplifies a thermally induced voltage. The side of the film facing the lens is coated with an organic coating to improve its absorptivity in the far infrared spectral range. This design results in a fine resolution (detection of small displacement at a longer distance), and a very small volume of a detector (Fig. 6-18B). Small detectors are especially useful for the installation in devices where overall dimensions are critical. For instance, one application is a light switch where the detector must be mounted into the wall plate of a switch.

Another example of how a curved faceted lens can produce a 360° vision is shown in Fig. 6-19. The sensor has a shape of two frustum cones joined at the bottoms. The upper half is a sensing element fabricated of either of a curved PVDF film with interdigitized electrodes, or multiple solid-state pyroelectric elements[1]. The bottom half of the sensor is fabricated of a faceted Fresnel lens, which focuses a thermal image of an intruder on the corresponding portion of the upper frustum cone where the sensing elements are positioned. Such a sensor can be located on a pole or a ceiling for the circumferential area protection.

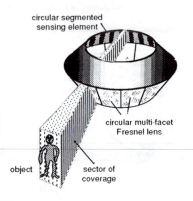

Fig. 6-19 A 360° far infrared motion detector

PIR sensor efficiency analysis

Regardless of the optical device employed, all modern PIR detectors operate on the same physical effect - pyroelectricity. To analyze the performance of such a sensor, first we must calculate the infrared power (flux) which is converted into an electric charge by the sensing element. The optical device focuses thermal radiation into a miniature thermal image on the surface of the sensor. The energy of an image is absorbed by the sensing element and is converted into heat. That heat, in turn, is converted by the pyroelectric crystalline element into a minute electric current.

To estimate a power level at the sensor's surface, let us make some assumptions. We assume that the moving object is a person whose effective surface area is b (Fig. 6-20), the temperature

---

[1] Naturally, the sensing element does not have to be pyroelectric. Any suitable and economically viable sensor (thermopyles, film thermistors, etc.) can be also employed. However, at the time of this writing the the pyroelectric sensor appears to be the most cost effective.

along this surface (T_b) is distributed uniformly and is expressed in K. The object is assumed to be a diffuse emitter (radiates uniformly within the hemisphere having a surface area of $A=2\pi L^2$). Also, we assume that the focusing device makes a sharp image of an object at any distance. For this calculation we select a lens which has a surface area a. The sensor's temperature in K is T_a - the same as that of ambient.

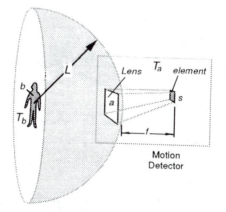

Fig. 6-20 Formation of a thermal image on the sensing element of the PIR motion detector

Total infrared power (flux) lost to surroundings from the object can be determined from the Stefan-Boltzmann law

$$\Phi = b\varepsilon_a\varepsilon_b\sigma(T_b^4 - T_a^4) \quad , \tag{6.5.3}$$

where σ is the Stefan-Boltzmann constant, ε_b and ε_a are the object and the surrounding emissivities, respectively. If the object is warmer than the surroundings (which is usually the case), this net infrared power is distributed toward an open space. Since the object is a diffusive emitter, we may consider that the same flux may be detected along an equidistant surface. In other words, the intensity of infrared power is distributed uniformly along the spherical surface having radius L.

Assuming that the surroundings and the object's surface are ideal emitters and absorbers ($\varepsilon_b=\varepsilon_a=1$) and the sensing element's emissivity is ε_s, the net radiative power density at distance L can be derived as

$$\phi = \frac{b}{2\pi L^2} \varepsilon_s\sigma(T_b^4 - T_a^4) \quad . \tag{6.5.4}$$

The lens efficiency (transmission coefficient) is γ, which may vary from 0 to 1 depending on the properties of the lens material and the lens design. After ignoring some minor nonlinearities, thermal power absorbed by the element can be expressed as

$$\Phi_s = a\gamma\phi \approx \frac{2\sigma\varepsilon_s}{\pi L^2} a\gamma b T_a^3 (T_b - T_a) \quad . \tag{6.5.5}$$

It is seen from Eq. (6.5.5) that the infrared flux which is focused by the lens on the surface of the sensing element is inversely proportional to the square distance from the object and directly proportional to the areas of the lens and the object. It is important to note that in the case of a multi-facet lens, the lens area a is related to a single facet only and not to the total lens area.

If the object is warmer than the sensor, the flux Φ_s is positive. If the object is cooler, the flux becomes negative, meaning it changes its direction: the heat goes from the sensor to the object. In reality, this may happen when a person walks into a warm room from the cold outside. The surface of his clothing will be cooler than the sensor and the flux will be negative. In the following discussion we will consider that the object is warmer than the sensor and the flux is positive.

A maximum operating distance for given conditions, can be determined by the noise level of the detector. For reliable discrimination, the worst case noise power must be at least 3-5 times smaller than that of the signal.

The pyroelectric sensor is a converter of thermal energy flow into electric charge. The energy flow essentially demands a presence of a thermal gradient across the sensing element. In the detector, the element of thickness h has the front side exposed to the lens, while the opposite side faces the detector's interior housing, which normally is at ambient temperature T_a. The front side of the sensor element is covered with a heat absorbing coating to increase its emissivity ε_s to the highest possible level, preferably close to unity. When thermal flux Φ_s is absorbed by the element's front side, the temperature goes up and heat starts propagating through the sensor toward its rear side. Thanks to the pyroelectric properties, electric charge is developing on the element surfaces in response to the heat flow.

Upon influx of the infrared radiation, the temperature of the sensor element increases (or decreases) with the rate which can be derived from the absorbed thermal power Φ_s and thermal capacity C of the element

$$\frac{dT}{dt} \approx \frac{\Phi_s}{C} \quad , \tag{6.5.6}$$

were t is time. This equation is valid during a relatively short interval, immediately after the sensor is exposed to the thermal flux, and can be used to evaluate the signal magnitude.

The electric current generated by the sensor can be found from the fundamental formula:

$$i = \frac{dQ}{dt} \quad ,$$

(6.5.7)

where Q is the electric charge developed by the pyroelectric sensor. This charge depends on the sensor's pyroelectric coefficient P sensor's area s and temperature change dT:

$$dQ = PsdT \quad .$$

(6.5.8)

Thermal capacity C can be derived through a specific heat c, area s, and thickness of the pyroelectric element h

$$C = csh \quad .$$

(6.5.9)

By substituting Eqs. (6.5.6), (6.5.8), and (6.5.9) into equation (6.5.7) we can evaluate the peak current which is generated by the sensor in response to the incident thermal flux

$$i = \frac{PsdT}{dt} = \frac{Ps\Phi_s}{csh} = \frac{P}{hc}\Phi_s \quad .$$

(6.5.10)

To establish the relationship between the current and the moving object, the flux from Eq. (6.5.5) has to be substituted into Eq. (6.5.10)

$$i = \frac{2P\sigma a\gamma}{\pi hc} bT_a^3 \frac{\Delta T}{L^2} \quad ,$$

(6.5.11)

where $\Delta T = (T_b - T_a)$.

There are several conclusions which can be drawn from Eq. (6.5.11). The first part of the equation (the first ratio) characterizes a detector while the rest relates to an object. The pyroelectric current i is directly proportional to the temperature difference (thermal contrast) between the object and its surroundings. It is also proportional to the surface area of the object which faces the detector. A contribution of the ambient temperature T_a is not that strong as it might appear from its third power. The ambient temperature must be entered in kelvin, hence its variations become relatively small with respect to the scale. The thinner the sensing element the more sensitive is the detector. The lens area also directly affects signal magnitude. On the other hand, pyroelectric current doesn't depend on the sensor's area as long as the lens focuses an entire image on a sensing element.

To evaluate Eq. (6.5.11) further, let us calculate voltage across the bias resistor. That voltage can be used as an indication of motions. We select a pyroelectric PVDF film sensor Kynar® with typical properties: $P=25$ μC/Km2, $c=2.4 \cdot 106$ j/m^{3}K, $h=25$ μm, lens area $a=1$ cm^2, $\gamma=0.6$, and the bias resistor $R=10^9\Omega$ (1GΩ). We will assume that the object's surface temperature is 27°C and surface area $b=0.1$m^2. The ambient temperature $T_a=20$°C. The output voltage is calculated from equation (6.5.11) as a function of distance L from the detector to the object and is shown in Fig. 16-21.

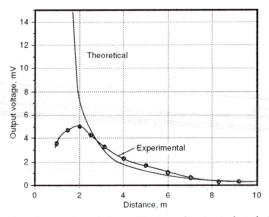

Fig. 6-21 Calculated and experimental amplitudes of output signals in the PIR detector

A graph for Fig. 6-21 was calculated under the assumption that the optical system provides a sharp image at all distances and that image is no larger that the sensing element area. In practice this is not always true, especially at shorter distances where the image is not only out of focus but also may overlap the out-of-phase parts of a symmetrical sensor. The reduction in the signal amplitude at shorter distances becomes apparent - the voltage doesn't go as high as in the calculated curve.

> For comparison, we can calculate what the output voltage would be if a thermopile sensor is used instead of the pyroelectric. A typical thermopile has a responsivity in the order 20 V/W. From Eq. (6.5.5) the infrared flux for a distance of 5 m, yields a signal amplitude from a thermopile on the order of $10\mu V$ - a value of about 100 times smaller than that produced by the pyroelectric film across the bias resistor. A low level signal requires a low noise amplifier. Combined with substantially higher cost for thermopiles, this makes their application to the motion detection quite impractical.

AFIR motion detector

Contrary to a passive motion detector which absorbs thermal radiation from a warmer object, an AFIR (Active Far Infrared) motion detector radiates heat waves *toward* the surroundings [10, 11]. The sensor's surface temperature (T_s) is maintained somewhat above ambient. The element is combined with a focusing system, very much like the PIR detector, however, the function of that system is inverse to that of passive detectors. A focusing part in the AFIR detector projects a thermal image of the warm sensing element into its surroundings. An electronic circuit supplies the element with electric power which is converted into heat. A portion of that power is emanated from the surface of the element in a form of far infrared radiation. The magnitude of the net radiation is a function of the surroundings which absorb the radiation. When an object of different from ambient temperature moves into the detector's field of view, the energy exchange between the sensor and the surroundings changes. In order to maintain the temperature of the sensor equal to a preset level under such changing conditions, the electronic circuit must readjust the amount of power which is provided to the sensing element. This readjustment results in a variable control signal which can be used to indicate the movement of an object within the field of view. The AFIR sensors have a significant advantage over the PIR detectors: immunity against to many interferences (such as RFI and microphonics).

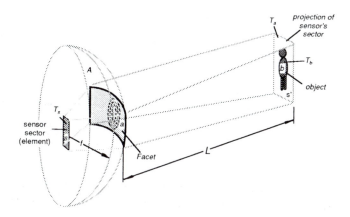

Fig. 6-22 The AFIR sensing element radiates heat which is projected onto the surroundings occupied by a moving object

AFIR sensor sensitivity analysis

Below, we are going to analyze a motion detector which consists of three essential components: a focusing Fresnel lens, an AFIR sensing element, and a threshold detector. Only one element of a dual AFIR sensor will be considered in the analysis. The sensor's surface is warmed up by the electric current passing through its built-in heating element. A part of the supplied energy in a form of an IR flux is emanated from the surface and is distributed by the lens to the surroundings (Fig. 6-22). Not all that radiation goes out of the detector toward the target (a moving object) and the surrounding objects. A part of it is absorbed by the detector's interior. If the lens is multi-faceted, then every facet is considered as a separate, independent lens which can focus a thermal flux on a specific area of the surrounding space where the movement is to be detected. The lens creates an image of the sensing surface on the surrounding objects, very much like a slide projector would make an image on a white screen. The lens efficiency is defined by the ratio of its facet surface area a to total area A of a hemisphere which can be imagined at focal distance f from the sensor

$$n = \gamma \frac{a}{A} = \gamma \frac{a}{2\pi f^2} \quad , \tag{6.5.12}$$

where γ is the lens transmission coefficient which depends on the type of the lens material, the thickness, and its surface reflectivity. A typical value of γ for a polyethylene Fresnel lens is about 0.6.

The lens distributes thermal radiation toward a space whose average temperature is either equal to ambient T_a or a part of the space is occupied by a target having surface area b and temperature T_b. The image of the sensor is situated at a distance L and its linear dimension is inversely proportional to focal length f. Area s' of the sensor's image can be expressed by a function

$$s' = s \frac{L^2}{f^2} \quad . \tag{6.5.13}$$

Similarly, that portion of the sensor's surface area b' which radiates an infrared flux toward the target is

$$b' = b \frac{f^2}{L^2} \quad . \tag{6.5.14}$$

A total thermal flux which is radiated by the detector (through a single facet of the lens) consists of two portions: that emanated toward a moving target Φ_2 and that emanated toward the static objects Φ_1

$$\Phi = \Phi_1 + \Phi_2 \quad . \tag{6.5.15}$$

A motion detector is not a precision measurement device, hence we can ignore a minor non-linearity in the Stefan-Boltzmann law. Then, the infrared flux emanated from the sensor to the surroundings is

$$\Phi_1 = 4\sigma n T_s^3 (T_s - T_b) \cdot \left(s - \frac{b f^2}{L^2} \right) \quad . \tag{6.5.16}$$

In the above equation, a value in the last parenthesis reflects the sensor's area reduced by that portion which radiates only toward a moving target. Similarly, with equation (6.5.15) in mind, the flux which is radiated toward a moving target can be expressed as

$$\Phi_2 = 4\sigma n T_s^3 (T_s - T_b) \frac{b f^2}{L^2} \quad . \tag{6.5.17}$$

Substituting Eq. (6.5.16) and (6.5.17) into (6.5.15) and after simple manipulations we arrive at

$$\Phi = 4\sigma n T_s^3 \left[sT_s - T_a(s - \frac{bf^2}{L^2}) - T_b\frac{bf^2}{L^2} \right] \quad . \tag{6.5.18}$$

The above equation is a mathematical model of the total infrared flux which is radiated from the AFIR sensor's element through a single lens facet toward the environment, a portion of which is occupied by a target. If the target is outside of the field of view ($b=0$), Eq. (6.5.18) can be simplified to represent the static background flux that is radiated toward the environment which contains no moving objects

$$\Phi_0 \approx 4\sigma n T_s^3 (T_s - T_a) \quad . \tag{6.5.19}$$

When a target moves into the filed of view, the radiated flux changes and a magnitude of its variable portion can be found from Eqs. (6.5.12), (6.5.18) and (6.5.19)

$$\Delta \approx \Phi - \Phi_0 = -4\sigma n T_s^3 \frac{bf^2}{L^2}(T_b - T_a) = -2\frac{ab}{\pi L^2}\gamma\sigma T_s^3 \Delta T \quad . \tag{6.5.20}$$

The negative sign for the warmer target ($T_b > T_a$) indicates that the total flux emanated from the detector is reduced when the target moves into the field of view. At normal room temperatures, the incremental flux radiated toward a moving adult person moving at a distance of about 5m is on the order of 1 μW. The above expression represents that portion of the infrared flux which is to be converted into an alternate electric signal. That signal can be used in the threshold circuit to indicate motions. Note that the signal doesn't depend on the focal length of the lens or the sensor's area. This is true if an entire target is covered by the projection of a sensing element.

The law of conservation of energy demands that electric power supplied to the sensor is equal to a combined loss power. The loss consists of two parts – the nonradiative heat and radiative flux. When the object moves into the field of view, the electric power supplied to the sensing element changes by the increment

$$\delta P = \frac{V_o^2}{R} - \frac{(V_o - \Delta V)^2}{R} \quad , \tag{6.5.21}$$

where V_o is the initial voltage across the heating element having resistance R, and ΔV is the voltage increment. The incremental electric power is equal to the incremental flux, so that $\Delta = \delta P$. Combining Eqs. (6.5.20) and (6.5.21) and after manipulations, the sensor's output alternate voltage amplitude is defined as

$$\Delta V \approx -\frac{R}{V_o}\frac{ab}{\pi L^2}\gamma\sigma T_s^3 \Delta T \quad . \tag{6.5.22}$$

Since T_s is in kelvin, its value doesn't make a significant contribution to the output signal. For instance, a rise in T_s by $10°$ results in a ΔV increase by about only 5%. Hence, T_s should be selected at a relatively low level to conserve energy. However, it always must be maintained higher than the ambient temperature to assure that heat flows from the sensor to the environment.

Making a negligibly small loss is essential for the detector's efficiency. As it was shown above, it is desirable to maintain T_s above ambient by a constant value Δt. Hence, the ambient temperature T_a should control T_s and, as a result, to maintain the loss P_o at a nearly constant level.

AFIR motion detector design

The Fourier law of thermal conduction states that the rate of heat loss is a function of thermal conductivity k of a material having cross-section area A and temperature gradient $\Delta t = T_s - T_a$ across its length Δx:

$$P_o = kA\frac{\Delta t}{\Delta x} \quad . \tag{6.5.23}$$

To compensate for P_o, the sensing element must contain a temperature sensor and possibly a heater. In one design, the sensing element is a thermistor operating in a self-heating mode, thus there is no need for an additional heater. In the other design, heating is provided by an separate *base* heater which is thermally coupled with the temperature sensor. As it was mentioned above, a function of the base heater is to keep the sensing element warmer than the ambient temperature. Thus, it must provide a compensation for the heat loss. The loss compensation concept is very simple. A motion detector power supply provides electric power W_o equal to thermal loss other than radiated to the surroundings[1]

$$W_o = P_o \quad . \tag{6.5.24}$$

However, there still is a need for another heater which would provide an additional heat equal to thermal loss variations Δ [Eq. (6.5.21)]. The second heater is called a *radiative* heater. The electric power developed by the base heater can be expressed through its resistance R_o:

$$W_o = E^2/R_o \quad , \tag{6.5.25}$$

where E is the power supply voltage. Combining Eqs. (6.5.23), (6.5.24), and (6.5.25) we can get an expression for a temperature difference between the sensing element and its surroundings

$$\Delta t = \frac{\Delta x}{kAR_o}E^2 \quad . \tag{6.5.26}$$

---

[1] P_o is equal to conductive and convective heat loss, and that stray-radiated inside the sensor's housing.

We can see that the temperature gradient Δt is not a function of T_a and depends only on the motion detector design and the component selection. Hence, the base heater will maintain a temperature of both the radiative heater and the temperature sensor above ambient by a constant value Δt.

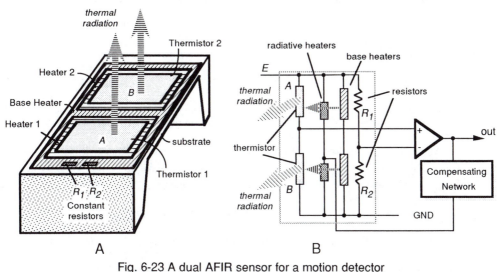

Fig. 6-23 A dual AFIR sensor for a motion detector
A: two symmetrical layers of resistive films are formed on a common substrate;
B: a control circuit provides constant voltage to the base heater and balances radiative heaters through a feedback control

An AFIR sensor can be fabricated in a form of a sandwich where layers of resistive and electrical isolating materials are deposited on a thin substrate. The substrate may be fabricated of a thin glass flake, ceramic or polymer film, or on a micromachined silicon membrane (Fig. 6-23A). In motion detectors, a dual AFIR sensor should be used almost exclusively to reduce common mode thermal and electrical interferences. A dual sensing element is comprised of two totally identical sensors, A and B, residing on a base heater which is common for both sensors. A thin layer of electrical isolation (not shown) is fabricated on the top of the base heater. Two fully symmetrical groups of layers are positioned above that isolator. These groups form two parts of a symmetrical sensor. The radiative heaters, 1 and 2, are made of a material having a relatively low temperature coefficient of resistance (TCR). Another layer of electrical isolation (not shown) is deposited on the top of both radiative heaters. Then, two temperature sensitive resistors (thermistors), 1 and 2, are deposited on the very top. These thermistors must be made of a material having a high value of TCR. Two additional fixed resistors, R_1 and R_2, may be deposited on the base heater for better thermal stability, however, this is not critical for the overall performance.

The base heaters provide identical heat to both parts of the sensor and maintain their temperature above ambient by a fixed value Δt. The electronic circuit (Fig. 6-23B) maintains both parts of the sensor in a thermal balance. A moving object causes thermal disbalance of two sensors and the circuit, in turn, provides a corrective signal to re-establish a thermally balanced state. The corrective signal to the radiative heaters are used as the output signal of the motion detector.

REFERENCES

1. Blumenkrantz, S. Personal and organizational security handbook. Government data publications, Washington, D.C., 1989

2. Ryser, P. and Pfister, G. Optical fire and security technology: sensor principles and detection intelligence. In: *Transducers'91. International conference on solid-state sensors and actuators. Digest of technical papers*, pp: 579-583, ©IEEE, 1991

3. Neukomm, P.A. Body-Mounted Antennas. Juris Druck + Verlag, Zürich, 1979

4. Calvin, N.M. Capacitance proximity sensor. *U.S. Patent* No. 4,345,167, Aug. 17, 1982

5. Long, D.J. Occupancy detector apparatus for automotive safety system. *U.S. Patent* No. 3,898,472. Aug. 5, 1975

6. Fraden, J. Apparatus and method for detecting movement of an object, *U.S. Patent* No. 5,019,804. May 28, 1991

7. Fraden, J. Motion discontinuance detection system and method. *U.S. Patent* No. 4,450,351. May 22, 1984

8. Fraden, J. Toy including motion-detecting means for activating same. *U.S. Patent* No. 4,479,329. Oct. 30, 1984

9. Fraden, J. Motion Detector, *U.S. Patent* No. 4,769,545, Sept. 6,1988

10. Fraden, J. Active Infrared Motion Detector and Method for Detecting Movement, *U.S. Patent* No. 4,896,039, Jan. 23, 1990

11. Fraden, J. Active Far Infrared Detectors. In: *7 International Converence on Temperature*, ©ISA, 1992

7
Velocity and Acceleration

Acceleration is a dynamic characteristic of an object, because according to Newton's 2nd law it essentially requires application of a force. In effect, displacement, velocity, and acceleration are all related — velocity is a first derivative of displacement and acceleration is the second derivative. However, in a noisy environment, taking derivatives may result in extremely high errors, even if complex and sophisticated signal conditioning circuits are employed. Therefore, velocity and acceleration are not derived from position or proximity detectors, but rather measured by special sensors. As a rule of thumb, in low-frequency applications (on the order of 1Hz), position and displacement measurements generally provide good accuracy. In intermediate-frequency applications (less than 1kHz), velocity measurement is usually favored. In measuring high-frequency motions with appreciable noise levels, acceleration measurement is preferred. However, a basic idea behind any sensor for transduction of velocity or acceleration is a measurement of a displacing object with respect to some reference object. That is, any such sensor must contain components which are sensitive to a displacement. Such components are described in Chapter 5, however, while being adapted for conversion of velocity and acceleration, rather than just displacement, their design and fabrication methods are specifically geared to the dynamic measurements.

7.1 ELECTROMAGNETIC VELOCITY SENSORS

Moving a magnet though a coil of wire will induce a voltage in the coil according to Faraday's law. This voltage is proportional to the magnet's velocity and the field strength [Eq. (3.4.5)]. Linear velocity transducers use this principle of magnetic induction, with a permanent magnet and a fixed geometry coil, so the output voltage of the coil is directly proportional to the magnet's relative velocity over its working range.

In the velocity sensor, both ends of the magnet are inside the coil. With a single coil, this would give a zero output (see Fig. 3-4.1B) because the voltage generated by one end of the magnet

-347-

would cancel the voltage generated by the other end. To overcome this limitation, the coil is divided into two sections.

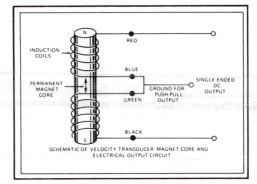

SCHEMATIC OF VELOCITY TRANSDUCER MAGNET CORE AND
ELECTRICAL OUTPUT CIRCUIT

Fig. 7-1 Operating principle of an electromagnetic velocity sensor
(Courtesy of Trans-Tek, Inc., Ellington, CT)

Table 7-1 Specification ranges of electromagnetic velocity sensors
(from Trans-Tek, Inc., Ellington, CT)

Characteristic	Value
Magnet core displacement, inches	0.5-24
Sensitivity, mV per inch/sec	35-500
Coil resistance, kΩ	2-45
Coil inductance, henry	0.06 - 7.5
Frequency response, Hz (at load >100·coil resistance)	500-1500
Weight, g	20-1500

The north pole of the magnet induces a current in one coil, while the south pole induces a current in the other coil (Fig. 7-1). The two coils are connected in a series-opposite direction to obtain an output proportional to the magnet's velocity. Maximum detectable velocity depends primarily on the input stages of the interface electronic circuit. Minimum detectable velocity depends on the noise floor, and especially of transmitted noise from nearby high ac current equipment. Typical specifications of an electromagnetic sensor are given in Table 7-1 and some practical assemblies are shown in Fig. 7-2. This design is very similar to LVDT position sensor (Section 5.5.1), except that LVDT is an active sensor with a moving ferromagnetic core, while the velocity sensor is a passive device with a moving permanent magnet. That is, this sensor is a current generating device which doesn't need an excitation signal.

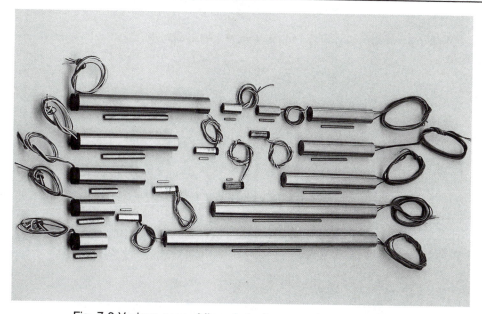

Fig. 7-2 Various assemblies of electromagnetic velocity sensors
(Courtesy of Trans-Tek, Inc.)

7.2 ACCELEROMETERS

Vibration is a dynamic mechanical phenomenon which involves periodic oscillatory motion around a reference position. In some cases (shock analysis, linear acceleration, etc.) the oscillating aspect may be missing, but the measurement and design of the sensor remains the same. An accelerometer can be specified as a single-degree-of-freedom device, which has some type of seismic mass, a spring-like supporting system, and a frame structure with damping properties (Fig. 3-18.1A).

A mathematical model of an accelerometer is represented by equation (3.18.9). To solve the equation, it is convenient to use Laplace transformation, which yields

$$Ms^2X(s) + bsX(s) + kX(s) = -MA(s) \quad , \qquad (7.1)$$

where $X(s)$ and $A(s)$ are the Laplace transforms of x(t) and d^2y/dt^2, respectively[1]. Solving the above for $X(s)$ we receive

$$X(s) = \frac{-MA(s)}{Ms^2 + bs + k} \quad . \qquad (7.2)$$

---

[1] d^2y/dt^2 is the input acceleration of the accelerometer body.

We introduce a conventional variable $\omega_0 = \sqrt{k/M}$, and $2\zeta\omega_0 = b/M$, then Eq. (7.2) can be expressed as:

$$X(s) = \frac{-A(s)}{s^2 + 2\zeta\omega_0 s + \omega_0^2} \ . \tag{7.3}$$

The value of ω_0 represents the accelerometer's angular natural frequency, and ζ is the normalized damping coefficient. Let us set

$$G(s) = \frac{-1}{s^2 + 2\zeta\omega_0 s + \omega_0^2} \ , \tag{7.4}$$

then, Eq. (7.3) becomes: $X(s) = G(s)A(s)$, and the solution can be expressed in terms of the inverse Laplace transform operator as

$$x(t) = \mathcal{L}^{-1}\{G(s)A(s)\} \ , \tag{7.5}$$

which from the convolution theorem for the Laplace transform can be expressed as

$$x(t) = \int_0^t g(t-\tau)a(\tau)d\tau \ , \tag{7.6}$$

where a is the time dependent impulse of the accelerometer body, and $g(t)$ is the inverse transform $\mathcal{L}^{-1}\{G(s)\}$. If we set $\omega = \omega_0\sqrt{1-\zeta^2}$, then the above has two solutions. One is for the underdamped mode ($\zeta < 1$)

$$x(t) = \int_0^t -\frac{1}{\omega}e^{-\zeta\omega_0(t-\tau)}\sin\omega(t-\tau)a(t)d\tau \ , \tag{7.7}$$

while for the overdamped mode ($\zeta > 1$)

$$x(t) = \int_0^t -\frac{1}{\omega}e^{-\zeta\omega_0(t-\tau)}\sinh\omega(t-\tau)a(t)d\tau \ , \tag{7.8}$$

where $\omega = \omega_0\sqrt{\zeta^2-1}$. The above solutions can be evaluated for different acceleration inputs applied to the accelerometer base. These can be found elsewhere [1].

A correctly designed, installed, and calibrated accelerometer should have one clearly identifiable resonant (natural) frequency, and a flat frequency response where the most accurate measurement can be made (Fig. 7-3). Within this flat region, as the vibrating frequency changes, the output of the sensor will correctly reflect the change without multiplying the signal by any variations in the frequency characteristic of the accelerometer. Viscous damping is used in many accelerometers to improve the useful frequency range by limiting effects of the resonant. As a damping medium, silicone oil is used quite often.

When calibrated, several characteristics of an accelerometer should be determined:

1. Sensitivity is the ratio of an electrical output to the mechanical input. It is usually expressed in terms of volts per unit of acceleration under the specified conditions. For instance, the sensitivity may specified as 1 V/g (unit of acceleration: $g = 9.80665$ m/s^2) at sea level, 45° lat. The sensitivity is typically measured at a single reference frequency of a sine-wave shape. In the U.S.A. it is 100 Hz, while in most European countries its is 160 Hz[1].

2. Frequency response is the outputs signal over a range of frequencies where the sensor should be operating. It is specified with respect to a reference frequency which is where the sensitivity is specified.

3. Resonant frequency in an undamped sensor shows as a clearly defined peak that can be 3-4dB higher than the response at the reference frequency. In a near-critically damped device the resonant may not be clearly visible, therefore, the phase shift is measured. At the resonant frequency, it is 180° of that at the reference frequency.

4. Zero stimulus output (for the capacitive and piezoresistive sensors) is specified for the position of the sensor where its sensitive (active) axis is perpendicular to Earth's gravity. That is, in the sensors which have a dc component in the output signal, the gravitational effect should be eliminated before the output at no mechanical input is determined.

5. Linearity of the accelerometer is specified over the dynamic range of the input signals.

When specifying an accelerometer for a particular application, one should answer a number questions, such as:

1. What is the anticipated magnitude of vibration or linear acceleration?

2. What is the operating temperature and how fast the ambient temperature may change?

3. What is the anticipated frequency range?

4. What linearity and accuracy are required?

5. What is the maximum tolerable size?

6. What kind of power supply is available?

7. Are any corrosive chemicals or high moisture present?

8. What is an anticipated overshock?

9. Are intense acoustic, electromagnetic, or electrostatic fields present?

10. Is the machinery grounded?

---

[1] These frequencies are chosen because they are removed from the power line frequencies and their harmonics.

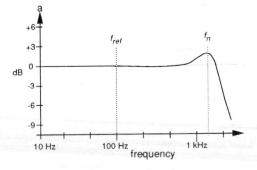

Fig. 7-3 A frequency response of an accelerometer
f_n is a natural frequency; f_{ref} is the reference frequency

7.2.1   Capacitive accelerometers

By definition, an accelerometer requires a component, whose movement lags behind that of the accelerometer's housing. This component is usually called either a seismic, or an internal mass. No matter what is the sensors' design or what is the conversion technique, an ultimate goal of the measurement is the detection of the mass displacement with respect to the housing. Hence, any suitable displacement sensor capable of measuring microscopic movements under strong vibrations can be used as an accelerometer. A capacitive displacement conversion is one of the proven and reliable methods. A capacitive acceleration sensor essentially contains at least two components, where the first is a "stationary" (i.e., connected to the housing) plate and the other is a plate attached to the internal mass. These plates form a capacitor whose value is a function of a distance d between the plates [Eq. (3.2.4)]. It is said, the capacitor value is modulated by the acceleration. A maximum displacement which is measured by the capacitive accelerometer rarely exceeds 20μm. Hence, such a small displacement requires a reliable compensation of drifts and interferences. This is usually accomplished by a differential technique, where an additional capacitor is formed in the same structure. The value of the second capacitor must be close to that of the first, and it should be subjected to changes with a 180°-phase shift. Then, an acceleration can be represented by a difference in values between the two capacitors.

Fig. 7-4A shows a cross-sectional diagram of a capacitive accelerometer where an internal mass is sandwiched between the upper cap and the base [2]. An entire sensor is micromachined from silicon. The mass is supported by four silicon springs (Fig. 7-4B). The upper plate and the base are separated from it by respective distances d_1 and d_2. All three parts are micromachined from a silicon wafer. Fig. 7-5 is a simplified circuit diagram for a capacitance-to-voltage converter which in many respects is similar to the circuit of Fig. 4-7.2.

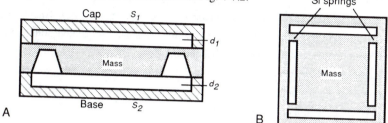

Fig. 7-4 A capacitive accelerometer with a differential capacitor
A: side cross-sectional view; B: top view of a seismic mass supported by four Si springs

A parallel plate capacitor C_{mc} between the mass and the cap electrodes has a plate area S_1. The plate spacing d_1 can be reduced by an amount Δ when the mass moves toward the upper plate. A second capacitor C_{mb} having a different plate area S_2 appears between the mass and the base. When mass moves toward the upper plate and away from the base, the spacing d_2 increases by Δ. The value of Δ is equal to the mechanical force F_m acting on the mass divided by the spring constant k of the silicon springs:

$$\Delta = \frac{F_m}{k} \ .$$

(7.9)

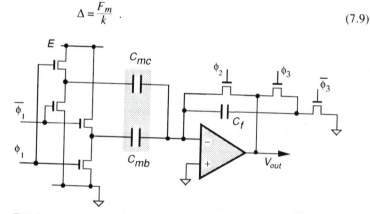

Fig. 7-5 A circuit diagram for a capacitance-to-voltage conversion
suitable for an integration on silicon

Strictly speaking, the accelerometer equivalent circuit is valid only when electrostatic forces do not affect the mass position, that is, when the capacitors depend linearly on F_m [3]. When an accelerometer serves as the input capacitors to a switched-capacitor summing amplifier, the output voltage depends on value of capacitors, and subsequently on force

$$V_{out} = 2E\frac{C_{mc} - C_{mb}}{C_f} \quad .$$
(7.10)

The above equation is true for small changes in sensor's capacitances. The accelerometer output is also a function of temperature and a capacitive mismatch. It is advisable that it is calibrated over an entire temperature range and an appropriate correction is made during the signal processing. Another effective method of assuring high stability is to design self-calibrating systems which makes use of electrostatic forces appearing in the accelerometer assembly when high voltage is applied to either a cap or a base electrode [2].

7.2.2 Piezoresistive accelerometers

As a sensing element a piezoresistive accelerometer incorporates strain gauges, which measure strain in mass-supporting springs. The strain can be directly correlated with the magnitude and rate of the mass displacement and, subsequently, with an acceleration. These devices can sense accelerations within a broad frequency range: from dc up to 13kHz. With a proper design, they can withstand overshock up to $10,000g$. Naturally, a dynamic range (span) is somewhat narrower ($\pm1,000g$ with error less than 1%). The overshock is a critical specification for many applications. However, piezoresistive accelerometers with discrete, epoxy-bonded strain gauges tend to have undesirable output temperature coefficients. Since they are manufactured separately, the gauges require individual thermal testing and parameter matching. This difficulty is virtually eliminated in modern sensors which use a micromachining technology of silicon wafers.

Table 7-2 Comparative characteristics of piezoresistive accelerometers

Characteristic	Endevco	Conventional
Die size (mm)	1.65 x 1.78	-
Range (g)	±1,000	±1,000
Sensitivity, typ. (mV/g)	0.2	0.1-0.25
Linearity (%)	1	1-3
Mounted resonant frequency (kHz)	65	25
Transverse sensitivity, max. (%)	3	3
Zero stimulus output (mV)	±25	±50
Operating temperature range (°C)	-54 to +135	-20 to +65
Shock survivability (g)	10,000	5,000
Weight (g)	0.8	1 to 5

An example of a wide dynamic range solid-state accelerometer is shown in Fig. 7-6. It was developed by Endevco/Allied Signal Aerospace Co. (Sunnyvale, CA). The microsensor is fabricated from three layers of silicon. The inner layer or the core consists of an internal mass, and the elastic hinge. The mass is suspended inside an etched rim on the hinge, which on either side has piezoelectric gauges. The gauges detect motion about the hinge. The outer two layers, the base and the lid, protect the moving parts from the external contamination. Both parts have recesses to allow the internal mass to move freely [4]. Several important features are incorporated into the sensor. One is that the sensitive axis lies in the plane of the silicon wafer, as opposed to many other designs where the axis is perpendicular to the wafer. A mechanical integrity and reliably are assured by the fabrication of all components of the sensor from a single silicon crystal.

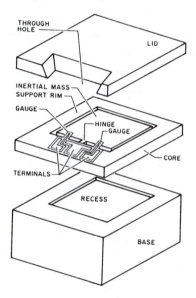

Fig. 7-6 Exposed view of a piezoresistive accelerometer
(Reprinted with permission from Sensors Magazine, ©1991]

When acceleration is applied along the sensitive axis, the internal mass rotates around the hinge. The gauges on both sides of the hinge allow rotation of the mass to create compressive stress on one gauge and tensile on the other. Since gauges are very short, even the small displacement produces large resistance changes. To trim the zero balance of the piezoresistive bridge, there are five trimming resistors positioned on the same crystal (not shown in the figure). A comparison of typical characteristics between some conventional and Endevco accelerometers are given in Table 7-2.

7.2.3   Piezoelectric accelerometers

The piezoelectric effect has a natural application in sensing vibration and acceleration. The effect is a direct conversion of mechanical energy into electrical (Section 3.6) in a crystalline material composed of electrical dipoles. These sensors operate from frequency as low as 2 Hz and up to about 5 kHz, they posses good off-axis noise rejection, high linearity, and a wide operating temperature range (up to 120°C). While quartz crystals are occasionally used as sensing elements, the most popular are ceramic piezoelectric materials, such as barium titanate, lead zirconite titanate (PZT), and lead metaniobite. A crystal is sandwiched between the case and the seismic mass which exerts on it the force proportional to the acceleration (Fig. 7-6). In miniature sensors, a silicon structure is usually employed. Since silicon does not possess piezoelectric properties, a thin film of lead titanate can be deposited on a micromachined silicon cantilever to fabricate an integral miniature sensor. For good frequency characteristics, a piezoelectric signal is amplified by a charge-to-voltage, or current-to-voltage converter which usually is built into the same housing as the piezoelectric crystal.

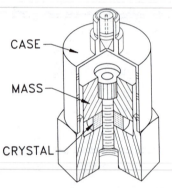

Fig. 7-6 A basic schematic of a piezoelectric accelerometer
An acceleration applied to the case moves it relative to the mass, which exerts a force on the crystal. The output is directly proportional to the acceleration or vibration level
(Reprinted with permission from Sensors Magazine, ©1992)

7.2.4   Thermal accelerometers

Since the basic idea behind an accelerometer is a sequential conversion of movement of seismic mass and measurement of its displacement, a fundamental formula of heat transfer can be used for that measurement [see Eq. (3.14.1)]. A thermal accelerometer, as any other accelerometer, contains a seismic mass which is suspended by a thin cantilever and positioned in close proximity with a heat sink, or between two heat sinks (Fig. 7-7) [5]. The mass and the cantilever structure are fabricated using a micromachined technology. The space between these components is filled with a thermally conductive gas. The mass is heated by a surface or imbedded heater to a defined temperature T_1. Under the no-acceleration conditions a thermal equilibrium is established between the mass and the heat sinks: the amounts of heat q_1 and q_2 conducted to the heat sinks through gas from the mass is a function of distances M_1 and M_2.

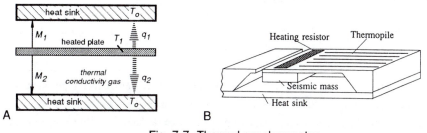

Fig. 7-7 Thermal accelerometer
A: a cross section of the heated part; B: an accelerometer design (shown without the roof)
(adapted from [5])

The temperature at any point in the cantilever beam supporting the seismic mass[1] depends on its distance from the support x and the gaps at the heat sinks. It can be found from

$$\frac{d^2T}{dx^2} - \lambda^2 T = 0 \quad , \tag{7.11}$$

where

$$\lambda = \sqrt{\frac{K_g(M_1 + M_2)}{K_{si}DM_1M_2}} \quad , \tag{7.12}$$

where K_g and K_{si} being thermal conductivities of gas and silicon respectively, and D is the thickness of a cantilever beam. For a boundary conditions, where the heat sink temperature is 0, a solution of the above equation for the temperature of the beam is

$$T(x) = \frac{P \text{singh}(\lambda x)}{WDK_{si}\lambda \cosh(\lambda L)} \quad , \tag{7.13}$$

where W and L is the width and length of the beam, and P is the thermal power. To measure that temperature, a temperature sensor can be deposited on the beam. It can be done by integrating sili-

---

[1] Here we assume steady-state conditions and neglect radiative and convective heat transfers.

cone diodes into the beam[1], or by forming serially connected thermocouples (a thermopile) on the beam surface. Eventually, the measured beam temperature in a form of an electrical signal is a measure of acceleration. The sensitivity of a thermal accelerometer (about 1% of change in the output signal per g) is somewhat smaller than that of the capacitive or piezoelectric types, however, it is much less susceptible to such interferences as ambient temperature or electromagnetic and electrostatic noise.

7.3 PIEZOELECTRIC CABLES

A piezoelectric effect is employed in a vibration sensor built with a mineral insulated cable. Such a cable generates an electric signal in its internal conductor when the outer surface of the cable is compressed. The piezoelectric Vibracoax™ cables[2] have been used in various experiments to monitor the vibration in compressor blades in turboshaft aircraft engines. Other applications include detection of insects in silos and automobile traffic analysis. In these applications, the cables are buried in the highway pavement, positioned perpendicular to the traffic. When properly installed, they last for at least five years [6]. The sensors are designed to be sensitive primarily to vertical forces. A piezoelectric cable consists of a solid insulated copper sheath having 3mm outer diameter, piezoelectric ceramic powder, and an inner copper core (Fig. 7-8A). The powder is tightly compressed between the outer sheath and the core. Usually, the cable is welded at one end and connected to a 50Ω extension cable at the other end.

Another method of fabrication of the piezoelectric cables is to use a PVDF polymer film as a component in the cable insulation (Fig. 7-8B). The PVDF can be made piezoelectric, thus giving the cable sensing properties. When a mechanical force is applied to the cable, the piezoelectric film is stressed, which results in the development of electric charges of the opposite polarities on it surfaces. The inner copper wire and the braided sheath serve as charge pick-up electrodes.

For the cable to possess piezoelectric properties, its sensing component (the ceramic powder or polymer film) must be poled during the manufacturing process. That is, the cable is warmed up to near the Curie temperature, and subjected to high voltage to orient ceramic dipoles in the powder, or polymer dipoles in the film, then cooled down while the high voltage is maintained. When the cable sensor is installed into the pavement (Fig. 7-9), its response should be calibrated, because the shape of the signal and its amplitude depend not only on the properties of the cable, but also on the type of the pavement and subgrade.

---

[1] See Chapter 16 for a description of a Si diode as a temperature sensor.
[2] Philips Electronic Instruments, Norcross, GA.

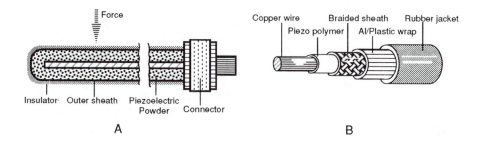

Fig. 7-8 Piezoelectric cable sensors

A - construction of Vibracoax; B - polymer film as a voltage generating component
(adapted from [7])

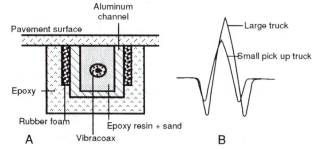

Fig. 7-9 Application of the piezoelectric cables in highway monitoring

A - Sensor installation in the pavement; B - shape of electrical response

REFERENCES

1. Articolo G. A. Shock impulse response of a force balance servo-accelerometer. *Sensors Expo West Proceedings*, 1989, © Helmers Publishing, Inc.

2. Sensor signal conditioning: an IC designer's perspective. *Sensors*, pp: 23-30. Nov. 1991

3. Allen, H., Terry, S. and De Bruin, D. Accelerometer System with self-testable features. *Sensors and Actuators*, 20, pp: 153-161, 1989

4. Suminto, J. T. A simple, high performance piezoresistive accelerometer. In: *Transducers'91. 1991 International conference on solid-state sensors and actuators. Digest of Technical Papers*. pp: 104-107, ©IEEE, 1991

5. Haritsuka, R., van Duyn, D.S., Otaredian, T., and de Vries, P. A novel accelerometer based on a silicon thermopile. In: *Transducers'91. International conference on solid-state sensors and actuators. Digest of technical papers*. pp: 420-423, ©IEEE, 1991

6. Bailleul, G. Vibracoax piezoelectric sensors for road traffic analysis. *Sensor Expo proceedings*, ©Helmers Publishing, Inc., 1991

7. Radice, P. F. Piezoelectric sensors and smart highways. In: *Sensors Expo Proceedings*, © Helmers Publishing, Inc., 1991

8
Force and Strain Sensors

The SI unit of force is derived from Newton's second law [Eq. (3.10.5)] and is one of the fundamental quantities of physics. The measurement of force is required in mechanical and civil engineering, for weighing objects, designing prosthesis, etc. Whenever pressure is measured, it requires the measurement of force. It could be said that force is measured when dealing with solids, while pressure - when dealing with fluids (i.e., liquids or gases). That is, force is considered when action is applied to a spot, and pressure is measured when force is distributed over a relatively large area.

Force sensors can be divided into two classes: quantitative and qualitative. A quantitive sensor actually measures the force and represents its value in terms of an electrical signal. Examples of these sensors are strain gauges and load cells. The qualitative sensors are threshold devices which are not concerned with good fidelity of representation of the force value. Their function is merely to indicate whether there is a sufficiently strong force is applied or not. That is, the output signal indicates when force magnitude exceeds a predetermined threshold level. An example of these detectors is a computer keyboard where a key makes a contact only when it is pressed sufficiently hard.

The various methods of sensing force can be categorized as follows [1]:

1. By balancing the unknown force against the gravitational force of a standard mass;

2. By measuring the acceleration of a known mass to which the force is applied;

3. By balancing the force against an electromagnetically developed force;

4. By converting the force to a fluid pressure and measuring that pressure;

5. By measuring the strain produced in a elastic member by the unknown force.

In the modern sensors the most commonly used method is *5*, while *3* and *4* are used occasionally.

In most sensors, force is not directly converted into an electric signal. Some intermediate steps are usually required. For instance, a force sensor can be fabricated by combining a position sensor and a force-to-displacement converter. The latter may be a simple coil spring, whose compression displacement x can be defined through the spring coefficient k and compressing force F as

$$x = kF \ . \tag{8.1}$$

The sensor shown in Fig. 8-1A is comprised of a spring and LVDT displacement sensor (Section 5.5.1). Within the linear range of the spring, the LVDT sensor produces voltage which is proportional to the applied force. A similar sensor can be constructed with other types of springs and pressure sensors, such as the one shown in Fig. 8-1B. The pressure sensor is combined with a fluid filled bellows which is subjected to force. The fluid-filled bellows functions as a force-to-pressure converter by distributing a localized force at its input over the sensing membrane of the pressure sensor.

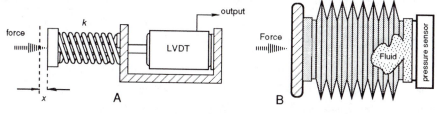

Fig. 8-1 A: Spring-loaded force sensor with LVDT
B: Force sensor with a pressure transducer

8.1 STRAIN GAUGES

A strain gauge is a resistive elastic sensor whose resistance is a function of applied strain (unit deformation). Since all materials resist to deformation, some force must be applied to cause deformation. Hence, resistance can be related to applied force. That relationship is generally called the *piezoresistive* effect (see Section 3.5) and is expressed through the gauge factor S_e of the conductor [Eq. (3.5.15)]:

$$\frac{dR}{R} = S_e e \ , \tag{8.2}$$

For many materials $S_e \approx 2$ with the exception of platinum for which $S_e \approx 6$ [2]. For small variations in resistance not exceeding 2% (which is usually the case), the resistance of the metallic wire is

$$R = R_0(1 + x) \; , \tag{8.3}$$

where R_0 is the resistance with no stress applied, and $x=S_e e$. For the semiconductive materials, the relationship depends on the doping concentration (Fig. 3-17.3B). Resistance decreases with compression and increases with tension. Characteristics of some resistance strain gauges are given in Table 8-1.

Table 8-1 Characteristics of some resistance strain gauges
(after [3])

Material	gauge factor, S_e	resistance, Ω	TCR ($°C^{-1} \cdot 10^{-6}$)	Notes
57% Cu – 43%Ni	2.0	100	10.8	S_e is constant over wide range of strain. For use under 260°C
Platinum alloys	4.0-6.0	50	2160	For high temperature use
Silicon	-100 to +150	200	90,000	High sensitivity, good for large strain measurements

A wire strain gauge is composed of a resistor bonded with an elastic carrier (backing). The backing, in turn, is applied to the object where stress or force should be measured. Obviously, that strain from the object must be reliably coupled to the gauge wire, while the wire must be electrically isolated from the object. The coefficient of thermal expansion of the backing should be matched to that of the wire. Many metals can be used to fabricate strain gauges. The most common materials are alloys constantan, nichrome, advance and karma. Typical resistances vary from 100 to several thousand ohms. To possess good sensitivity, the sensor should have long longitudinal and short transverse segments (Fig. 8-2), so that transverse sensitivity is no more than a couple of percent of the longitudinal. The gauges may be arranged in many ways to measure strains in different axes. Typically, they are connected into Wheatstone bridge circuits (Section 4.7). It should be noted, that semiconductive strain gauges are quite sensitive to temperature variations. Therefore, interface circuits or the gauges must contain temperature compensating networks.

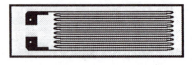

Fig. 8-2 Wire strain gauge bonded on elastic backing

8.2 TACTILE SENSORS

The tactile sensors are a special class of force or pressure transducers which are characterized by small thickness. This makes the sensors useful in the applications where force or pressure can be developed between two surfaces being in close proximity to one another. Examples include robotics, where tactile sensors can be positioned on the "fingertips" of a mechanical actuator to provide a feedback upon developing a contact with an object - very much like tactile sensors work in human skin. They can be used to fabricate "touch screen" displays, keyboards, and other devices where a physical contact has to be sensed. A very broad area of applications is in the biomedical field where tactile sensors can be used in dentistry for the crown or bridge occlusion investigation, in studies of forces developed by a human foot during locomotion. They can be installed in artificial knees for the balancing of the prosthesis operation, etc. In mechanical and civil engineering the sensors can be used to study forces developed by fastening devices.

Several methods can be used to fabricate tactile sensors. Some of them require a formation of a thin layer of a material which is responsive to strain. A simple tactile sensor producing an "on-off" output can be formed with two leaves of foil and a spacer (Fig. 8-3). The spacer has round (or any other suitable shape) holes. One leaf is grounded and the other is connected to a pull-up resistor. A multiplexer can be used if more than one sensing area is required. When an external force is applied to the upper conductor over the opening in the spacer layer, the conductor flexes and upon reaching the lower conductor, makes an electric contact, grounding by that the pull-up resistor. The output signal becomes zero indicating the applied force. The upper and lower conducting leaves can be fabricated by a silk-screen printing of conductive ink on the backing material, like Mylar® or polypropylene.

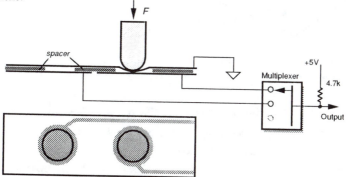

Fig. 8-3 Membrane switch as a tactile sensor

Good tactile sensors can be designed with piezoelectric films, such as polyvinylidene fluoride (PVDF) used in active or passive modes. An active ultrasonic coupling touch sensor with the piezoelectric films is illustrated in Fig. 8-4 where three films are laminated together (the sensors also has additional protective layers which are not shown in the figure). The upper and the bottom films are PVDF, while the center film is for the acoustic coupling between the other two. The softness of the center film determines sensitivity and the operating range of the sensor. The bottom piezoelectric film is driven by an ac voltage from an oscillator. This excitation signal results in mechanical contractions of the film which are coupled to the compression film and, in turn, to the upper piezoelectric film, which acts as a receiver. Since piezoelectricity is a reversible phenomenon, the upper film produces alternating voltage upon being subjected to mechanical vibrations from the compression film. These oscillations are amplified and fed into a synchronous demodulator. The demodulator is sensitive to both the amplitude and the phase of the received signal. When compressing force F is applied to the upper film, mechanical coupling between the three-layer assembly changes. This affects the amplitude and the phase of the received signal. These changes are recognized by the demodulator and appear at its output as a variable voltage.

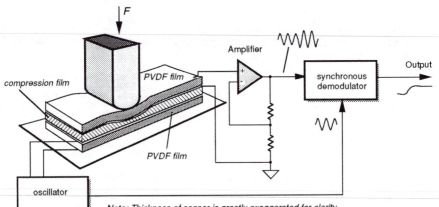

Fig. 8-4 An active piezoelectric tactile sensor

Within certain limits, the output signal linearly depends on the force. If 25μm PVDF films are laminated with a 40 mμ silicone rubber compression film, the thickness of an entire assembly (including protective layers) doesn't exceed 200μm. The PVDF film electrodes may be fabricated with a cell-like pattern on either the transmitting or receiving side. This would allow us to use electronic multiplexing of the cells to achieve spatial recognition of applied stimuli. The sensor also can be used to measure small displacements. Its accuracy is better than ±2μm over a few millimeter range. The advantages of this sensor is in its simplicity and a dc response, that is, in the ability to recognize static forces.

A piezoelectric tactile sensor can be fabricated with the PVDF film strips imbedded into a rubber skin (Fig. 8-5A). This sensor is passive, that its, its output signal is generated by the piezoelectric film without the need for an excitation signal. As a result, it produces a response proportional to the rate of stress, rather than to the stress magnitude. A design of this sensor is geared to robotic applications where it is desirable to sense sliding motions causing fast vibrations. The piezoelectric sensor is directly interfaced with a rubber skin, thus the electric signal produced by the strips reflect movements of the elastic rubber which results from the friction forces.

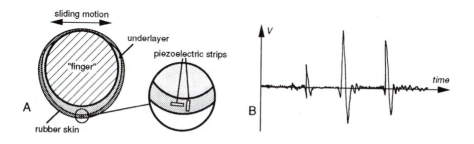

Fig. 8-5 Tactile sensor with a piezoelectric film for detecting sliding forces
A: a cross sectional view; B: a typical response
(adapted from [4])

The sensor is built on a rigid structure (a robot's "finger") which has a foamy compliant underlayer (1 mm thick), around which a silicon rubber "skin" is wrapped. It is also possible to use a fluid underlayer for better smooth surface tracking. Because the sensing strips are located at some depth beneath the skin surface, and because the piezoelectric film responds differently in different directions, a signal magnitude is not the same for movements in any direction. The sensor responds with a bipolar signal (Fig. 8-5B) to surface discontinuity or bumps as low as 50μm high.

A resistive tactile sensor can be fabricated by using materials whose electrical resistance is a function of strain. The sensor incorporates a force sensitive resistor (FSR) whose resistance varies with applied pressure. Such materials are conductive elastomers or pressure sensitive inks. A conductive elastomer is fabricated of silicone rubber, polyurethane and other compounds which are impregnated with conductive particles or fibers. For instance, conductive rubber can be fabricated by using carbon powder as an impregnating material. Operating principles of elastomeric tactile sensors are based either on varying the contact area when the elastomer is squeezed between two conductive plates (Fig. 8-6A) or in changing the thickness. When the external force varies, the contact area at the interface between the pusher and the elastomer changes, resulting in a reduction of electrical resistance (Fig. 8-6B).

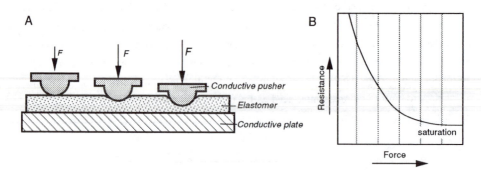

Fig. 8-6   FSR tactile sensor
A: a through-thickness application with an elastomer; B: transfer function

At a certain pressure, the contact area reaches its maximum and the transfer function (Fig. 8-6B) goes to saturation. This sensor is quite useful in some applications, however, for robotics and medical use its disadvantage is in relatively large thickness, which at best can be made on the order of 1mm. A much thinner sensor can be fabricated with a semiconducting polymer whose resistance varies with pressure. A design of the sensor resembles a membrane switch (Fig. 8-7) [5]. Compared with a strain gauge, the FSR has a much wider dynamic range: typically 3 decades of resistance change over a 0-3kg force range and much lower accuracy (typically ±10%). However, in many applications, where an accurate force measurement is not required, a very low cost of the sensor makes it an attractive alternative. A typical thickness of a FSR polymer sensor is on the range of 0.25mm (0.010").

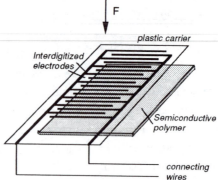

Fig. 8-7 Tactile sensor with a polymer FSR

Miniature tactile sensors are especially in high demand in robotics, where good spatial resolution, high sensitivity, and a wide dynamic range are required. A plastic deformation in silicon can be used for the fabrication of a threshold tactile sensor with a mechanical hysteresis. In one design [6], the expansion of trapped gas in a sealed cavity formed by wafer bonding is used to plastically deform a thin silicon membrane bonded over the cavity, creating a spherically shaped cap. The structure shown in Fig. 8-8 is fabricated by a micromachining technology of a silicon wafer. At normal room temperature and above a critical force, the upper electrode will buckle downward, making contact with the lower electrode.

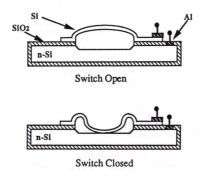

Fig. 8-8 Micromachined silicon threshold switch with trapped gas
(from [6])

Experiments have shown that the switch has hysteresis of about 2 psi of pressure with a closing action near 13psi. The closing resistance of the switch is on the order of 10kΩ, which for the micropower circuits is usually low enough.

In another design, a vacuum, instead of pressurized gas, is used in a microcavity. This sensor shown in Fig. 8-9 [7] has a silicon vacuum configuration, with a cold field emission cathode and a movable diaphragm anode. The cathode is a sharp silicon tip. When a positive potential difference is applied between the tip and the anode, an electric field is generated, which allows electrons to tunnel from inside the cathode to the vacuum, if the field exceeds $5 \cdot 10^7$V/cm [8]. The field strength at the tip and quantity of electrons emitted (emission current) are controlled by the anode potential. When an external force is applied, the anode deflects downward, thus changing the field and the emission current.

The emission current can be expressed through the anode voltage V as

$$I = V^2 a \cdot \exp(-\frac{b}{\beta V}) \quad , \tag{8.4}$$

where a and b are constants, and β is the tip geometry factor, which depends on the distance between the anode and the cathode. To achieve a better sensitivity, the tip is fabricated with a radius of curvature of about 0.02 μm.

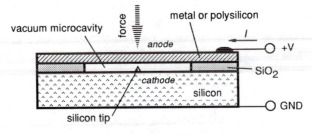

Fig. 8-9 Schematic of a vacuum diode force sensor
(adapted from [7])

8.3   QUARTZ FORCE SENSOR

A thin quartz crystal can be adapted to measure force (and pressure) over a relatively narrow range from 0 to 1.5 kg, however with a good linearity and over 11-bit resolution. A basic idea behind the sensor's operation is that certain cuts of quartz crystal, when used as resonators in electronic oscillators, shift the resonant frequency upon being mechanically loaded. To fabricate the sensor, a rectangular plate is cut of the crystal where only one edge is parallel to the x axis, and the face of the plate is cut at the angle of approximately $\Theta=35°$ with respect to the z axis. This cut is commonly known as AT-cut plate (Fig. 8-10A).

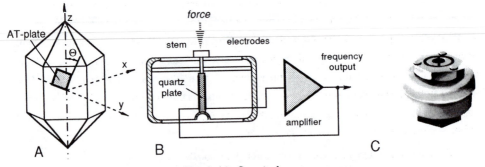

Fig. 8-10 Quartz force sensor
A: AT-cut of a quartz crystal; B: structure of the sensor; C: is the outside appearance
(Courtesy of Quartzcell, Santa Barbara, CA)

The plate is given surface electrodes for utilizing a piezoelectric effect (see Fig. 3-6.2), which are connected in a positive feedback of an oscillator (Fig. 8-10B). A quartz crystal oscillates at a fundamental frequency f_o (unloaded) which shifts at loading by [9]

$$\Delta f = F \frac{K f_o^2 n}{D} \quad , \tag{8.5}$$

where F is the applied force, K is a constant, n is the number of the overtone mode, and D is the size of the crystal. To compensate for frequency variations due to temperature effects, a double crystal can be employed, where one half is used for a temperature compensation. Each resonator is connected into its own oscillating circuit and the resulting frequencies are subtracted, thus negating a temperature effect. A commercial force sensor is shown in Fig. 8-10C.

8.4 PIEZOELECTRIC FORCE SENSORS

A simple piezoelectric sensor is shown in Fig. 8-11. This sensor is used in a passive mode. That is, it directly converts mechanical stress into an electrical signal. This however, only makes it sensitive to changing stimuli only and insensitive to a constant force. The sensor consists of three layers where the PVDF film is laminated between a backing material (for instance, silicone rubber) and a pushing layer.

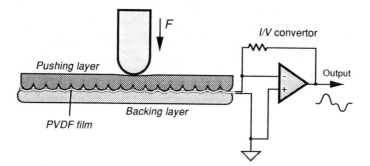

Fig. 8-11 Piezoelectric force rate sensor

The pushing layer is fabricated of a plastic film (for instance, Mylar®) whose side facing the PVDF film is preformed to have a corrugated surface. Upon touching the sensor, the PVDF film is stressed by the groves of the pusher. This results in a generation by the film of electric charge. The charge flows out of the film through a current-to-voltage (I/V) converter which produces variable output voltage. The amplitude of that voltage within certain limits is proportional to the applied force. A modified version of such a sensor was built for a medical application where minute movements of a sleeping infant had to be monitored in order to detect cessation of its breathing [10]. The sensor was placed under the mattress in a crib. A body of a normally breathing infant slightly shifts with each inhale and exhale due to a moving diaphragm. This results in a displacement of the body's center of gravity which is detected by the sensor.

REFERENCES

1. Doebelin, E. O. Measurement systems: applications and design. McGraw-Hill, New York, 1966

2. Pallás-Areny, R. and Webster, J. G. Sensors and signal conditioning, John Wiley & Sons, Inc., 1991

3. Holman J. P. Experimental methods for engineers. McGraw-Hill Book Co., New York, 1978

4. Howe, R. T. Surface micromachining for microsensors and microactuators. *J. Vac. Sci. Technol. B*, vol. 6, No. 6, American Institute of Physics, pp: 1809-1813, Nov.-Dec. 1988

5. Yates, B. A keyboard controlled joystick using force sensing resistor. *Sensors*, pp: 38-39, April 1991

6. Huff, M.A., Nikolich, A.D., and Schmidt, M.A. A threshold pressure switch utilizing plastic deformation of silicon. In: *Transducers'91. International conference on solid-state sensors and actuators. Digest of technical papers*, pp: 177-180, ©IEEE, 1991

7. Jiang, J.C., White, R.C. and Allen, P.K. Microcavity vacuum tube pressure sensor for robotic tactile sensing. In: *Transducers'91. International conference on solid-state sensors and actuators. Digest of technical papers*, pp: 238-240, IEEE, 1991

8. Brodie, I. Physical considerations in vacuum microelectronics devices. In: *IEEE trans. Electron. Devices*, vol. 36, p: 2641, 1989

9. Corbett, J.P. Quatz steady-state force and pressure sensor. In: *Sensors Expo West Proceedings*, paper 304A-1, © Helmers Publishing, Inc., Peterborough, NH, 1991

10. Fraden, J. Cardio-Respiration Transducer, U.S. Patent No. 4,509,527, April 9, 1985

9

Pressure Sensors

9.1 CONCEPTS OF PRESSURE

The pressure concept was primarily based on the pioneering work of Evangelista Torricelli who for a short time was a student of Galileo [1]. During his experiments with mercury filled dishes, in 1643, he realized that the atmosphere exerts pressure on earth. Another great experimenter Blaise Pascal, in 1647, conducted an experiment with the help of his brother-in-law, Perier, on the top of the mountain *Puy de Dome* and at its base. He observed that pressure exerted on the column of mercury depends on elevation. He named a mercury-in-vacuum instrument they used in the experiment the *barometer*. In 1660, Robert Boyle stated his famous relationship: "*The product of the measures of pressure and volume is constant for a given mass of air at fixed temperature*". In 1738 Daniel Bernoulli developed an impact theory of gas pressure to the point where Boyle's law could be deducted analytically. Bernoulli also anticipated the Charles-Gay-Lussac law by stating that pressure is increased by heating gas at a constant volume. For a detailed description of gas and fluid dynamics a reader should be referred to one of the many books on the fundamentals of physics. Below, we briefly summarize the basics which are essential for design and use of pressure sensors.

In general terms, matter can be classified into solids and fluids. The word *fluid* describes something which can flow. That includes liquids and gases. The distinction between liquids and gases are not quite definite. By varying pressure it is possible to change liquid into gas and vice versa. It is impossible to apply pressure to fluid in any direction except normal to its surface. At any angle, except 90°, fluid will just slide over, or flow. Therefore, any force applied to fluid is tangential and the pressure exerted on boundaries is normal to the surface. For a fluid at rest, pressure can be defined as the force F exerted perpendicularly on a unit area A of a boundary surface [2]:

$$p = \frac{dF}{dA} \, . \tag{9.1}$$

Pressure is basically a mechanical concept that can be fully described in terms of the primary dimensions of mass, length, and time. It is a familiar fact that pressure is strongly influenced by the position within the boundaries, however at a given position it is quite independent of direction. We note the expected variations in pressure with elevation

$$dp = -wdh \, , \tag{9.2}$$

where w is the specific weight of the medium, and h represents the vertical height.

Pressure is unaffected by the shape of the confining boundaries. Thus a great variety of pressure sensors can be designed without a concern of shape and dimensions. If pressure is applied to one of the sides of the surface confining fluid or gas, pressure is transferred to the entire surface without diminishing in value.

Kinetic theory of gases states that pressure can be viewed as a measure of the total kinetic energy of the molecules

$$p = \frac{2}{3}\frac{KE}{V} = \frac{1}{3}\rho C^2 = NRT \, , \tag{9.3}$$

where KE is the kinetic energy, V is the volume, C^2 is an average value of the square of the molecular velocities, ρ is the density, N is the number of molecules per unit volume, R is a specific gas constant and T is the absolute temperature.

Eq. (9.3) suggests that pressure and density of compressible fluids (gases) are linearly related. The increase in pressure results in the proportional increase in density. For example, at $0°C$ and 1atm pressure, air has a density of $1.3kg/m^3$, while at the same temperature and 50atm of pressure its density is $65kg/m^3$ which is 50 times higher. Contrary, for liquids the density varies very little over ranges of pressure and temperature. For instance, water at $0°C$ and 1atm has a density of $1000kg/m^3$, while at $0°C$ and 50atm its density is $1002kg/m^3$, and at $100°C$ and 1atm its density is $958kg/m^3$.

Whether gas pressure is above or below pressure of ambient air we speak about overpressure or vacuum. Pressure is called relative when it is measured with respect to ambient pressure. It is called absolute when it is measured with respect to a vacuum at 0 pressure. The pressure of a medium may be static when it is referred to fluid at rest, or dynamic when it is referred to kinetic energy of a moving fluid.

9.2 UNITS OF PRESSURE

The SI unit of pressure is the *pascal*: $1Pa=1N/m^2$. That is, one pascal is equal to one newton force uniformly distributed over 1 square meter of surface. Sometimes, in technical systems, the *atmosphere* is used which is denoted 1atm. One atmosphere is the pressure exerted on 1 square cm by a column of water having height of 1 meter at a temperature of $+4°C$ and normal gravitational acceleration. 1Pa may be converted into other units by use of the following relationships (see also Tables 1-10)

$$1\ Pa = 1.45 \cdot 10^{-4}\ lb/in^2 = 9.869 \cdot 10^{-6}\ atm = 7.5 \cdot 10^{-4}\ cm\text{-}Hg \ .$$

For practical estimation, it is useful to remember that 0.1 mm H_2O is roughly equal to 1 Pa. In industry, another unit of pressure is often used. It is defined as pressure exerted by 1 mm column of mercury at $0°C$ at normal atmospheric pressure and normal gravity. This unit is named after Torricelli and is called the torr. The ideal pressure of the Earth atmosphere is 760 torr and is called the *physical atmosphere*

$$1\ atm = 760\ torr = 101325\ Pa.$$

The US customary system of units defines pressure as a pound per square inch (lbs/sq) or psi. Conversion into SI systems is the following

$$1\ psi = 6.89 \cdot 10^3 Pa = 0.0703\ atm \ .$$

A pressure sensor operating principle is based on the conversion of a result of the pressure exertion on a sensitive element into an electrical signal. Virtually in all cases, pressure results in the displacement of an element, having a defined surface area. Thus, a pressure measurement may be reduced to a measurement of a displacement or force, which results from a displacement. Thus, we recommend that the reader also familiarizes oneself with displacement sensors covered in Chap. 5 and force sensors of Chap. 8.

9.3 MERCURY PRESSURE SENSOR

A simple yet efficient sensor is based on communicating vessels principle (Fig. 9-1). Its prime use is for the measurement of gas pressure. A U-shaped wire is immersed into mercury which shorts its resistance in proportion with the height of mercury in each column. The resistors are connected into a Wheatstone bridge circuit which remains in balance as long as the differential pressure in the tube is zero. Pressure is applied to one of the arms of the tube and disbalances the bridge which results in the output signal. The higher the pressure in the left tube the higher is resistance of the corresponding arm and the lower is the resistance of the opposite arm. The output voltage is proportional to a difference in resistances ΔR of the wire arms which are not shunted by mercury:

$$V_{out} = V\frac{\Delta R}{R} = V\beta\Delta p \ . \tag{9.4}$$

The sensor can be directly calibrated in units of torr. While being simple, this sensor suffers from several drawbacks, such as necessity of precision leveling, susceptibility to shocks and vibration, and contamination of gas by mercury vapors.

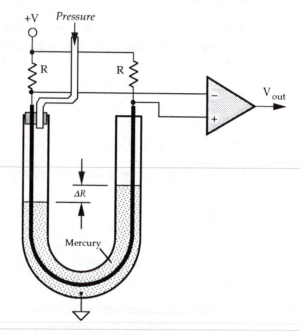

Fig. 9-1 Mercury-filled U-shaped sensor for measuring gas pressure

9.4 BELLOWS, MEMBRANES, AND THIN PLATES

In pressure sensors, a sensing element is a mechanical device which undergoes structural changes under strain. Historically, such devices were bourdon tubes (C-shaped, twisted and helical), corrugated [3] and catenary diaphragms, capsules, bellows, barrel tubes, and other components whose shape was changing under pressure.

A bellows (Fig. 9-2) is intended for the conversion of pressure into a linear displacement which can be measured by an appropriate sensor. Thus, bellows performs a first step in the conversion of pressure into an electrical signal. It is characterized by a relatively large surface area and, therefore, by a large displacement at low pressures. The stiffness of seamless metallic bellows is proportional to the Young's modulus of the material and inversely proportional to the outside diameter and to the number of convolutions of the bellows. Stiffness also increases with roughly the third power of the wall thickness.

Fig. 9.2 Steel bellows for a pressure transducer (x2 magnification)
(fabricated by Servometer Corp., Cedar Grove, NJ)

A popular example of pressure conversion into a linear deflection is a diaphragm in an aneroid barometer. A deflecting device always forms at least one wall of a pressure chamber and is coupled to a strain sensor (for instance, a strain gauge) which converts deflection into electrical signals. Nowadays, a great majority of pressure sensors are fabricated with silicon membranes by using a micromachining technology.

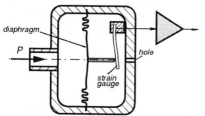

Fig. 9-3 Metal corrugated diaphragm
ɔr conversion of pressure into linear deflection

A membrane is a thin diaphragm under radial tension S which is measured in N/m (Fig. 9-4). The stiffness to bending forces can be neglected as the thickness of the membrane is much smaller as compared with its radius (at least 200 times smaller). When pressure is applied to one side of a membrane, it shapes spherically, like a soap bubble. At low pressure p differences, the center deflection z_m and the stress σ_m[1] are quasi-linear functions of pressure

$$z_{max} = \frac{r^2 p}{4S} \quad , \tag{9.5}$$

$$\sigma_{max} \approx \frac{S}{g} \quad , \tag{9.6}$$

where r is the membrane radius and g is the thickness. Stress is generally uniform over the membrane area.

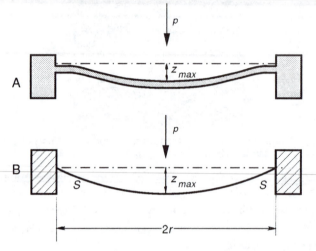

Fig. 9-4 Thin plate (A) and membrane (B) under pressure p

For the membrane, the lowest natural frequency can be calculated from equation [4]

$$f_o = \frac{1.2}{\pi r} \sqrt{\frac{S}{\rho g}} \quad , \tag{9.7}$$

where ρ is the membrane material density.

If the thickness of the membrane is not negligibly small (r/g ratio is 100 or less), the membrane is called a *thin plate* (Fig. 9-4). If the plate is compressed between some kind of clamping rings, it exhibits a noticeable hysteresis due to friction between the thin plate and the clamping rings. A much better arrangement is a one-piece structure where the plate and the supporting components are fabricated of a single bulk of material.

---

[1] stress is measured in $\frac{N}{m^2}$.

For a plate, the maximum deflection is also linearly related to pressure

$$z_{max} = \frac{3(1 - v^2)r^4 p}{16Eg^3},$$

$$(9.8)$$

where E is Young's modulus (N/m^2) and v is Poisson's ratio. The maximum stress at the circumference is also a linear function of pressure:

$$\sigma_{max} \approx \frac{3r^2 p}{4g^2}.$$

$$(9.9)$$

The above equations suggest that a pressure sensor can be designed by exploiting membrane and thin plate deflections. The next question is: what physical effect to use for the conversion of deflection into an electrical signal. There are several options which we discuss below.

9.5 PIEZORESISTIVE SENSORS

To make a pressure sensor, two essential components are required. They are the plate (membrane) having known area A and a detector which responds to applied force F [Eq. (9.1)]. Both these components can be fabricated of silicon. A silicon-diaphragm pressure sensor consists of a thin silicon diaphragm as an elastic material [5] and piezoresistive gauge resistors made by diffusive impurities into the diaphragm. Thanks to single crystal silicon superior elastic characteristics, virtually no creep and no hysteresis occur, even under strong static pressure. The gauge factor of silicon is many times stronger than that of thin metal conductors [6]. It is customary to fabricate strain gauge resistors connected as the Wheatstone bridge. The full scale output of such a circuit is on the order of several hundred millivolts, thus a signal conditioner is required for bringing the output to an acceptable format. Further, silicon resistors exhibit quite strong temperature sensitivity, therefore, a conditioning circuit should include temperature compensation.

When stress is applied to a semiconductor resistor, having initial resistance R piezoresistive effect results in change in the resistance ΔR [7]

$$\frac{\Delta R}{R} = \pi_l \sigma_l + \pi_t \sigma_t,$$

$$(9.10)$$

where π_l and π_t are the piezoresistive coefficients in a longitudinal and transverse directions, respectively. Stresses in longitudinal and transverse directions are designated σ_l and σ_t. The π-coefficients depend on the orientation of resistors on the silicon crystal. Thus, for p-type diffused resistor arranged in the <110> direction or an n-type silicon square diaphragm with (100) surface orientation as shown in Fig. 9-5, the coefficients are approximately denoted as [7]

$$\pi_1 = -\pi_t = \frac{1}{2}\pi_{44} \ . \tag{9.11}$$

A change in resistivity is proportional to applied stress and, subsequently, to applied pressure. The resistors positioned on the diaphragm in such a manner as to have the longitudinal and transverse coefficients of the opposite polarities, therefore resistors change in the opposite directions:

$$\frac{\Delta R_1}{R_1} = -\frac{\Delta R_2}{R_2} = \frac{1}{2}\pi_{44}(\sigma_{1y} - \sigma_{1x}) \ . \tag{9.12}$$

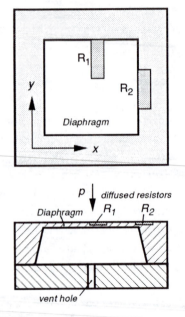

Fig. 9-5 Position of piezoresistors on a silicon diaphragm

When connecting R_1 and R_2 in a half-bridge circuit, and exciting the bridge with E, the output voltage V_{out} is

$$V_{out} = \frac{1}{4}E\ \pi_{44}(\sigma_{1y} - \sigma_{1x}) \ . \tag{9.13}$$

As a result, pressure sensitivity a_p and temperature sensitivity of the circuit b_T can be found by taking partial derivatives:

$$a_p = \frac{1}{E}\frac{\partial V_{out}}{\partial p} = \frac{\pi_{44}}{4}\frac{\partial(\sigma_{1y} - \sigma_{1x})}{\partial p} \quad , \tag{9.14}$$

$$b_T = \frac{1}{a_p}\frac{\partial a_p}{\partial T} = \frac{1}{\pi_{44}}\frac{\partial \pi_{44}}{\partial T} \quad . \tag{9.15}$$

Since $\partial\pi_{44}/\partial T$ has a negative value, the temperature coefficient of sensitivity is negative, that is, sensitivity decreases at higher temperatures.

There are several methods of fabrication which can be used for the silicon pressure sensor processing. In one method [8], the starting material is n-type silicon substrate with (100) surface orientation. Piezoresistors with $3\cdot10^{18}$ cm^{-3} surface-impurity concentration are fabricated using a boron ion implantation. One of them (R_1) is parallel and the other is perpendicular to the <110> diaphragm orientation. Other peripheral components, like resistors and pn-junctions used for temperature compensation are also fabricated during the same implantation process as that for the piezoresistors. They are positioned in a thick-rim area surrounding the diaphragm. Thus, they are insensitive to pressure applied to the diaphragm.

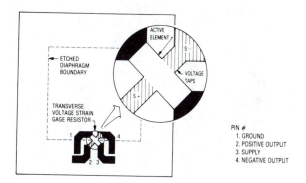

**Fig. 9-6 Basic uncompensated piezoresistive element
of Motorola MPX pressure sensor**
(Copyright of Motorola, Inc. Used with permission)

Another approach of stress sensing was used in Motorola MPX pressure sensor chip shown in Fig. 9-6 The piezoresistive element, which constitutes a strain gauge, is ion implanted on a thin silicon diaphragm. Excitation current is passed longitudinally through the resistor's taps 1 and 3, and the pressure that stresses the diaphragm is applied at a right angle to the current flow. The stress establishes a transverse electric field in the resistor that is sensed as voltage at taps 2 and 4. The single-element transverse voltage strain gauge can be viewed as the mechanical analog of a Hall effect device (Section 3.8). Using a single element eliminates the need to closely match the four stress and temperature sensitive resistors that form a Wheatstone bridge design. At the same time, it greatly simplifies the additional circuitry necessary to accomplish calibration and temperature compensation. Nevertheless, the single element strain gauge electrically is analogous to the bridge circuit. Its balance (offset) does not depend on matched resistors, as it would be in a conventional bridge, but instead on how well the transverse voltage taps are aligned. This alignment is accomplished in a single photolithographic step, making it easier to control. To make the designer's job easier, Motorola incorporated into the chip not only temperature compensation and calibration components, but signal conditioning and interfacing elements as well (Fig. 9-7).

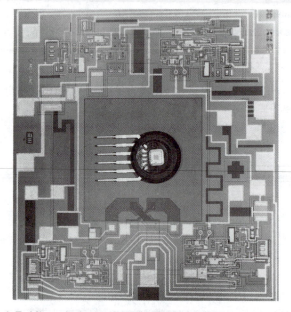

Fig. 9-7 Microphotograph of the integrated pressures sensor
(a chip in a plastic carrier is superimposed on the diaphragm)
Note the piezoresistive element at the bottom side of the diaphragm
(Copyright of Motorola, Inc. Used with permission)

A thin diaphragm with $1mm^2$ area size may be formed by using one of the commonly used silicon etching solutions, for instance hydrazine-water ($N_2H_4 \cdot H_2O$) anisotropic etchant. A SiO_2 or Si_3N_4 layer serves as an etch mask and the protective layer on the bottom side of the wafer. The etching time is about $1.7\mu m/min$ at 90°C in reflux solution. The final diaphragm thickness is achieved at about $30\mu m$.

Another method of diaphragm fabrication is based on the so-called silicon fusion bonding (SFB) where single crystal silicon wafers can be reliably bonded with near-perfect interfaces without the use of intermediate layers [9]. This technique allows the making of very small sensors which find use in catheter-tip transducers for medical *in vivo* measurements. Total chip area may be as much as 8 times smaller than that of the conventional silicon-diaphragm pressure sensor. The sensor consists of two parts - the bottom and the top wafers (Fig. 9-8A). The bottom constraint wafer (substrate) is first anisotropically etched with a square hole which has desirable dimensions of the diaphragm. The bottom wafer has thickness about 0.5 mm and the diaphragm has side dimensions of 250 μm, so the anisotropic etch forms a pyramidal cavity with a depth of about 175μm. The next step is SFB bonding to a top wafer consisting of a *p*-type substrate with an *n*-type epi layer. The thickness of the epi layer corresponds to the desired final thickness of the diaphragm. Then, the bulk of the top wafer is removed by a controlled-etch process, leaving a bonded-on single crystal layer of silicon which forms the sensor's diaphragm. Next, resistors are ion implanted, contact vias are etched, and is deposited and etched to a desired pattern. In the final step, the constrain wafer is ground and polished back to the desired thickness of the device; about 140 μm. Despite the act that the dimensions of the SFB chip are about half of those of the conventional chip, their pressure sensitivities are identical. A comparison of conventional and SFB technology is shown in Fig. 9-8B. For the same diaphragm dimensions and the same overall thickness of the chip, the SFB device is about 50% smaller.

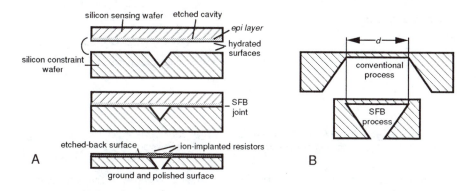

Fig. 9-8 Silicon fusion bonding method of a silicon membrane fabrications
A: productions steps; B: comparison of an SFB chip size with a conventionally fabricated diaphragm

Pressure sensors are usually available in three basic configurations that permit measurement of *absolute*, *differential*, and *gauge* pressures. Absolute pressure, such a barometric pressure, is measured with respect to a reference vacuum chamber. The chamber may be either external, or it can be built directly into the sensor (Fig. 9-9A). A differential pressure, such as the pressure drop in a pressure-differential flowmeter, is measured by applying pressure to opposite sides of the diaphragm simultaneously. Gauge pressure is measured with respect to some kind of reference pressure. An example is a blood pressure measurement which is done with respect to atmospheric pressure. Thus, gauge pressure is a special case of a differential pressure. Diaphragm and strain gauge designs are the same for all three configurations, while the packaging makes them different. For example, to make a differential or gauge sensor, a silicon die is positioned inside the chamber (Fig. 9-9B) which has two openings at both sides of the die. To protect them from harsh environment, the interior of the housing is filled with a silicone gel which isolates the die surface and wire bonds, while allowing the pressure signal to be coupled to the silicon diaphragm. A differential sensor may be incorporated into various porting holders (Fig. 9-10). Certain applications, such as a hot water hammer, corrosive fluids, and load cells, require physical isolation and hydraulic coupling to the chip-carrier package. It can be done with additional diaphragms and bellows. In either case, silicon oil, such as Dow Corning DS200, can be used to fill the air cavity so that system frequency response is not degraded.

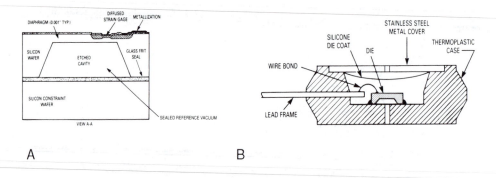

A   B

Fig. 9-9 Absolute (A) and differential (B) pressure sensor packagings
(Copyright of Motorola, Inc. Used with permission)

All silicon-based sensors are characterized by temperature dependence. Temperature coefficient of sensitivity b_T as defined by Eq. (9.15) is usually negative and for the accurate pressure sensing it must be compensated for. Typical methods of temperature compensation of bridge circuits are covered in Section 4.7.3. Without the compensation, sensor's output voltage may look like the one shown in Fig. 9-11A for three different temperatures.

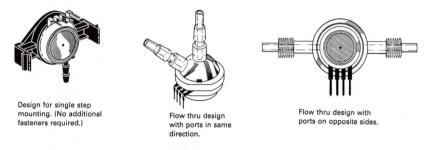

Design for single step
mounting. (No additional
fasteners required.)

Flow thru design
with ports in same
direction.

Flow thru design with
ports on opposite sides.

Fig. 9-10 Examples of differential pressure packagings
(Copyright of Motorola, Inc. Used with permission)

In many applications, a simple yet efficient temperature compensation can be accomplished by adding to the sensor either a series or parallel temperature stable resistor. By selecting an appropriate value of the resistor, the sensor's output can be tailored to the desirable operating range (Fig. 9-11B). Whenever a better temperature correction over a broad range is required, more complex compensation circuits with temperature detectors can be employed. A viable alternative is a software compensation where the temperature of the pressure transducer is measured by an imbedded temperature sensor. Both data from the pressure and temperature sensors relayed to the processing circuit where numerical compensation is digitally performed.

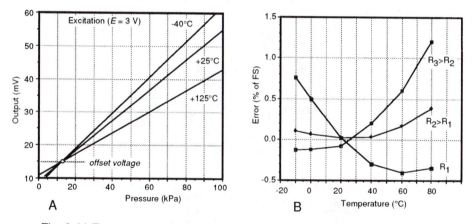

Fig. 9-11 Temperature characteristics of a piezoresistive pressure sensor
A: transfer function at three different temperatures;
B: full-scale errors for three values of compensating resistors

9.6 CAPACITIVE SENSORS

A silicon diaphragm can be used with another pressure-to-electric output conversion process: in a capacitive sensor. Here, the diaphragm displacement modulates capacitance with respect to the reference plate (backplate). This conversion is especially effective for the low pressure sensors. An entire sensor can be fabricated from a solid piece of silicon, thus maximizing its operational stability. The diaphragm can be designed to produce up to 25% capacitance change over the full range which makes these sensors candidates for direct digitization (see Section 4.5). While a piezoresistive diaphragm should be designed to maximize stress at its edges, the capacitive diaphragm utilizes a displacement of its central portion. These diaphragms can be protected against overpressure by including mechanical stops close to either side of the diaphragm (for a differential pressure sensor). Unfortunately, in the piezoresistive diaphragms, the same protection is not quite effective because of small operational displacements. As a result, the piezoresistive sensors typically have burst pressures of about ten times the full scale rating, while capacitive sensors with overpressure stops can handle a thousand times the rated full scale pressure. This is especially important for the low pressure applications, where relatively high pressure pulses can occur.

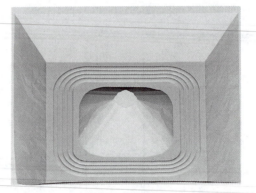

Fig. 9-12 Backside view of the corrugated and bossed diaphragm
for a capacitive pressure sensor
(Courtesy of Monolithic Sensors, Inc. Rolling Meadows, IL)

While designing a capacitive pressure sensor, for good linearity it is important to maintain flatness of the diaphragm. Traditionally, these sensors are linear only over the displacements which are much less then their thickness. One way to improve the linear range is to make a diaphragm with groves and corrugations by applying a micromachining technology. In addition, Monolithic Sensors, Inc. has added a boss to maintain a parallel plate action of the capacitor (Fig. 9-12) The combination of the corrugations and the boss lead to a nonlinearity of 0.01% full scale. The diaphragm is sandwiched between two other pieces of silicon. The one on the top serves as a mechanical overpressure stop. The one on the bottom contains the CMOS circuitry, the backplate of the capacitor, and serves as an overpressure stop in the other direction. The complete assembly is shown in Fig. 9-13A. Planar diaphragms are generally considered more sensitive than the corrugated diaphragms with the same size and thickness. However, in the presence of the in-plane tensile stresses, the corrugations serve to release some of the stresses, thus resulting in better sensitivity and linearity (Fig. 9-13B).

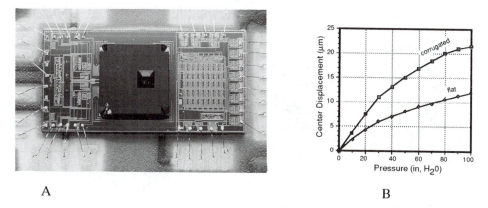

A   B

Fig. 9-13 A complete assembly of a capacitive diaphragm
and interface electronics (A); central deflection of a flat and corrugated
diaphragms of the same sizes under the in-plate tensile stresses (B)
(Courtesy of Monolithic Sensors, Inc. Rolling Meadows, IL)

9.7 VRP SENSORS

When measuring small pressures, deflection of a thin plate or a diaphragm can be very small. In fact, it can be so small that use of strain gauges attached to or imbedded into the diaphragm becomes impractical due to the low output signal. One possible answer to the problem may be a capacitive sensor where a diaphragm deflection is measured by its relative position to a reference base rather than by the internal strain in the material. Such sensors were described above. Another solution which is especially useful for very low pressures is a magnetic sensor. A *variable reluctance pressure* (VRP) sensor uses a magnetically conductive diaphragm to modulate the magnetic resistance of a differential transformer. The operation of the sensor is very close to that of the magnetic proximity detectors described in Chapter 5. Fig. 9-14A illustrates a basic idea behind the magnetic flux modulation. The assembly of an E-shaped core and a coil produces a magnetic flux whose field lines travel through the core, the air gap and the diaphragm. The permeability of the E-core magnetic material is at least 1000 times higher than that of the air gap [10], and, subsequently, its magnetic resistance is lower than the resistance of air. Since the magnetic resistance of the air gap is much higher than the resistance of the core, it is the gap which determines the inductance of the core-coil assembly. When the diaphragm deflects, the air gap increases or decreases depending on the direction of a deflection, thus causing the modulation of the inductance.

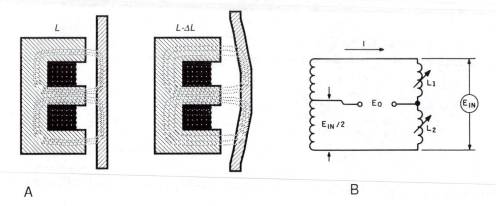

A   B

Fig. 9-14 Variable reluctance pressure sensor
A: basic principle of operation; B: an equivalent circuit

To fabricate a pressure sensor, a magnetically permeable diaphragm is sandwiched between two halves of the shell (Fig. 9-15). Each half incorporates a E-core/coil assembly. The coils are encapsulated in a hard compound to maintain maximum stability under even very high pressure. Thin pressure cavities are formed at both sides of the diaphragm. The thickness of the diaphragm defines a full scale operating range, however, under most of circumstances, total deflection doesn't exceed 25-30µm which makes this device very sensitive to low pressures. Further, due to thin pressure cavities, the membrane is physically prevented from excessive deflection under the overpressure conditions. This makes VRP sensors inherently safe devices. When excited by an ac current, a magnetic flux is produced in each core and the air gaps by the diaphragm. Thus, the sensors contain two inductances and can therefore be thought of as half of a variable-reluctance bridge where each inductance forms one arm of the bridge (Fig. 9-14B). As a differential pressure across the diaphragm is applied, the diaphragm deflects, one side decreasing and the other increasing, and the air gap reluctances in the electromagnetic circuit change proportionally to the differential pressure applied. A full scale pressure on the diaphragm, while being very small, will produce a large output signal that is easily differentiated from noise.

VRP sensor's output is proportional to the reluctance in each arm of the inductive Wheatstone bridge that uses the equivalent inductive reactances $x_{1,2}$ as the active elements. The inductance of a coil is determined by the number of turns and the geometry of the coil. When a magnetically permeable material is introduced into the field flux, it forms a low resistance path attracting magnetic field. This alters the coil's self-inductance. The inductance of the circuit, and subsequently its reactance, is inversely proportional to the magnetic reluctance, that is $x_{1,2} = k/d$ where k is a constant and d is the gap size. When the bridge is excited by a carrier, the output signal across the bridge becomes amplitude-modulated by the applied pressure. The amplitude is proportional to the bridge imbalance, and the phase of the output signal changes with the direction of the imbalance. The ac signal can be demodulated to produce a dc response.

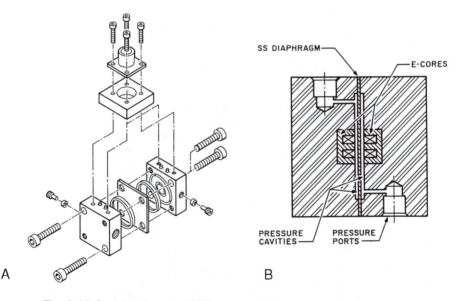

Fig. 9-15 Construction of a VRP sensor for low pressure measurements
A: assembly of the sensor; B: double E-core at both sides of the cavity
(Reprinted with permission. Sensors Magazine, ©1991)

9.8 OPTOELECTRONIC SENSORS

When measuring low level pressures or, contrary, when thick membranes are required to enable a broad dynamic range, a diaphragm displacement may be too small to assure a sufficient resolution and accuracy. Besides, most of piezoresistive sensors, and some capacitive, are quite temperature sensitive, that requires an additional thermal compensation. An optical readout has several advantages over other technologies, namely, a simple encapsulation, small temperature effects, high resolution and accuracy. Especially promising are the optoelectronic sensors operating with the light interference phenomena [11]. A simplified circuit of such a sensor is shown in Fig. 9-16.

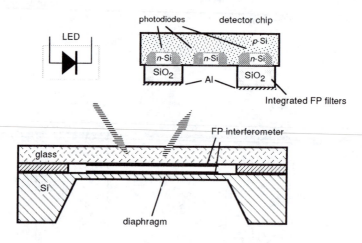

Fig. 9-16 Schematic of an optoelectronic pressure sensor
operating on the interference phenomenon
(adapted from [12])

The sensor consists of the following essential components: a passive optical pressure chip with a membrane etched in silicon, a light emitting diode (LED), and a detector chip [12]. A pressure chip is similar to a capacitive pressure sensor as described above, except that a capacitor is replaced by an optical cavity forming a Fabry-Perot (FP) interferometer [13] measuring the deflection of the diaphragm. A back-etched single-crystal diaphragm on a silicon chip is covered with a thin metallic layer, and a glass plate with a metallic layer on its backside. The glass is separated from the silicon chip by two spacers at a distance w. Two metallic layers form a variable-gap FP interferometer with a pressure sensitive movable mirror (on the membrane) and a plain-parallel stationary fixed half-transparent mirror (on the glass). A detector chip contains three *pn*-junction photodiodes. Two of them are covered with integrated optical FP filters of slightly different thicknesses. The filters are formed as first surface silicon mirrors, coated with a layer of SiO_2 and thin metal (Al) mirrors on their surfaces. An operating principle of the sensor is based on the measurement of a wavelength modulation of the reflected and transmitted light depending on the width of the FP cavity. The reflection and transmission from the cavity is almost a periodic function in the inverse wavelength, $1/\lambda$, of the light with a period equal to $1/2w$. Since w is a linear function of the applied pressure, the reflected light is wavelength modulated.

The detector chip works as a demodulator and generates electrical signals representing the applied pressure. It performs an optical comparison of the sensing cavity of the pressure sensor with a virtual cavity formed by the height difference between two FP filters. If both cavities are the same, the detector generates the maximum photocurrent, and, when the pressure changes, the photocurrent is cosine modulated with a period defined by a half the mean wavelength of the light source. The photodiode without the FP filter serves as a reference diode, which monitors the total light intensity arriving at the detector. Its output signal is used for the ratiometric processing of the information. Since the output of the sensor is inherently nonlinear, a linearization by a microprocessor is generally required. Similar optical pressure sensors can be designed with fiber-optics, which makes them especially useful for remote sensing where radio frequency interferences present serious problem [14].

REFERENCES

1. Benedict, R. P. Fundamentals of temperature, pressure, and flow measurements. 3rd ed., John Wiley & Sons, New York, 1984

2. Plandts, L. Essentials of fluid dynamics. Hafner, New York, 1952

3. Di Giovanni, M. Flat and corrugated diaphragm design handbook. Marcel Dekker, Inc., New York, 1982

4. Neubert, H. K. P. Instrument transducers. An introduction to their performance and design. 2nd ed., Clarendon Press. Oxford, 1975

5. Clark, S. K. and Wise, K. D. Pressure sensitivity in anisotropically etched thin-diaphragm pressure sensor. *IEEE Trans. Electron Devices*, vol. ED-26, pp: 1887-96, Dec. 1979

6. Tufte, O. N., Chapman, P.W. and Long, D. Silicon diffused-element piezoresistive diaphragm. *J. Appl. Phys.*, vol. 33, pp: 3322-27, Nov. 1962

7. Kurtz, A. D. and Gravel, C. L. Semiconductor transducers using transverse and shear piezoresistance. *Proc. 22nd ISA Conf.*, No. P4-1 PHYMMID-67, Sept. 1967

8. Tanigawa, H., Ishihara, T., Hirata, M., Suzuki K. MOS integrated silicon pressure sensor. *IEEE Trans. Electron Devices*, vol. ED-32, no. 7, pp: 1191-95, July 1985

9. Petersen, K., Barth, P., Poydock, J., Brown, J., Mallon, Jr., J. and Bryzek, J. Silicon fusion bonding for pressure sensors. *Rec. of the IEEE Solid-state sensor and actuator workshop*, pp: 144-147, 1988

10. Proud, R. VRP transducers for low-pressure measurement. Sensors. Feb., pp. 20-22 (1991)

11. Wolthuis, R., A., Mitchell, G.L., Saaski, E., Hratl, J.C., and Afromowitz, M.A. Development of medical pressure and temperature sensors employing optical spectral modulation. In: *IEEE Trans. on Biomed. Eng.*, vol. 38, No. 10, pp: 974-981, Oct. 1991

12. Hälg, B. A silicon pressure sensor with an interferometric optical readout. In: *Transducers'91. International conference on solid-state sensors and actuators. Digest of technical papers*, pp: 682-684, IEEE, 1991

13. Vaughan, J.M. The Fabry-Perot interferometers. Bristol, Adam Hilger, 1989

14. Saaski, E.W., Hartl, J.C. and Mitchell, G.L. A fiber optic sensing system based on spectral modulation. Paper #86-2803, ©ISA, 1989

10

Flow Sensors

10.1 BASICS OF FLOW DYNAMICS

One of the fundamentals of physics is that mass is a conserved quantity. It can't be created or destroyed. In the absence of sources or sinks of mass, its quantity remains constant regardless of boundaries. However, if there is influx or outflow of mass through the boundaries, the sum of influx and efflux must be zero. When both are measured over the same interval of time, mass entering the system (M_{in}) is equal to mass leaving the system (M_{out}). Therefore [1],

$$\frac{dM_{in}}{dt} = \frac{dM_{out}}{dt} \; . \tag{10.1}$$

In mechanical engineering, moving media whose flow is measured are liquids (water, oil, solvents, gasoline, etc.), air, gases (oxygen, nitrogen, CO, CO_2, methane CH_4, water vapor, etc.).

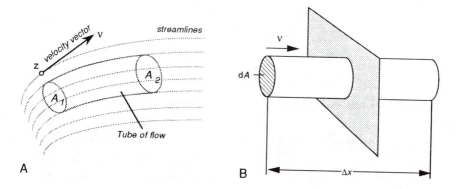

Fig. 10-1 Tube of flow (A) and flow of a medium through a plane (B)

In a steady flow, the velocity at a given point is constant in time. We can draw a streamline through every point in a moving medium (Fig. 10-1A). In steady flow, the line distribution is time independent. A velocity vector is tangent to a stream line in every point z. Any boundaries of flow which envelop a bundle of streamlines is called a *tube of flow*. Since the boundary of such a tube consists of streamlines, no fluid (gas) can cross the boundary of a tube of flow and the tube behaves something like a pipe of some shape. The flowing medium can enter such a pipe at one end, having cross section A_1 and exit at the other through cross section A_2. The velocity of a moving material inside a tube of flow will in general have different magnitudes at different points along the tube.

The volume of moving medium passing a given plane (Fig. 10-1B) in a specified time interval Δt is

$$\Lambda = \frac{V}{\Delta t} = \int \frac{\Delta x dA}{\Delta t} = \int v dA \ , \tag{10.2}$$

where v is the velocity of moving medium which must be integrated over area A and Δx is the displacement of volume V. Fig. 10.2 shows that the velocity of liquid or gas in a pipe may vary over the cross section. It is often convenient to define an average velocity

$$v_a = \frac{\int v dA}{A} \tag{10.3}$$

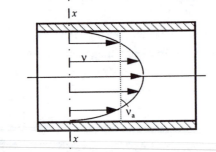

Fig. 10-2 Profile of velocity of flow in a pipe

When measuring the velocity by a sensor whose dimensions are substantially smaller than the pipe size, one should be aware of a possibility of erroneous detection of either too low or too high velocity, while the average velocity, v_a, is somewhere in-between. A product of the average velocity and a cross-sectional area is called *flux* or *flow rate*. Its SI unit is m³/sec. The US customary system unit is ft³/sec. The flux can be found by rearranging equation (10.3)

$$Av_a = \int v dA \ . \tag{10.4}$$

What a flow sensor usually measures is v_a. Thus, to determine a flow rate, a cross-section area of tube of flow A must be known, otherwise the measurement is meaningless.

The measurement of flow is rarely conducted for the determination of a displacement of volume. Usually, what is needed is to determine the flow of mass rather than volume. Of course, when dealing with virtually incompressible fluids (water, oil, etc.), either volume or mass can be used. A relationship between mass and volume for a incompressible material is through density ρ

$$M = \rho V \ . \tag{10.5}$$

The densities of some materials are given in Table 3-14. The rate of mass flow is defined as

$$\frac{dM}{dt} = \rho A \ \overline{v} \tag{10.6}$$

SI unit for mass flow is kg/sec while the US customary system unit is lb/sec. For a compressible medium (gas) either mass flow or volume flow at a given pressure should be specified.

There is a great variety of sensors which can measure flow velocity by determining the rate of displacement either mass, or volume. Whatever sensor is used, inherent difficulties of the measurement make the process a complicated procedure. It is necessary to take into consideration many of the natural characteristics of the medium, its surroundings, barrel and pipe shapes and materials, medium temperature and pressure, etc. When selecting any particular sensor for the flow measurement it is advisable to consult with the manufacturer's specifications and very carefully considering the application recommendations for a particular sensor. In this book we do not cover such traditional flow measurement systems as turbine type meters. It is of interest to us to consider sensors without moving components which introduce either no or little restriction into the flow.

10.2 PRESSURE GRADIENT TECHNIQUE

A fundamental equation in fluid mechanics is *Bernoulli equation* which is strictly applicable only to steady flow of nonviscous, incompressible medium

$$p + \rho(\tfrac{1}{2}v_a^2 + gy) = \text{const} \ , \tag{10.7}$$

where p is the pressure in a tube of flow, $g = 9.80665 \text{ m/s}^2 = 32.174 \text{ ft/s}^2$ is the gravity constant, and y is the height of medium displacement. Bernoulli's equation allows us to find fluid velocity by measuring pressures along the flow.

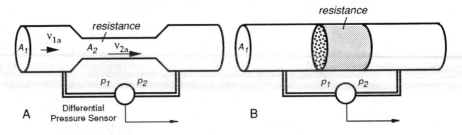

Fig. 10-3 Two types of flow resistors: a narrow channel (A) and a porous plug (B)

The pressure gradient technique (of flow measurement) essentially requires an introduction a flow resistance. Measuring the pressure gradient across a known resistor allows to calculate a flow rate. The concept is analogous to Ohm's law: voltage (pressure) across a fixed resistor is proportional to current (flow). In practice, the restricting elements which cause flow resistances are orifices, porous plugs, and Venturi tubes (tapered profile pipes). Fig. 10-3 shows two types of flow resistors. In the first case it is a narrow in the channel, while in the other case there is a porous plug which somewhat restricts the medium flow. A differential pressure sensor is positioned across the resistor. When moving mass enters the higher resistance area, its velocity increases in proportion to the resistance increase:

$$v_{1a} = v_{2a}R \quad . \tag{10.8}$$

The Bernoulli equation defines differential pressure as[1]

$$\Delta p = p_1 - p_2 = \frac{\rho}{2}(v_{2a}^2 - v_{1a}^2) = k\frac{\rho}{2}v_{2a}^2(1-R^2) \tag{10.9}$$

where k is the correction coefficient which is required because the actual pressure p_2 is slightly lower than the theoretically calculated. From equation (10.9) the average velocity can be calculated as

$$v_{2a} = \frac{1}{\sqrt{k(1-R^2)}}\sqrt{\frac{2}{\rho}\Delta p} \quad . \tag{10.10}$$

---

[1] It is assumed that both pressure measurements are made at the same height ($y = 0$) which is usually the case.

To determine the mass flow rate per unit time, for a incompressible medium, the Eq. (10.10) is simplified to

$$q = \xi A_2 \sqrt{\Delta p} \ , \tag{10.11}$$

where ξ is a coefficient which is determined through calibration. The calibration must be done with a specified liquid or gas over an entire operating temperature range. It follows from the above that the pressure gradient technique essentially requires the use of either one differential pressure sensor or two absolute sensors. If a linear representation of the output signal is required, a square root-extraction must be used. The root-extraction can be performed in a microprocessor by using one of the conventional computation techniques. An advantage of the pressure gradient method is in the absence of moving components and use of standard pressure sensors which are readily available. A disadvantage is in the restriction of flow by resistive devices.

10.3 THERMAL TRANSPORT SENSORS

A good method to measure flow would be by somehow marking the flowing medium and detecting the movement of the mark. For example, a mark can be a floating object which can move with the medium while being stationary with respect to the medium. The time which it would take the object to move with the flow from one position to another could be used for the calculation of the flow rate. Such an object may be a float, radioactive element, or die which changes optical properties (for instance, color) of flowing medium. Also, the mark can be a different gas or liquid whose concentration and rate of dilution can be detectable by appropriate sensors.

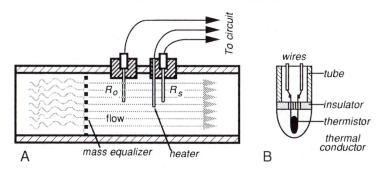

Fig. 10-4 Thermoanemometer
A: a basic two-sensor design;   B: cross-sectional view of a temperature detector

In medicine, a die dilution method of flow measurement is used for studies in hemodynamics. In most instances, however, placing any foreign material into the flowing medium is either impractical or forbidden for some other reasons. An alternative would be to change some physical properties of the moving medium and to detect the rate of displacement of a changed portion or rate of its dilution. Usually, the best physical property which can be easily modified without causing undesirable effects, is temperature.

Fig. 10-4A shows a sensor which is called a thermoanemometer. It is comprised of three small tubes immersed into a moving medium. Two tubes contain temperature detectors R_o and R_s. The detectors are thermally coupled to the medium and are thermally isolated from the structural elements and the pipe where the flow is measured. In between two detectors, a heating element is positioned. Both detectors are connected to electrical wires through tiny conductors to minimize thermal loss through conduction (Fig. 10-4B). The sensor operates as follows. The first temperature detector R_o measures the temperature of the flowing medium. The heater warms up the medium and the elevated temperature is measured by the second temperature detector R_s. In a still medium, heat would be dissipated from the heater through media to both detectors. In a medium with a zero flow, heat moves out from the heater mainly by thermal conduction and gravitational convection. Since the heater is positioned closer to the R_s detector, that detector will register higher temperature. When the medium flows, heat dissipation increases due to forced convection. The higher the rate of flow the higher the heat dissipation and the lower temperature will be registered by the R_s detector. Heat loss is measured and converted into the flow rate of medium.

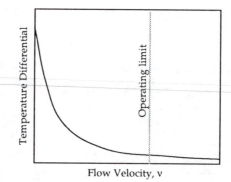

Fig. 10-5 Transfer function of a thermoanemometer

A fundamental relationship of the thermoanemometry is based on King's law [2]

$$\Delta Q = kl \left(1 + \sqrt{\frac{2\pi \rho c d v}{k}}\right)(T_s - T_o) , \qquad (10.12)$$

where k and c are the thermal conductivity and specific heat of a medium at a given pressure, ρ is the density of the medium, l and d are the length and diameter of the sensor, T_s is the surface temperature of the sensor, T_o is the temperature of the medium away from the sensor, and v is the velocity of the medium. Collis and Williams experimentally proved [3] that King's theoretical law needs some correction. For a cylindrical sensor with $l/d \gg 1$ a modified King's equation yields the velocity of the medium:

$$v = \frac{K}{\rho}\left(\frac{dQ}{dt} \frac{1}{T_s - T_o}\right)^{1.87} , \qquad (10.13)$$

where K is the calibration constant. It follows from the above that to measure a flow, a temperature gradient between the sensor and the moving medium, and dissipated heat must be measured. Then velocity of the fluid or gas becomes although nonlinear, but a quite definitive function of thermal loss (Fig. 10-5).

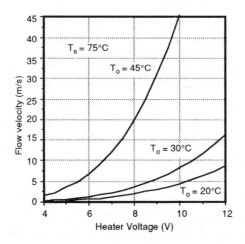

Fig. 10-6 Calibration curves for a self-heating sensor in a thermoanemometer for three different levels of heat

To maintain the R_s detector at T_s, and for assuring a sufficient thermal gradient with respect to T_o, heat loss must be compensated for by supplying the appropriate power to the heating element. Also, we may consider a flow sensor without a separate heating element. In such a sensor, the R_s detector operates in a *self-heating mode*. That is, the electric current passing through its resistance generates enough Joule heat to elevate its temperature to T_s. At that temperature, the second detector has resistance R_s. Assuming that conductive heat loss through connecting wires and sensor's enveloping tube is negligibly small, the law of conservation of energy demands that electric power W is equal to thermal loss to flowing medium

$$W = \frac{dQ}{dt}.$$ (10.14)

On the other hand, the electric power through a heating resistance is in square relationship with voltage e across the heating element:

$$W = \frac{e^2}{R_s}.$$ (10.15)

Eqs. (10.13 to 10.15) yield a relationship between the voltage across the self-heating detector and the velocity of flow

$$v_{2a} = \frac{K}{\rho} \left(\frac{e^2}{R_s} \frac{1}{T_s - T_o} \right)^{1.87}.$$ (10.16)

Fig. 10-6 shows an example of a calibrating curve for a flow sensor using a self-heating thermistor ($T_s = 75°C$) operating in air whose temperature varies from 20 to 45°C. The thermistor temperature was maintained constant over an entire range of T_o temperatures[1]. It should be emphasized, that T_s must always be selected higher than the highest temperature of the flowing medium.

Formula (10.13) suggests that two methods of measurement are possible. In the first method, the voltage and resistance of a heating element is maintained constant, while the temperature differential (T_s-T_o) is used as the output signal. In the second method, the temperature differential is maintained constant by a control circuit which regulates the heater's voltage e. In the latter case, e is the output signal. This method is often preferable for use in the miniature sensors where self-heating temperature detectors are employed. A self-heating sensor (it can be either a RTD or thermistor) operates at high excitation currents. That current serves two purposes - it measures the resistance of a detector to determine its temperature, and provides Joule heat. Fig. 10-7 illustrates the design of a sensor with a separate heater and Fig. 10-8A shows that both temperature detectors (heated and reference) can be connected in a bridge circuit. At very low flow velocities, the bridge is imbalanced and the output signal is high. When the flow rate increases, the heated detector cools down and its temperature comes closer to that of a reference detector, lowering the output voltage. Fig. 10-8B illustrates that the sensor's response is different for various fluids and gases. A sensor manufacturer usually provides calibration curves for any

---

[1] This can be accomplished by using a self-balancing resistive bridge. See for example [4]

particular medium, however, whenever precision measurement is required, on-site calibration is recommended.

Fig. 10-7 Thermal flow sensor
Note that the heater is positioned downstream closer to the second temperature sensor (Courtesy of Fluid Components, Inc., San Marcos, CA)

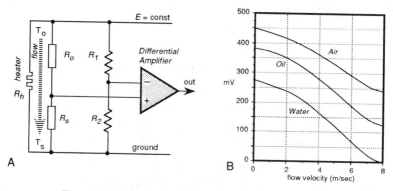

A

B

Fig. 10-8 Bridge circuit for a thermal flowmeter (A); the sensor responses for different fluids (B)

For accurate temperature measurements in a flowmeter, any type of temperature detectors can be used - resistive, semiconductor, optical, etc. (Chap. 17). Nowadays, however, the majority of manufacturers use resistive sensors. In industry and scientific measurements, RTDs are the prime choice as they assure higher linearity, predictable response, and long term stability over broader temperature range. In medicine, thermistors are often preferred thanks to their higher sensitivity. Whenever a resistive temperature sensor is employed, especially for a remote sensing, a four-wire measurement technique should be seriously considered. The technique is a solution for a problem arising from a finite resistance of connecting wires which may be a substantial source of error. See Section 4.8.2 for the description of a four-wire method.

A sensor's design determines its operating limits. At a certain velocity, the molecules of a moving medium while passing near a heater do not have sufficient time to absorb enough thermal energy for developing a temperature differential between two detectors. Since the differential is in the denominator of formula (10.13), at high velocities computational error becomes unacceptably large and accuracy drops dramatically. The upper operating limits for the thermal transport sensors usually are determined experimentally. For instance, under normal atmospheric pressure and room temperature (about 20°C), the maximum air velocity which can be detected by a thermal transport sensor is in the range of 60 m/sec (200foot/sec).

While designing thermal flow sensors, it is important to assure that the medium moves through the detectors without turbulence in a nonlaminar well mixed flow. The sensor is often supplied with mixing grids or turbulence breakers which sometimes are called *mass equalizers* (Fig. 10-4).

The pressure and temperature of a moving medium, especially of gases, make a strong contribution to the accuracy of a volume rate calculation. It is interesting to note that for the mass flow meters, pressure makes very little effect on the measurement as the increase in pressure results in a proportional increase in mass.

A data processing system for the thermal transport sensing must receive at least three variable input signals: a flowing medium temperature, a temperature differential, and a heating power signal. These signals are multiplexed, converted into digital form and processed by a computer to calculate characteristics of flow. Data are usually displayed in velocities (m/s or ft/s), volume rates (m^3/s or ft^3/s), or mass rate (kg/s or lb/s).

Thermal transport flowmeters are far more sensitive than other types and have a broad dynamic range. They can be employed to measure very minute gas or liquid displacements as well as fast and strong currents. Major advantages of these sensors are the absence of moving components and an ability to measure very low flow rates. "Paddle wheel", hinged vane, and pressure differential sensors have low and inaccurate outputs at low rates. If a small diameter of tubing is required, as in automotive, aeronautic, medical, and biological applications, sensors with moving components become mechanically impractical. In these applications, thermal transport sensors are indispensable.

10.4 ULTRASONIC SENSORS

Flow can be measured by employing ultrasonic waves. The main idea behind the principle is the detection of frequency or phase shift caused by flowing medium. One possible implementation is based on the Doppler effect (see Section 6.2 for the description of the Doppler effect), while the other relies on the detection of the increase or decrease in effective ultrasound velocity in the medium. Effective velocity of sound in a moving medium is equal to the velocity of sound relative to the medium plus the velocity of the medium with respect to the source of the sound. Thus, a sound wave propagating upstream will have a smaller effective velocity, while the sound propagating downstream will have a higher effective velocity. Since the difference between the two velocities is exactly twice the velocity of the medium, measuring the upstream-downstream velocity difference allows us to determine the velocity of the flow.

Fig. 10-9A shows two ultrasonic generators positioned at opposite sides of a tube of flow. Piezoelectric crystals are usually employed for that purpose. Each crystal can be used for either the generation of the ultrasonic waves (motor mode), or for receiving the ultrasonic waves (generator mode). In other words, the same crystal, when needed, acts as a "speaker" or a "microphone".

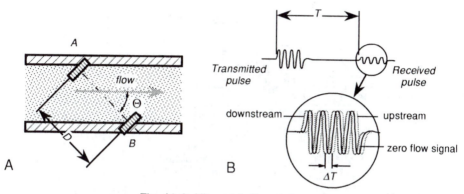

Fig. 10-9 Ultrasonic flowmeter
A: Position of transmitter-receiver crystals in the flow; B: Wave forms in the circuit

Two crystals are separated by distance D and positioned at angle Θ with respect to flow. Also, it is possible to place small crystals right inside the tube along the flow. That case corresponds to $\Theta=0$. The transit time of sound between two transducers A and B can be found through the average fluid velocity v_c

$$T = \frac{D}{c \pm v_c \cos\Theta} \ , \tag{10.17}$$

where c is the velocity of sound in the fluid. The plus/minus signs refer to the downstream/upstream directions, respectively. The velocity v_c is the flow velocity averaged along the path of the ultrasound. Gessner [5] has shown that for laminar flow $v_c = 4v_a/3$, and for turbulent flow, $v_c=1.07v_a$, where v_a is the flow averaged over the cross sectional area. By taking the difference between the downstream and upstream velocities we find [6]

$$\Delta T = \frac{2Dv_c \cos\Theta}{c^2 + v_c \cos^2\Theta} \approx \frac{2Dv_c \cos\Theta}{c^2} \ , \tag{10.18}$$

which is true for the most practical cases when $c \gg v_c \cos\Theta$. To improve the signal-to-noise ratio, the transit time is often measured for both up- and downstream directions. That is, each piezoelectric crystal at one time works as a transmitter and at the other time as a receiver. This can be accomplished by a selector (Fig. 10-10) which is clocked by a relatively slow sampling rate (400Hz in the example). The sinusoidal ultrasonic waves (about 3MHz) are transmitted as bursts with the same slow clock rate (400Hz). A received sinusoidal burst is delayed from the transmitted one by time T which is modulated by the flow. This time is detected by a transit time detector, then, the time difference in both directions is recovered by a synchronous detector. Such a system can achieve a quite good accuracy, with a zero-drift as small as $5 \cdot 10^{-3}$m/s^2 over the 4-hour period.

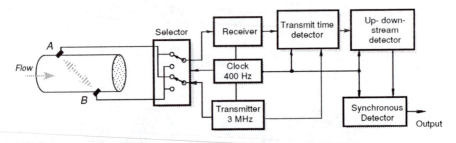

Fig. 10-10 Block diagram of an ultrasonic flowmeter
with alternating transmitter and receiver

An alternative way to measure flow with the ultrasonic sensors is to detect a phase difference in transmitted and received pulses in the up- and downstream directions. The phase differential can be derived from Eq. (10.18)

$$\Delta f = \frac{4\pi f D v_c \cos\Theta}{c^2} , \tag{10.19}$$

where f is the ultrasonic frequency. It is clear that the sensitivity is better with the increase in the frequency, however, at higher frequencies one should expect stronger sound attenuation in the system, which may cause reduction in the signal-to-noise ratio.

For the Doppler flow measurements, continuos ultrasonic waves can be used. Fig. 10-11 shows a flowmeter with a transmitter-receiver assembly positioned inside the flowing stream. Like in a Doppler radio receiver, transmitted and received frequencies are mixed in a non-linear circuit (a mixer). The output low frequency differential harmonics are selected by a bandpass filter. That differential is defined as

$$\Delta f = f_s - f_r \approx \pm \frac{2 f_s v}{c} , \tag{10.20}$$

where f_s and f_r are the frequencies in the transmitting and receiving crystals respectively, and the plus/minus signs indicate different directions of flow. An important conclusion from the above equation is that the differential frequency is directly proportional to the flow velocity. Obviously, the crystals must have much smaller sizes than the clearance of the tube of flow. Hence, the measured velocity is not the average but rather a localized velocity of flow. In practical systems, it is desirable to calibrate ultrasonic sensors with actual fluids over the useful temperature range, so that contribution of a fluid viscosity is accounted for.

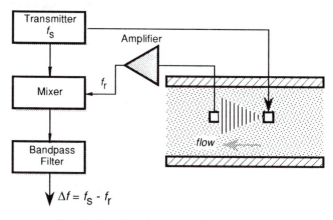

Fig. 10-11 Ultrasonic Doppler flowmeter

An ultrasonic piezoelectric sensors/transducers can be fabricated of small ceramic discs encapsulated into a flowmeter body. The surface of the crystal can be protected by a suitable material, for instance, silicone rubber. An obvious advantage of an ultrasonic sensor is in its ability to measure flow without a direct contact with the fluid.

10.5 ELECTROMAGNETIC SENSORS

The electromagnetic flow sensors are useful for measuring the movement of conductive liquids. The operating principle is based on the discovery of Faraday and Henry (see Section 3.4) of the electromagnetic induction. When a conductive media, wire, for instance, or for this particular purpose, flowing conductive liquid crosses the magnetic flux lines, e.m.f. is generated in the conductor. The value of e.m.f. is proportional to velocity of moving conductor [Eq. (3.4.5)]. Fig. 10-12 illustrates a tube of flow positioned into magnetic field B. There are two electrodes incorporated into a tube to pick up e.m.f. induced in the liquid. The magnitude of the e.m.f. is defined by

$$v = e - e' = 2aBv \ ,$$
(10.21)

where a is the radius of the tube of flow, and v is the velocity of flow.

By solving Maxwell's equations, it can be shown that for a typical case when the fluid velocity is nonuniform within the cross-sectional area but remains symmetrical about the tube axis (axisymmetrical), the e.m.f generated is the same as that given above, except that v is replaced by the average velocity, v_a [Eq. (10.3)]:

$$v_a = \frac{1}{\pi a^2} \int_0^a 2\pi vr \, dr \ ,$$
(10.22)

where r is the distance from the center of the tube. Eq. (10.21) can be expressed in terms of the volumetric flow rate

$$v = \frac{2\Lambda B}{\pi a} \ .$$
(10.23)

It follows from the above equation that the voltage registered across the pick-up electrodes is independent of the flow profile or fluid conductivity. For a given tube geometry and the magnetic flux, it depends only on the instantaneous volumetric flow rate.

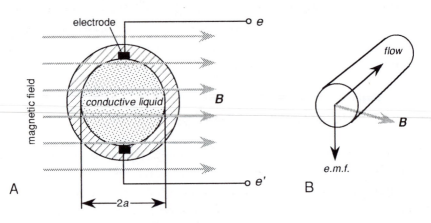

Fig. 10-12 Principle of electromagnetic flowmeter
A: position of electrodes is perpendicular to the magnetic field; B: relationships between flow and electrical and magnetic vectors

There are two general methods of inducing voltage in the pick-up electrodes. The first is a dc method where the magnetic flux density is constant and induced voltage is a dc or slow changing signal. One problem associated with this method is a polarization of the electrodes due to small but unidirectional current passing through their surface. The other problem is a low frequency noise which makes it difficult to detect small flow rates.

Another and far better method of excitation is with an alternating magnetic field, which causes appearance of an ac voltage across the electrodes (Fig. 10-13). Naturally, the frequency of the magnetic field should meet a condition of the Nyquist rate. That is, it must be at least two times higher than the highest frequency of flow rate spectrum variations. In practice, the excitation frequency is selected in the range between 100 and 1000Hz.

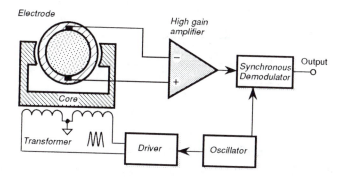

Fig. 10-13 Electromagnetic flowmeter
with synchronous (phase sensitive) demodulator

10.6 MICROFLOW SENSORS

In some applications, such as process control in precise semiconductor manufacturing, chemical and pharmaceutical industries, and biomedical engineering, miniaturized gas flow sensors are encountered with an increasing frequency. Most of them operate on the method of a thermal transport (see Section 10.3) and are fabricated from a silicon crystal by using a micromachining technology. Many of the microflow sensors use a thermopile as a temperature sensor [7], however, the thermoelectric coefficient [Eq. (3.9.5)] of standard elements used in the IC processing (silicon and aluminum), is smaller than that of conventional thermocouples by factors ranging from 10 to 100. Thus, a resulting output signal may be very small, which requires amplification by amplifiers integrated directly into the sensor. A cantilever design of a microflow sensor is shown in Fig. 10-14. The thickness of the cantilever may be as low as 2μm. It is fabricated in the form of a sandwich consisting of layers of field oxide, CVD oxide, and nitrate [8]. The cantilever sensor is heated by an imbedded resistor with a rate of 26K per 1mW of applied electric power, and a typical transfer function of the flow sensor has a negative slope of about \F(4 mV,m/s).

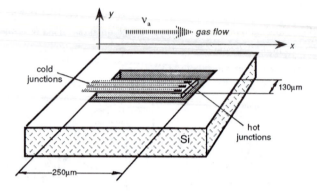

Fig. 10-14 Micromachined gas flow sensor

The heat is removed from the sensor by three means: conductance L_b through the cantilever beam, gas flow $h(v)$ and thermal radiation, which is governed by the Stefan-Boltzmann law

$$P = L_b(T_s - T_b) + h(v)(T_s - T_b) + a\sigma\varepsilon(T^4_s - T^4_b) , \tag{10.24}$$

where σ is the Stefan-Boltzmann constant, a is the area along which the beam-to-gas heat transfer occurs, ε is surface emissivity, and v is the gas velocity. From the principles of energy and particle conservation we deduce a generalized heat transport equation governing the temperature distribution $T(x,y)$ in the gas flowing near the sensor's surface

$$\frac{\partial^2 T}{\partial x^2} + \frac{\partial^2 T}{\partial y^2} = \frac{vnc_p}{k_g}\cdot\frac{\partial T}{\partial x} \qquad \text{for } y>0 , \tag{10.25}$$

where n is the gas density, c_p is the molecular gas capacity, and k_g is the thermal conductivity of gas. It can be shown that solution of this equation for the boundary condition of vanishing thermal gradient far off the surface is [8]

$$\Delta V = B \cdot \left(\frac{1}{\sqrt{\mu^2 + 1}} - 1 \right), \qquad (10.26)$$

where V is the input voltage, B is a constant, and $\mu = \frac{Lvnc_p}{2\pi k_g}$, and L is the gas-sensor contact length. This solution coincides very well with the experimental data.

Another design of a thermal transport microsensor is shown in Fig. 10-15A [9] where titanium films having a thickness of 0.1μm serve as both the temperature sensors and the heaters. The films are sandwiched between two layers of SiO_2. Titanium was used because of its high TCR (temperature coefficient of resistance) and excellent adhesion to SiO_2. Two microheaters are suspended with four silicon girders at a distance of 20 μm from one another. The Ti film resistance is about 2 kΩ. Fig. 10-5.2B shows a simplified circuit diagram for the sensor, which exhibits an almost linear relationship between the flow and output voltage ΔV.

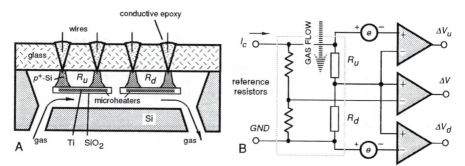

Fig. 10-15 Gas microflow sensor with self-heating titanium resistors
A - sensor design; B - interface circuit: R_u and R_d are resistances
of the up- and downstream heaters respectively
(adapted from [8])

A microflow sensor can be constructed by utilizing a capacitive pressure sensor [10] shown in Fig. 10-16. An operating principle of the sensor is based on a pressure gradient technique as described in Section 10.2. The sensor was fabricated using silicon micromachining and defused boron etch-stops to define the structure. The gas enters the sensor's housing at pressure P_1 through the inlet, and the same pressure is established around the silicon plate, including the outer side of the etched membrane. The gas flows into the microsensor's cavity through a narrow channel having a relatively high pressure resistance. As a result, pressure P_2 inside the cavity is lower than P_1, thus creating a pressure differential across the membrane. Therefore, the flow rate can be calculated from Eq. (10.10).

The pressure differential is measured by a capacitive pressure sensor, which is composed of a thin, stress compensated, p^{++} boron-doped silicon membrane suspended above a metal plate. Pressure differential changes capacitance C_x between the metal plate and the silicon structure with a resolution of 1mTorr/1fF with a full pressure of about 4torr. An overall resolution of the sensor is near 14-15 bits and the accuracy of pressure measurement about 9-10 bits. At approximately twice the full scale pressure differential, the membrane touches the metal plate, hence a dielectric layer is required to prevent an electric short, while the substrate glass plate protects the membrane from rupturing. A capacitance measurement circuit (see Fig. 4-5.2) is integrated with the silicon plate using standard CMOS technology.

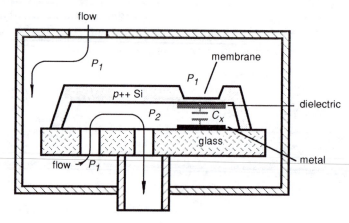

Fig. 10-16 Structure of a gas microflow sensor utilizing capacitive pressure sensor
(adapted from [10])

10.7 BREEZE SENSOR

In some applications, it is desirable just to merely detect a change in the air (or any other gas for that matter) movement, rather then to measure its flow rate quantitatively. This task can be accomplished by a breeze sensor, which produces an output transient whenever the velocity of the gas flow happens to change. One example of such a device is a piezoelectric breeze sensor produced by Nippon Ceramic, Japan. A sensor contains a pair of the piezoelectric (or pyroelectric) elements[1], where one is exposed to ambient air and the other is protected by the encapsulating resin

---

[1] In this sensor, the crystalline element which is poled during the manufacturing process is the same as used in piezo- or pyroelectric sensors. However, the operating principle of the breeze sensor is neither related to mechanical stress nor heat flow. Nevertheless, for the simplicity of the description, we will use the term *piezoelectric*.

coating. Two sensors are required for a differential compensation of variations in ambient temperature. The elements are connected in a series-opposed circuit, that is, whenever both of them generate the same electric charge, the resulting voltage across the bias resistor R_b (Fig. 10-6.1A) is essentially zero. Both elements, the bias resistor, and the JFET voltage follower are encapsulated into a TO-5 metal housing with vents for exposing the S_1 element to the gas movement (Fig. 10-6.1B).

An operating principle of the sensor is illustrated in Fig. 10-6.2. When air flow is either absent or is very steady, the charge across the piezoelectric element is balanced. An element internal electric dipoles, which are oriented during the poling process (Section 3.6), are balanced by both the free carriers inside the material, and by the charged floating air molecules at the element's surface. In the result, voltage across the piezoelectric elements S_1 and S_2 is zero, which results in zero output voltage V_{out}.

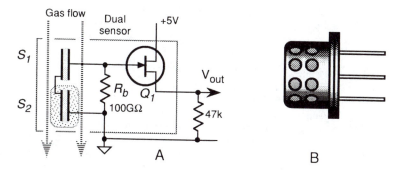

Fig. 10-6.1 Piezoelectric breeze sensor
A: a circuit diagram; B: a packaging in a TO-5 can

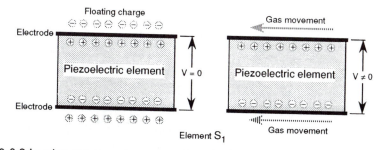

Fig. 10-6.2 In a breeze sensor, gas movement strips off electric charges from the surface of a piezoelectric element

When the gas flow across both S_1 surfaces changes (S_2 surfaces are protected by resin), moving gas molecules strip off the floating charges from the element. This results in the appearance of voltage across the element's electrodes, because the internally poled dipoles are not balanced any more by the outside floating charges. The voltage is repeated by the JFET follower which serves as an impedance converter, and appears as a transient in the output terminal.

REFERENCES

1. Benedict, R. P. Fundamentals of temperature, pressure, and flow measurements. 3rd ed., John Wiley & Sons, New York, 1984

2. King, L.V. On the convention of heat from small cylinders in a stream of fluid. *Phil. Trans. Roy. Soc.*, A214, 373, 1914

3. Collis, D. C., Williams, M. J. Two-dimensional convection from heated wires at low Reynolds' numbers. *J. Fluid Mech.* vol. 6, p: 357, 1959

4. Williams, J. Thermal techniques in measurement and control circuitry. AN5, In: *Linear application handbook*, Linear Technology Corp., 1990

5. Gessner, U. The performance of the ultrasonic flowmeter in complex velocity profiles. IEEE Trans. *Bio-Med. Eng. MBE-16*, pp: 139-142, April 1969

6. Cobbold, R.S.C. Transducers for biomedical measurements. John Wiley & Sons, 1974

7. Van Herwaarden, A.W. and Sarro, P.M. Thermal sensors based on the Seebeck effect. *Sensors and actuators*, No. 10, pp:321-346, 1986

8. Wachutka, G., Lenggenhager, R., Moser, D, and Baltes, H. Analytical 2D-model of CMOS micromachined gas flow sensors. In: *Transducers'91. International conference on solid-state sensors and actuators. Digest of technical papers.* ©IEEE, 1991

9. Esashi, M. Micro flow sensor and integrated magnetic oxygen sensor using it. In: In: *Transducers'91. International conference on solid-state sensors and actuators. Digest of technical papers.* IEEE, 1991

10. Cho, S.T., Wise, K.D. A high performance microflowmeter with built-in self test. In: *Transducers'91. International conference on solid-state sensors and actuators. Digest of technical papers*, pp: 400-403, IEEE, 1991

11

Acoustic Sensors

The fundamentals of acoustics are given in Section 3.12. Here we will discuss sensors which are generally called microphones. In essence, a microphone is a pressure transducer adapted for the transduction of sound waves over a broad spectral range. The microphones differ by their sensitivity, directional characteristics, frequency bandwidth, dynamic range, sizes, etc. Also, their designs are quite different depending on the media from which sound waves are sensed. For the perception of air waves or vibrations in solids, the sensor is called a microphone, while for the operation in liquids, it is called a hydrophone. The main difference between a pressure sensor and an acoustic sensor is that latter does not need to measure constant or very slow changing pressures. Its operating frequency range usually starts at several hertz (or as low as tens of millihertz for some applications), while the upper operating frequency limit is quite high - up to several megahertz for the ultrasonic applications.

Since acoustic waves are mechanical pressure waves, any microphone or hydrophone has the same basic structure as a pressure sensor: it is comprised of a moving diaphragm and a displacement sensor which converts the diaphragm's deflections into an electrical signal. That is, all microphones or hydrophones differ by the design of these two essential components. Also, they may include some additional parts such as mufflers, focusing reflectors, etc., however, in this chapter we will review only the sensing parts of some of the most interesting, from our point of view, acoustic sensors.

11.1 RESISTIVE MICROPHONES

In old times, resistive pressure converters were used quite extensively in microphones. The converter consisted of a semiconductive powder (usually graphite) whose bulk resistivity was sensitive to pressure. Nowadays we would say that the powder possessed piezoresistive properties. However, these early devices had quite a limited dynamic range, poor frequency response, and a high noise floor. Presently, the same piezoresistive principle can be employed in the micromachined sensors, where stress sensitive resistors are integral parts of a silicon diaphragm (Section 9.5).

11.2   CONDENSER MICROPHONES

If a parallel-plate capacitor is given an electric charge q, voltage across its plates is governed by Eq. (3.2.3). On the other hand, according to Eq. (3.2.4) the capacitance depends on distance d between the plates. Thus solving these two equations for voltage we arrive to

$$V = q \frac{d}{\varepsilon_0 A} \ , \tag{11.1}$$

where $\varepsilon_0 = 8.8542 \cdot 10\text{-}12$ C^2/Nm^2 is the permitivity constant (Section 3.1). The above equation is the basis for operation of the condenser microphones, which is the other way to say "capacitive" microphones. Thus, a capacitive microphone linearly converts a distance between the plates into electrical voltage which can be further amplified. The device essentially requires a source of an electric charge q whose magnitude directly determines the microphone sensitivity. The charge can be provided either from an external power supply having a voltage in the range from 20 to 200 V, or from an internal source of an electric charge. This is accomplished by a built-in electret layer which is a polarized dielectric crystal.

Presently, many condenser microphones are fabricated with silicon diaphragms which serve two purposes: to convert acoustic pressure into displacement, and to act as a moving plate of a capacitor. Some promising designs are described in [1-3]. To achieve high sensitivity, a bias voltage should be as large as possible, resulting in a large static deflection of the diaphragm, which may result in a reduced shock resistivity and lower dynamic range. Besides, if the air gap between the diaphragm and the backplate is very small, the acoustic resistance of the air gap will reduce the mechanical sensitivity of the microphone at higher frequencies. For instance, at an air gap of 2μm, an upper cutoff frequency of only 2kHz has been measured [1].

One way to improve the characteristics of a condenser microphone is to use a mechanical feedback from the output of the amplifier to the diaphragm [4]. Fig. 11-1A shows a circuit diagram and Fig. 11-1B is a drawing of interdigitized electrodes of the microphone. The electrodes serve different purposes - one is for the conversion of a diaphragm displacement into voltage at the input of the amplifier A_1 while the other electrode is for converting feedback voltage V_a into a mechanical deflection by means of electrostatic force. The mechanical feedback clearly improves linearity and the frequency range of the microphone, however, it significantly reduces the deflection which results in a lower sensitivity.

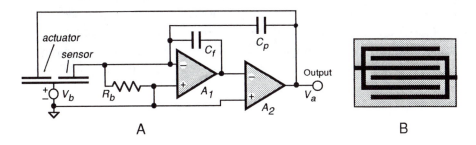

Fig. 11-1 Condenser microphone with a mechanical feedback
A: a circuit diagram; B: interdigitized electrodes on the diaphragm
(adapted from [4])

11.3 FIBER-OPTIC MICROPHONE

Direct acoustic measurements in hostile environments, such as in turbojets or rocket engines, require sensors which can withstand high heat and strong vibrations. The acoustic measurements under such hard conditions are required for computational fluid dynamics (CFD) code validation, structural acoustic tests, and jet noise abatement. For such applications, a fiber-optic interferometric microphone can be quite suitable. One such design [5] is comprised of a single-mode temperature insensitive Michelson interferometer, and a reflective plate diaphragm (see Fig. 3-16.7). The interferometer monitors the plate deflection which is directly related to the acoustic pressure. The sensor is water cooled to provide thermal protection for the optical materials and to stabilize the mechanical properties of the diaphragm.

To provide an effect of interference between the incoming and outgoing light beams, two fibers are fused together and cleaved at the minimum tapered region (Fig. 11-2). The fibers are incorporated into a stainless steel tube which is water cooled. The internal space in the tube is filled with epoxy, while the end of the tube is polished until the optical fibers are observed. Next, aluminum is selectively deposited to one of the fused fiber core ends to make its surface mirror reflective. This fiber serves as a reference arm of the microphone. The other fiber core is left open and serves as the sensing arm. Temperature insensitivity is obtained by the close proximity of the reference and sensing arms of the assembly.

Light from a laser source (a laser diode operating near 1.3μm wavelength) enters one of the cores and propagates toward the fused end, where it is coupled to the other fiber core. When reaching the end of the core, light in the reference core is reflected from the aluminum mirror toward the input and output sides of the sensor. The portion of light which goes toward the input is lost and makes no effect on the measurement, while the portion which goes to the output, strikes the detector's surface. That portion of light which travels to the right in the sensing core, exits the fiber and strikes the copper diaphragm. Part of the light is reflected from the diaphragm back toward the sensing fiber and propagates to the output end, along with the reference light. Depending on the position of the diaphragm, the phase of the reflected light will vary, thus becoming different from the phase of the reference light.

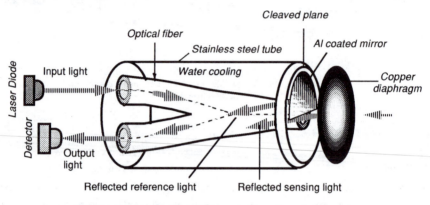

Fig. 11-2 Fiber optic interferometric microphone
Movement of copper diaphragm is converted into light intensity in the detector

While traveling together to the output detector, the reference and sensing lights interfere with one another, resulting in the light intensity modulation. Therefore, the microphone converts the diaphragm displacement into a light intensity. Theoretically, the signal-to-noise ratio in such a sensor is obtainable on the order of 70-80dB, thus resulting in an average minimum detectable diaphragm displacement of 1Å (10^{-10}m).

Fig. 11-3 shows a typical plot of the optical intensity in the detector versus the phase for the interference patterns. To assure a linear transfer function, the operating point should be selected near the middle of the intensity, where the slope is the highest and the linearity is the best. The slope and the operating point may be changed by adjusting the wavelength of the laser diode. It is important for the deflection to stay within a quarter of the operating wavelength to maintain a proportional input.

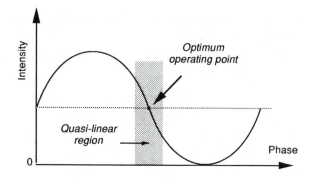

Fig. 11-3 Intensity plot as function of a reflected light phase

The diaphragm is fabricated from a 0.05 mm foil with a 1.25mm diameter. Copper is selected for the diaphragm due to its good thermal conductivity and relatively low modulus of elasticity. The latter feature allows us to use a thicker diaphragm which provides better heat removal while maintaining a usable natural frequency and deflection. A pressure of 1.4kPa produces a maximum center deflection of 39nm (390Å) which is well within a 1/4 of the operating wavelength (1300nm). The maximum acoustic frequency which can be transferred with the optical microphone is limited to about 100 kHz which is well above the desired working range needed for the structural acoustic testing.

11.4   PIEZOELECTRIC MICROPHONES

The piezoelectric effect can be used for the design of simple microphones, as it is a direct conversion of a mechanical stress developed in a crystalline material into an electric charge. The most frequently used material for the sensor is a piezoelectric ceramic, which can operate up to a very high frequency limit. This is the reason, why piezoelectric sensors are used for transduction of ultrasonic waves (Section 5.9).

11.5   ELECTRET MICROPHONES

An electret is a close relative to piezoelectric and pyroelectric materials. In effect, they are all electrets with enhanced either piezoelectric or pyroelectric properties. An electret is a permanently electrically polarized crystalline dielectric material. The first application of electrets to microphones and earphones where described in 1928 [6]. An electret microphone is an electrostatic transducer consisting of a metallized electret diaphragm and backplate separated from the diaphragm by an air gap (Fig. 11-4).

The upper metallization and a metal backplate are connected through a resistor R, voltage V across which can be amplified and used as an output signal. Since the electret is permanently electrically polarized dielectric, charge density σ_1 on its surface is constant and sets in the air gap an electric field E_1. When acoustic wave impinges on the diaphragm, the latter deflects downward reducing the air gap thickness s_1 for a value of Δs. Under open-circuit conditions, the amplitude of a variable portion of the output voltage becomes

$$V = \frac{s \cdot \Delta s}{\varepsilon_0(s + \varepsilon s_1)} \; . \qquad (11.2)$$

Thus, the deflected diaphragm generate voltage across the electrodes. That voltage is in phase with the diaphragm deflection. If the sensor has a capacitance C the above equation should be written as

$$V = \frac{s \cdot \Delta s}{\varepsilon_0(s + \varepsilon s_1)} \frac{2\pi f RC}{\sqrt{1 + (2\pi f RC)^2}} \; , \qquad (11.3)$$

where f is frequency of sonic waves.

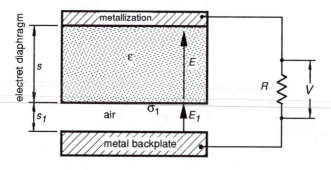

Fig. 11-4 General structure of an electret microphone
Thicknesses of layers are exaggerated for clarity
(after [7])

If the restoring forces are due to elasticity of the air cavities behind the diaphragm (effective thickness is s_o) and the tension T of the membrane, its displacement Δs to a sound pressure Δp assuming negligible losses is given by [8]

$$\Delta s = \frac{\Delta p}{(\gamma p_o/s_o) + (8\pi T/A)} ,$$ (11.4)

where γ is the specific heat ratio, p_o the atmospheric pressure, and A is the membrane area. If we define the electret microphone sensitivity as $\delta_m = \Delta V/\Delta p$, then below resonant it can be expressed as [7]

$$\delta_m = \frac{s s_o \sigma_1}{\varepsilon_0(s + \varepsilon s_1)\gamma p_o} .$$ (11.5)

It is seen that the sensitivity doesn't depend on area. If mass of the membrane is M, then the resonant frequency is defined by

$$f_r = \frac{1}{2\pi}\sqrt{\frac{p_o}{s_o M}} .$$ (11.6)

This frequency should be selected well above the upper frequency of the microphone's operating range.

The electret microphone differs from other similar sensors in the sense that it doesn't require a dc bias voltage. For a comparable design dimensions and sensitivity, a condenser microphone would require well over 100V bias. The mechanical tension of the membrane is generally kept at a relatively low value (about 10Nm^{-1}), so that the restoring force is determined by the air gap compressibility. A membrane may be fabricated of Teflon FEP which is permanently charged by an electron beam to give it electret properties. The temperature coefficient of sensitivity of the electret microphones are in the range of 0.03dB/°C in the temperature range from -10 to +50°C [9].

Foil-electret (diaphragm) microphones have more desirable features than any other microphone types. Among them is very wide frequency range from 10^{-3}Hz and up to hundreds of MHz. They also feature a flat frequency response (within ±1dB), low harmonic distortion, low vibration sensitivity, good impulse response, and insensitivity to magnetic fields. Sensitivities of electret microphones are in the range of few mV/μbar.

For operation in the infrasonic range, an electret microphone requires a miniature pressure equalization hole on the backplate. When used in the ultrasonic range, the electret is often given an additional bias (like a condenser microphone) in addition to its own polarization.

REFERENCES

1. Hohm, D. and Hess, G. A subminiature condenser microphone with silicon nitrite membrane and silicon back plate. In: *J. Acoust. Soc. Am.* vol. 85, pp: 476-480, Jan. 1989

2. Bergqvist, J. and Rudolf, F. A new condenser microphone in silicon. In: *Sensors and actuators,* vol. A21-A23, pp: 123-125, June 1990

3. Sprenkels, A.J., Groothengel, R.A., Verloop and A.J., Bergveld, P. Development of an electret microphone in silicon. *Sensor and actuators*, vol. 17, no. 3&4, pp: 509-512, 1989

4. van der Donk, A.G.H., Sprenkels, A.J., Olthuis, W., and Bergveld, P. Preliminary results of a silicon condenser microphone with internal feedback. In: *Transducers'91. International conference on solid-state sensors and actuators. Digest of technical papers*, pp. 262-265, IEEE, 1991

5. Hellbaum, R.F. *et al.* An experimental fiber optic microphone for measurement of acoustic pressure levels in hostile environments. In: *Sensors Expo Proceedings,* Helmers Publishing, Inc., 1991

6. Nishikawa, S. and Nukijama, S. *Proc. Imp. Acad.* Tokyo, vol. 4, p: 290, 1928

7. Sessler, G.M., ed. Electrets. Springer-Verlag. Berlin, 1980

8. Morse, P.M. Vibration and Sound. McGraw-Hill, New York, 1948

9. Griese, H.J. Paper Q29, *Proc. 9th Int. Conf. Acoust.*, Madrid, 1977

12

Humidity and Moisture Sensors

12.1 CONCEPT OF HUMIDITY

The water content in surrounding air is an important factor for the well-being of humans and animals. The level of comfort is determined by a combination of two factors: relative humidity and ambient temperature. You may be quite comfortable at -30°C in Siberia, where the air is usually very dry in winter, and feel quite miserable in Cleveland near lake Erie at 0°C, where air may contain substantial amount of moisture[1]. Humidity is an important factor for operating certain equipment, for instance, high impedance electronic circuits, electrostatic sensitive components, high voltage devices, fine mechanisms, etc. A rule of thumb is to assure a relative humidity near 50% at normal room temperature (20-25°C). This may vary from as low as 38% for class-10 clean rooms to 60% in hospital operating rooms.

Moisture is the ingredient common to most manufactured goods and processed material. It can be said that a significant amount of the U.S. GNP (Gross National Product) is moisture [1]. Humidity can be measured by the instruments called *hygrometers*. The first hygrometer was invented by Sir John Leslie (1766-1832) [2]. To detect moisture contents, a sensor in a hygrometer must be selective to water, and its internal properties should be modulated by the water concentration. Generally, sensors for moisture, humidity and dew temperature can be capacitive, conductive, oscillating, or optical. The optical sensors for gases detect dewpoint temperature, while the optical hygrometers for organic solvents employ absorptivity of near infrared (NIR) light in the spectral range from 1.9µm to 2.7µm [3] (see Fig. 13-19).

There are many ways to express moisture and humidity, often depending on industry or the particular application. Moisture of gases sometimes is expressed in pounds of water vapor per million cubic feet of gas. Moisture in liquids and solids is generally given as a percentage of water per

---

[1] Naturally, here we disregard other comfort factors, such as political or economical.

total mass (wet weight basis), but may be given on a dry weight basis. Moisture in liquids with low water miscibility is usually expressed as parts per million by weight (PPM_w).

The term *moisture* generally refers to the water content of any material, but for practical reasons, it is applied only to liquids and solids, while the term *humidity* is reserved for the water vapor content in gases. Below we provide some useful definitions [1].

Moisture - the amount of water contained in a liquid or solid by absorption or adsorption which can be removed without altering its chemical properties.

Mixing ratio (humidity ratio) r - the mass of water vapor per unit mass of dry gas.

Absolute humidity (mass concentration or density of water vapor) - the mass m of water vapor per unit volume v of wet gas: $d_w=m/v$. In other words, absolute humidity is the density of water vapor component. It can be measured, for example, by passing a measured quantity of air through a moisture-absorbing substance (such as silica-gel) which is weighed before and after the absorption. Absolute humidity is expressed in grams per cubic meter, or in grains per cubic foot. Since this measure is also a function of atmospheric pressure, it is not generally useful in engineering practice.

Relative humidity (RH) is the ratio of the actual vapor pressure of the air at any temperature, to the maximum of saturation vapor pressure at the same temperature. Relative humidity in percents is defined as

$$H = 100\frac{P_w}{P_s} \quad , \tag{12.1}$$

where P_w is the partial pressure of water vapor and P_s is the pressure of saturated water vapor at a given temperature. The value of H expresses the vapor content as a percentage of the concentration required to cause the vapor saturation, that is, the formation of water droplets (dew) at that temperature. An alternative way to present RH is as a ratio of the mole fraction of water vapor in a space to the mole fraction of water vapor in the space at saturation.

The value of P_w together with partial pressure of dry air P_a is equal to pressure in the enclosure, or to the atmospheric pressure P_{atm} if the enclosure is open to the atmosphere:

$$P_w + P_a = P_{atm} \quad . \tag{12.2}$$

At temperatures above the boiling point, water pressure could displace all other gases in the enclosure. The atmosphere would then consist entirely of superheated steam. In this case, $P_w=P_{atm}$. At temperatures above 100°C, RH is a misleading indicator of moisture content because at these temperatures P_s is always more the P_{atm}, and maximum RH never can reach 100%. Thus, at normal atmospheric pressure and temperature of 100°C, the maximum RH is 100%, while at 200°C it is only 6%. Above 374°C, saturation pressures are not thermodynamically specified.

Dewpoint temperature - the temperature at which the partial pressure of the water vapor present, would be at its maximum, or saturated vapor condition, with respect to equilibrium with a plain surface of ice, or the temperature to which the gas-water vapor mixture must be cooled isobarically (at constant pressures) to induce frost or ice (assuming no prior condensation). The *dewpoint* is the temperature at which relative humidity is 100%.

Relative humidity displays an inverse relationship with absolute temperature. Dewpoint temperature is usually measured with a chilled mirror (see below). However, below 0°C dewpoint, the measurement becomes uncertain as moisture eventually freezes and a crystal lattice growth will slowly occur, much like a snowflake. Nevertheless, moisture can exist for prolonged time below 0°C in a liquid phase, depending on such variables as molecular agitation, rate of convection, sample gas temperature, contaminations, etc.

12.2 CAPACITIVE SENSORS

An air filled capacitor may serve as a relative humidity sensor because moisture in the atmosphere changes air electrical permitivity according to the following equation [4]

$$\kappa = 1 + \frac{211}{T}\left(P + \frac{48P_s}{T}H\right)10^{-6} , \qquad (12.3)$$

where T is the absolute temperature in kelvin, P is the pressure of moist air in mm Hg, P_s is the pressure of saturated water-vapor at temperature T, mm Hg, H is the relative humidity in %. Eq. (12.3) shows that the dielectric constant of moist air, and, therefore, the capacitance, is proportional to the relative humidity.

Instead of air, the space between the capacitor plates can be filled with an appropriate isolator whose dielectric constant changes significantly upon being subjected to humidity. The capacitive sensor may be formed of a hygroscopic polymer film with metallized electrodes deposited on the opposite sides. In one design [6] the dielectric was composed of a hygrophilic polymer thin film (8-12 μm thick) made of cellulose acetate butyrate and the dimetylephtalate as plasticizer. A size of the film sensor is 12x12 mm. The 8-mm-diameter gold porous disk electrodes (200Å thick) were deposited on the polymer by the vacuum deposition. The film was suspended by a holder and the electrodes were connected to the terminals. The capacitance of such a sensor is approximately proportional to relative humidity H:

$$C_h \approx C_o(1 + \alpha_h H) \ , \tag{12.4}$$

where C_o is the capacitance at $H=0$.

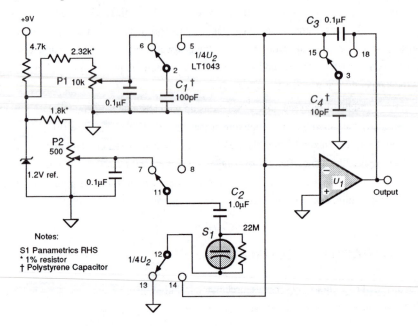

Fig. 12-1 Simplified circuit for measuring humidity with a capacitive sensor
(adapted from [5])

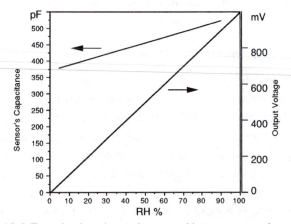

Fig. 12-2 Transfer functions of a capacitive sensor and a system

For the use with capacitive sensors, a 2% accuracy in the range from 5% to 90% RH can be achieved with a simple circuit shown in Fig. 12-1. The sensor and the circuit transfer characteristics are in Fig. 12-2. The sensor's nominal capacitance at 75% RH is 500pF. It has a quasi-linear transfer function with the offset at zero humidity of about 370pF and the slope of 1.7pF/% RH. The circuit effectively performs two functions: makes a capacitance-to-voltage conversion and subtracts the offset capacitance to produce an output voltage with zero intercept. A heart of the circuit is a self-clocking analog switch LT1043 which multiplexes several capacitors at the summing junction (virtual ground) of the operational amplifier U_1. The capacitor C_1 is for the offset capacitance subtraction, while the capacitor C_2 is connected in series with the capacitive sensor S_1. The average voltage across the sensor must be zero; otherwise, electrochemical migration could damage it permanently. Nonpolarized capacitor C_2 protects the sensor against building up any dc charge. Trimpot P_2 adjusts the amount of charge delivered to the sensor and P_1 trims the offset charge which is subtracted from the sensor. The net charge is integrated with the help of the feedback capacitor C_3. Capacitor C_4 maintains dc output when the summing junction is disconnected from the sensor.

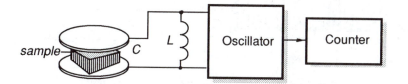

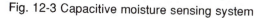

Fig. 12-3 Capacitive moisture sensing system

A similar technique can be used for measuring moisture in material samples [7]. Fig. 12-3 shows a block diagram of the capacitive measurement system where the dielectric constant of the sample changes frequency of the oscillator. This method of moisture measurement is quite useful in the process control of the pharmaceutical products. Dielectric constants of most of the medical tablets is quite low (between 2.0 and 5.0) as compared with that of water (Fig. 3-2.4). The sampled material is placed between two test plates which form a capacitor connected into an LC-oscillating circuit. The frequency is measured and related to the moisture. The best way to reduce variations attributed to environmental conditions, such as temperature and room humidity, is a use of a differential technique[1]. That is, the frequency shift $\Delta f = f_o - f_1$ is calculated, where f_o and f_1 are frequencies produced by the empty container and that filled with the sampled material, re-

---

[1] An example is Model 4586 Moisture Tester produced by Forte's Technology, Inc., Norwood, MA.

spectively. The method has some limitations, for instance, its accuracy is poor when measuring moistures below 0.5%, the sample must be clean of foreign particles having relatively high dielectric constants - examples are metal and plastic objects, a packing density, and a fixed sample geometry must be maintained.

A thin film capacitive humidity sensor can be fabricated on a silicon substrate [8]. A layer of SiO_2 3,000Å thick is grown on an *n*-Si substrate (Fig. 12-B). Two metal electrodes are deposited on the SiO_2 layer. They are made of aluminum, chromium, or phosphorous doped polysilicon (LPCVD). The electrode thickness is in the range from 2000 to 5000Å. The electrodes are shaped in an interdigitized patter as shown in Fig. 12-4A. To provide additional temperature compensation, two temperature sensitive resistors are formed on the same substrate. The top of the sensor is coated with a dielectric layer. For this layer, several materials can be used, such as chemically vapor-deposited SiO_2 (CVD PSG) or phosphorosilicate glass (CVD PSG). The thickness of the layer is in the range from 300 to 4000Å.

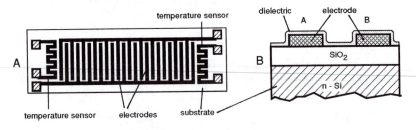

Fig. 12-4 A capacitive thin-film humidity sensor
A: interdigitized electrodes form capacitor plates;
B: cross section of the sensor

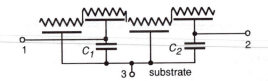

Fig. 12-5 A simplified equivalent electric circuit of a capacitive thin-film humidity sensor

A simplified equivalent electrical circuit is shown in Fig. 12-5. Each element of the circuit represents a *RC*-transmission line [9]. When the relative humidity increases, the distributed surface resistance drops and the equivalent capacitance between the terminals 1 and 2 grows. The capacitance is frequency dependent, hence for the low humidity range measurement frequency should be selected near 100Hz, while for the higher humidities it is in the range between 1 and 10kHz.

12.3 CONDUCTIVITY SENSORS

A general concept of a conductive hygrometric sensor is shown in Fig. 12-6. The sensor contains a material of relatively low resistivity which changes significantly under varying humidity conditions. The material is deposited on the top of two interdigitized electrodes to provide a large contact area. When water molecules are absorbed by the upper layer, resistivity between the electrodes changes which can be measured by an electronic circuit. The first such sensor was developed by F. W. Dunmore in 1935 of a hygroscopic film consisting of 2%-5% aqueous solution of LiCl [10]. Another example of a conductive humidity sensor is the so-called "Pope element" which contains a polystyrene film treated with sulfuric acid to obtain the desired surface-resistivity characteristics.

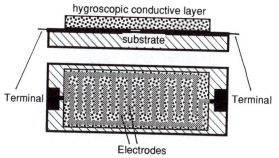

hygroscopic conductive layer

substrate

Terminal

Terminal

Electrodes

Fig. 12-6 A composition of a conductive humidity sensor

Other promising materials for the fabrication of a film in a conductivity sensor are solid polyelectrolytes because their electrical conductivity varies with humidity. Long term stability and repeatability of these compounds, while generally not too great, can be significantly improved by using the interpenetrating polymer networks and carriers, and supporting media. When measured at 1kHz, an experimental sample of such a film has demonstrated a change in impedance from $10M\Omega$ to 100Ω while RH was changing from 0% to 90% [11].

A solid-state humidity sensor can be fabricated on a silicon substrate (Fig. 12-7A). The silicone must be of a high conductance [12] which provides an electrical path from the aluminum electrode vacuum deposited on its surface. An oxide layer is formed on the top of the conductive aluminum layer, and on the top of that, another electrode is formed. The aluminum layer is anodized in a manner to form a porous oxide surface. The average cross-sectional dimension of pores is sufficient to allow penetration by water molecules. The upper electrode is made in a form of porous gold which is permeable to gas and, at the same time, can provide electrical contact. Electrical connections are made to the gold and silicon layers. Aluminum oxide (AL_2O_3), like numerous other materials, readily sorb water when in contact with a gas mixture containing water in the vapor phase. The amount of sorption is proportional to the water vapor partial pressure and inversely proportional to the absolute temperature. Aluminum oxide is a dielectric material. Its dielectric constant and surface resistivity are modified by the physisorption of water. For this reason, this material lends itself as a humidity sensing compound.

Fig. 12-7B shows an electrical equivalent circuit of the sensor [13]. The values of R_1 and C_1 depend on the Al_2O_3 average pore sizes and density. These components of resistance and capacitance vary with the number of water molecules that penetrate the pores and adhere to the surface. R_2 and C_2 represent the resistance and capacitance components of the bulk oxide material between the pores and are therefore unaffected by moisture. C_3 is an equivalent series capacitance term as determined by the measurement of the total resistance components in a dry atmosphere at very low frequencies. The sensor's resistance becomes very large ($>10^8\Omega$) as the frequency approaches dc. Thus, the measurement of humidity involves the measurement of the sensor's impedance. The residual of non-humidity dependent resistance and capacitance terms that exist in a typical sensor shunt the humidity dependent variables, thus causing the continuous reduction in slope (sensitivity) as the humidity is lowered, which, in turn, reduces the accuracy at lower humidities. Since temperature is a factor in humidity measurement, the sensor usually combines in the same package a humidity sensor, a thermistor, and a reference capacitance, which is protected against humidity influence and has a low temperature coefficient.

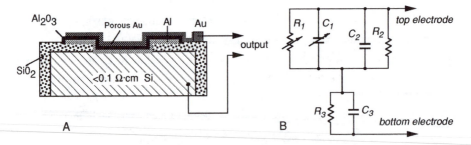

Fig. 12-7 A: Structure of Al_2O_3 thin film moisture sensor
B: Simplified equivalent circuit of the sensor
R_1 and C_1 moisture dependent variable terms; R_2 and C_2 shunting terms of bulk oxide between pores (unaffected by moisture); R_3 and C_3 series terms below pores (unaffected by moisture)

12.4 THERMAL CONDUCTIVITY SENSOR

Using thermal conductivity of gas to measure humidity can be accomplished by used a thermistor based sensor (Fig. 12-8A) [14]. Two tiny thermistors (R_{t1} and R_{t2}) are supported by thin wires to minimize thermal conductivity loss to the housing. The left thermistor is exposed to the outside gas through small venting holes, while the right thermistor is hermetically sealed in dry air. Both thermistors are connected into a bridge circuit (R_1 and R_2), which is powered by voltage $+E$. The thermistors develop self-heating due to the passage of electric current. Their temperatures rise up to 170°C over the ambient temperature. Initially, the bridge is balanced in dry air to establish a zero reference point. The output of this sensor gradually increases as absolute humidity rises from zero. At about 150 g/m^3 it reaches the saturation and then decreases with a polarity change at about 345g/m^3 (Fig. 12-8B).

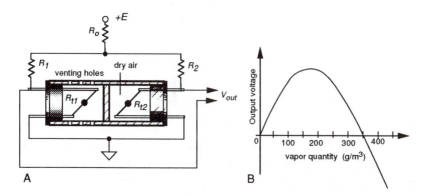

Fig. 12-8 Absolute humidity sensor with self-heating thermistors
A - design and electrical connection; B - output voltage

12.5 OPTICAL HYGROMETER

Most of the humidity sensors exhibit some repeatability problems, especially hysteresis with a typical value from 0.5% to 1%RH. In precision process control, this may be a limiting factor, therefore indirect methods of humidity measurements should be considered. The most efficient method is a calculation of absolute or relative humidity through dewpoint temperature. As was indicated above, the dewpoint is the temperature at which liquid and vapor phases of water (or any fluid for that matter) are in equilibrium. The temperature at which the vapor and solid phases are in equilibrium is called the *frostpoint*. At the dewpoint, only one value of saturation vapor pressure exist. Hence, absolute humidity can be measured from this temperature as long as the pressure is known. The optimum method of moisture measurement by which the minimum hysteresis effects are realized require the use of optical hygrometry. The cost of an optical hygrometer is considerably greater, but if the benefit of tracking low level moisture enhances product yield and quality, the cost is easily justified.

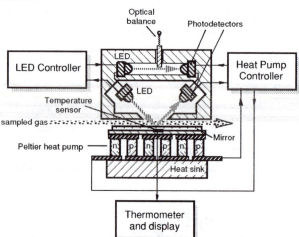

Fig. 12-9 Chilled mirror dewpoint sensor with an optical bridge

The basic idea behind the optical hygrometer is the use of a mirror whose surface temperature is precisely regulated by a thermoelectric heat pump. The mirror temperature is controlled at a threshold of formation of dew. Sampled air is pumped over the mirror surface and, if the mirror temperature crosses a dewpoint, releases moisture in the form of water droplets. The reflective properties of the mirror change at water condensation because water droplets scatter light rays. This can be detected by an appropriate photodetector. Fig. 12-9 shows a simplified block diagram of a chilled mirror hygrometer. It is comprised of a heat pump operating on a Peltier effect. The pump removes heat from a thin mirrored surface which has an imbedded temperature sensor. That sensor is part of a digital thermometer which displays the temperature of the mirror. The hygrometer's circuit is of a differential type, where the top optocoupler, a light emitting diode (LED) and a photodetector, are used for the compensation of drifts, while the bottom optocoupler is for measuring the mirror reflectivity. The sensor's symmetry can be balanced by a wedged optical balance inserted into the light path of the upper optocoupler. The lower optocoupler is positioned at 45° angle with respect to the mirror. Above the dewpoint, the mirror is dry and its reflectivity is the highest. The heat pump controller lowers the temperature of the mirror through the heat pump. At the moment of the water condensation, the mirror reflectivity drops abruptly, which caused reduction in a photocurrent in the photodetector. The photodetector signals pass to the controller to regulate electric current through the heatpump to maintain its surface temperature at the level of a dewpoint, where no additional condensation or evaporation from the mirror surface occurs. Actually, water molecules are continuously being trapped and are escaping from the surface, but the average net level of the condensate density does not change once equilibrium is established.

Since the sensed temperature of the mirrored surface precisely determines the actual prevailing dewpoint, this is considered the moisture's most fundamental and accurate method of measurement. Hysteresis is virtually eliminated and sensitivity is near 0.03°C DP (dewpoint). From the dewpoint, all moisture parameters such as %RH, vapor pressure, etc. are obtainable once the prevailing temperature and pressure are known.

There are several problems associated with the method. One is a relatively high cost, the other is a potential mirror contamination, and the third is a relatively high power consumption by the heat pump. Contamination problems can be virtually eliminated with use of particle filters and a special technique which deliberately cools the mirror well below the dewpoint to cause excessive condensation, with the following fast rewarming. This flashes the contaminants keeping the mirror clean [15].

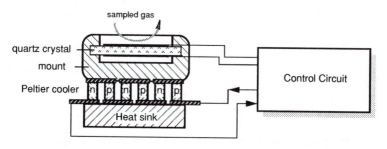

Fig. 12-10 An oscillating humidity sensor

12.6 OSCILLATING HYGROMETER

The idea behind the oscillating hygrometer is similar to that behind the optical chilled mirror sensor. The difference is that the measurement of the dewpoint is made not by the optical reflectivity of the surface, but rather by detecting the changing mass of the chilled plate. The chilled plate is fabricated of a thin quartz crystal, that is a part of an oscillating circuit. This implies the other name for the sensor: *the piezoelectric hygrometer*, because the quartz plate oscillation is based on the piezoelectric effect. A quartz crystal is thermally coupled to the Peltier cooler (see Section 3.9) which controls the temperature of the crystal with a high degree of accuracy (Fig. 12-10). When the temperature drops to that of a dewpoint, a film of water vapor deposits on the exposed surface of the quartz crystal. Since the mass of the crystal changes, the resonant frequency of the oscillator shifts from f_o to f_1. The new frequency f_1 corresponds to a given thickness of the water layer. The frequency shift controls current through the Peltier cooler, thus changing the temperature of the quartz crystal to stabilize at the dewpoint temperature. The major difficulty in designing the piezo-electric hygrometer is in providing an adequate thermal coupling between the cooler and the crystal, while maintaining small size of the crystal at a minimum mechanical loading [16].

REFERENCES

1. Quinn, F. C. The most common problem of moisture/humidity measurement and control. In: Moisture and Humidity, *Proc. of the 1985 Int. Symp. on Moisture and Humidity.* , pp: 1-5 Washington, DC, ©ISA, 1985

2. Carter, E. F., ed. Dictionary of inventions and discoveries. Crane, Russak and Co., N.Y., © 1966 by F. Muller.

3. Baughman E. H. and Mayes, D. NIR applications to process analysis. *American Laboratory*, vol. 21, No.10, pp: 54-58, 1989

4. Lea. Notes on the stability of LC oscillators. *J. Inst. Elec. Eng.*, vol. 92, pt. II, 1945

5. Conditioner circuit, AN3, Linear Technology, Inc., Appl. Handbook, 1990

6. Sashida, T. and Sakaino, Y. An interchangeable humidity sensor for an industrial hygrometer. In: Moisture and Humidity. *Proc. of the Intern. Symp. on Moisture and Humidity*, Washington, DC, April 15-18, 1985

7. Carr-Brion, K. Moisture sensors in process control. Elsevier Applied Science Publishers, New York, 1986

8. Jachowicz, R. S. and Dumania, P. Evaluation of thin-film humidity sensor type MCP-MOS. In: *Moisture and Humidity. Proc. of the Intern. Symp. on Moisture and Humidity*, Washington, DC, April 15-18, 1985

9. Jachowicz, R. S. and Senturia, S.D. A thin film humidity sensor. *Sensors and Actuators*, vol. 2, 1981, 1982

10. Norton, H. N. Handbook of transducers. Prentice Hall, Englewood Cliffs, NJ, 1989

11. Sakai, Y., Sadaoka, Y., Matsuguchi, M., and Hirayama, K. Water resistive humidity sensor composed of interpenetrating polymer networks of hydrophilic and hydrophobic methacrylate. In: *Transducers'91. International conference on solid-state sensors and actuators. Digest of technical papers*, pp: 562-565, IEEE, 1991

12. Fong, V. Al_2O_3 moisture sensor chip for inclusion in microcircuit package and the new MIL standard for moisture content. In: *Moisture and Humidity, Proceedings of the 1985 Int. Symp. on Moisture and Humidity.* Washington, pp: 345-357, DC, ISA, 1985

13. Harding, Jr., J. C. Overcoming limitations inherent to aluminum oxide humidity sensors. In: *Moisture and Humidity, Proceedings of the 1985 Int. Symp. on Moisture and Humidity.* Washington, DC, pp: 367-378, ISA, 1985

14. Miura, T. Thermistor humidity sensor for absolute humidity measurements and their applications. In: *Moisture and Humidity. Proc. of the Intern. Symp. on Moisture and Humidity*, Washington, DC, April 15-18, 1985

15. Harding, Jr. J. C. A chilled mirror dewpoint sensor/psychrometric transmitter for energy monitoring and control systems. In: *Moisture and Humidity. Proc. of the Intern. Symp. on Moisture and Humidity*, Washington, DC, April 15-18, 1985

16. Porlier, C. Chilled piezoelectric hygrometer: sensor interface design. In: *Sensors Expo proceedings*, 107B-7, Helmers Publishing, Inc., 1991

13

Light Detectors

13.1   INTRODUCTION

Detectors of electromagnetic radiation in the spectral range from ultraviolet to infrared are called light detectors. From the standpoint of a sensor designer, absorption of photons by a sensing material may result either in a quantum or thermal response. Therefore, all light detectors are divided into two major groups that are called *quantum* and *thermal*. The quantum detectors operate from the ultraviolet to mid infrared spectral ranges, while thermal detectors are most useful in the mid and far infrared spectral range where their efficiency at room temperatures exceeds that of the quantum detectors.

Quantum detectors (photovoltaic and photoconductive devices) relay on the interaction of individual photons with a crystalline lattice of semiconductor materials. Their operations are based on the photoeffect that was discovered by A. Einstein, and brought him the Nobel Prize. In 1905, he made a remarkable assumption about the nature of light, that at least under certain circumstances, its energy was concentrated into localized bundles, later named photons. The energy of a single photon is given by

$$E = h\nu ,\tag{13.1}$$

where ν is the frequency of light and $h=6.63 \cdot 10^{-34}$J·s (or $4.13 \cdot 10^{-15}$eV·s) is Plank's constant derived on the basis of the wave theory of light. When a photon strikes a surface of a conductor, it may result in the generation of a free electron. Part (ϕ) of the photon energy E is used to detach the electron from the surface, while the other part gives it kinetic energy. The photoelectric effect can be described as

$$h\nu = \phi + K_m ,\tag{13.2}$$

where ϕ is called the *work function* of the emitting surface and K_m is the maximum kinetic energy of the electron upon its exiting the surface. The similar processes occur when a semiconductor *pn*-junction is subjected to radiant energy: the photon transfers its energy to an electron and, if the energy is sufficiently high, the electron may become mobile which results in an electric current.

The periodic lattice of crystalline materials establishes allowed energy bands for electrons that exist within that solid. The energy of any electron within the pure material must be confined to one of these energy bands which may be separated by gaps or ranges of forbidden energies.

If light of a proper wavelength (sufficiently high energy of photons) strikes a semiconductor crystal, the concentration of charge carriers (electrons and holes) in the crystal increases, which manifests in the increased conductivity of a crystal

$$\sigma = e(\mu_e n + \mu_h p) \ , \tag{13.3}$$

where e is the electron charge, μ_e is the electron mobility, μ_h is the hole mobility, and n and p are the respective concentrations of electrons and holes.

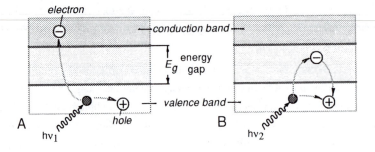

Fig. 13-1 Photoeffect in a semiconductor for high (A) and low (B) energy photons

Fig. 13-1A shows energy bands of a semiconductor material, where E_g is the magnitude in *eV* of the forbidden band gap. The lower band is called the valence band, which corresponds to those electrons that are bound to specific lattice sites within the crystal. In the case of silicon or germanium, they are parts of the covalent bonding which constitute the interatomic forces within the crystal. The next higher-lying band is called the conduction band and represents electrons that are free to migrate through the crystal. Electrons in this band contribute to the electrical conductivity of the material. The two bands are separated by the band gap, the size of which determines the whether the material is classified as a semiconductor or an isolator. The number of electrons within the crystal is just adequate to completely fill all available sites within the valence band. In the absence of thermal excitation, both isolators and semiconductors would therefore have a configuration in which the valence band is completely full, and the conduction band completely empty. Under these circumstances, neither would theoretically show any electrical conductivity. In a metal, the highest occupied energy band is not completely full, Therefore, electrons can easily migrate throughout the material because they need achieve only a small incremental energy to the above occupied states. Metals, therefore, are always characterized by very high electrical conductivity. In isolators or semiconductors, on the other hand, the electron must first cross the energy band gap in order to reach the conduction band and the conductivity is therefore many orders of magnitude lower. For isolators, the band gap is usually 5 eV or more, whereas for semiconductors, the gap is considerably less (Table 13-1).

Table 13-1 Band gaps and longest wavelengths for various semiconductors (after [1])

Material	Band gap (eV)	Longest wavelength (μm)
ZnS	3.6	0.345
CdS	2.41	0.52
CdSe	1.8	0.69
CdTe	1.5	0.83
Si	1.12	1.10
Ge	0.67	1.85
PbS	0.37	3.35
InAs	0.35	3.54
Te	0.33	3.75
PbTe	0.3	4.13
PbSe	0.27	4.58
InSb	0.18	6.90

When the photon of frequency v_1 strikes the crystal, its energy is high enough to separate the electron from its site in the valence band and push it through the band gap into a conduction band at a higher energy level. In that band, the electron is free to serve as a current carrier. The deficiency of an electron in the valence band creates a hole which also serves as a current carrier. This is manifested in the reduction of specific resistivity of the material. On the other hand, Fig. 13-1B shows that a photon of lower frequency v_2 doesn't have sufficient energy to push the electron through the band gap. The energy is released without creating current carriers.

The energy gap serves as a threshold, below which the material is not light sensitive. However, the threshold is not abrupt. Throughout the photon-excitation process, the law of conservation of momentum applies. The momentum and density of hole-electron sites are higher at the center of both the valence and conduction bands, and fall to zero at the upper and lower ends of the bands. Therefore, the probability of an excited valence-band electron finding a site of like momentum in the conduction band is greater at the center of the bands, and the lowest at the ends of the bands. Therefore, the response of a material to photon energy increases from E_g gradually to its maximum and then falls back to zero at the energy corresponding to the difference between the bottom of the valence band and the top of the conduction band. A typical spectral response of a semiconductive material is shown in Fig. 13-2. The light response of a bulk material can be altered by adding various impurities. They can be used to reshape and shift a spectral response of the material. All devices that directly convert photons of electromagnetic radiation into charge carriers are called *quantum detectors* which are generally produced in a form of photodiodes, phototransistors, and photoresistors.

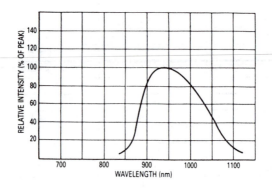

Fig. 13-2 Spectral response of an infrared photodiode

When comparing the characteristics of different photodetectors, the following specifications usually should be considered:

NEP (noise equivalent power) is the amount of light equivalent to the intrinsic noise level of the detector. Stated differently, it is the light level required to obtain a signal-to-noise ratio equal to unity. Since the noise level is proportional to the square root of the bandwidth, the *NEP* is expressed in units of $W/\sqrt{Hz}$

$$NEP = \frac{\text{noise current } (A/\sqrt{Hz})}{\text{Radiant sensitivity at } \lambda_p \ (A/W)}. \qquad (13.4)$$

$D*$ *(D-star)* refers to the *detectivity* of a detector's sensitive area of 1cm^2 and a noise bandwidth of 1Hz

$$D* = \frac{\sqrt{\text{area (cm}^2)}}{NEP} \ .$$ (13.5)

Detectivity is another way to measure the sensor's signal-to-noise ratio. Detectivity is not uniform over the spectral range for operating frequencies, therefore, the chopping frequency and the spectral content must be also specified. The detectivity is expressed in units of cm·$\sqrt{\text{Hz}}$/W. It can be said, that the higher the value of D* the better the detector.

IR cutoff wavelength (λ_c) represents the long wavelength limit of spectral response and often is listed as the wavelength at which the detectivity drops by 10% of the peak value.

Maximum current is specified for photoconductive detectors (such as HgCdTe) which operate at constant currents. The operating current never should exceed the maximum limit.

Maximum reverse voltage is specified for Ge and Si photodiodes and photoconductive cells. Exceeding this voltage can cause the breakdown and severe deterioration of the sensor's performance.

Radiant responsivity is the ratio of the output photocurrent (or output voltage) divided by the incident radiant power at a given wavelength, expressed in A/W or V/W.

Field of view (FOV) is the angular measure of the volume of space where the sensor can respond to the source of radiation.

Junction capacitance (C_j) is similar to the capacitance of a parallel plate capacitor. It should be considered whenever a high speed response is required. The value of C_j drops with reverse bias and is higher for the larger diode areas.

13.2 PHOTODIODES

Photodiodes are semiconductive optical sensors, which if broadly defined, may even include solar batteries. However, here we consider only the information aspect of these devices rather than the power conversion. In a simple way, the operation of a photodiode can be described as follows. If a *pn*-junction is forward biased (positive side of a battery is connected to the *p* side) and is exposed to light of proper frequency, the current increase will be very small with respect to a dark current. If the junction is reverse biased (Fig. 13-3), the current will increase quite noticeably. Impinging photons create electron-hole pairs on both sides of the junction. When electrons enter the conduction band they starts flowing toward the positive side of the battery. Correspondingly, the created holes flow to the negative terminal, meaning that photocurrent i_p flows in the network. Under dark conditions, leakage current i_o is independent of applied voltage and mainly is the result of thermal generation of charge carriers. Thus, a reverse-biased photodiode electrical equivalent circuit (Fig. 13-4A) contains two current sources and a RC network.

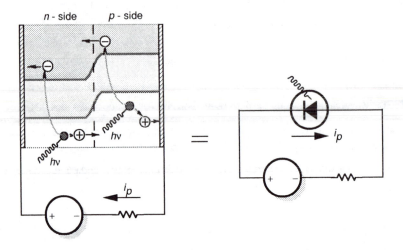

Fig. 13-3 Structure of a photodiode

The process of optical detection involves the conversion of optical energy (in the form of photons) into an electrical signal (in the form of electrons). If the probability that a photon of energy hv will produce an electron in a detection is η, then the average rate of production of electrons $<r>$ for an incident beam of optical power P is given by [2]:

$$<r> = \frac{\eta P}{hv} \tag{13.6}$$

The production of electrons due to the incident photons at constant rate $<r>$ is randomly distributed in time and obeys Poisson statistics, so that the probability of the production of m electrons in some measurement interval τ is given by

$$p(m,\tau) = (<r>\tau)^m \frac{1}{m!} e^{-<r>\tau} \tag{13.7}$$

The statistics involved with optical detection are very important in the determination of minimum detectable signal levels and hence the ultimate sensitivity of the sensors. At this point, however, we just note that the electrical current is proportional to the optical power incident on the detector:

$$i = <r>e = \frac{\eta e P}{h\nu} \; ,\tag{13.8}$$

where e is the charge of an electron. A change in input power ΔP (due to intensity modulation in a sensor, for instance) results in the output current Δi. Since power is proportional to squared current, the detector's electrical power output varies quadratically with input optical power, making it a "square-law" detector.

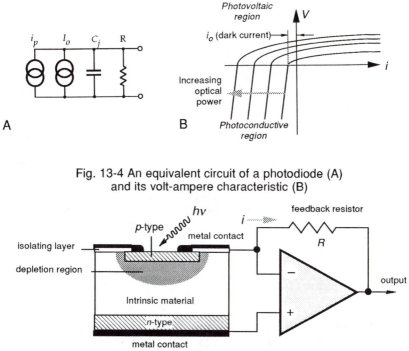

Fig. 13-4 An equivalent circuit of a photodiode (A)
and its volt-ampere characteristic (B)

Fig. 13-5 Structure of a PIN photodiode connected to a current-to-voltage converter

The voltage-to-current response of a typical photodiode is shown in Fig. 13-4B. If we attach a high input impedance voltmeter to the diode (corresponds to the case when $i=0$), we will observe that with increasing optical power, the voltage changes in a quite nonlinear fashion. In fact, variations are logarithmic. For the short circuit conditions ($V=0$), that is when the diode is connected to a current-to-voltage converter (Fig. 4-2.12), current varies linearly with the optical power. The current-to-voltage response of the photodiode is given by [3]

$$i = i_o(e^{eV/k_bT} - 1) - i_s \ , \qquad (13.9)$$

where i_o is a reverse "dark current" which is attributed to the thermal generation of electron-hole pairs, i_s is the current due to the detected optical signal, k_b is Boltzmann constant, and T is the absolute temperature. Combining Eqs. (13.8) and (13.9) yields

$$i = i_o(e^{eV/k_bT} - 1) - \frac{\eta eP}{h\nu} \ , \qquad (13.10)$$

which is the overall characteristic of a photodiode. An efficiency of the direct conversion of optical power into electric power is quite low. Typically, it is in the range of 5%-10%, however, in 1992 it was reported that some experimental photocells were able to reach an efficiency as high as 25%. In the sensor technologies, however, the photocells are generally not used. Instead, an additional high resistivity intrinsic layer is present between p and n types of the material, which is called a PIN photodiode (Fig. 13-5). The depth to which a photon can penetrate a photodiode is a function of its wavelength which is reflected in a spectral response of a sensor (Fig. 13-2).

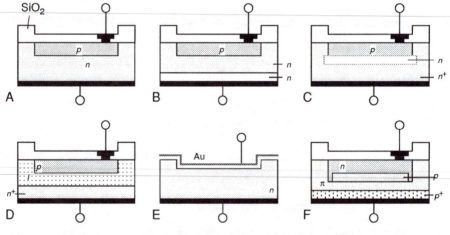

Fig. 13-6 Simplified structures of six types of photodiodes

Besides very popular PIN diodes, several other types of photodiodes are used for sensing light. In general, depending on the function and construction, all photodiodes may be classified as follows:

1. The *PN photodiodes* may include a SiO$_2$ layer on the outer surface (Fig. 13-6A). This yields a low level dark current. To fabricate a high-speed version of the diode, a depletion layer is

increased thus reducing the junction capacitance (Fig. 13-6B). To make the diode more sensitive to UV, a *p* layer can be made extra thin. A version of the planar diffusion type is pnn^+ diode (Fig. 13-6C) which has a lower sensitivity to infrared and higher sensitivity at shorter wavelengths. This is due primarily to a thick layer of a low-resistance n^+ silicon to bring the $nn+$ boundary closer to depletion layer.

2. The *PIN photodiodes* (Fig. 13-6D) are an improved version of low-capacitance planar diffusion diodes. It uses an extra high-resistance *I* layer between the *p* and *n* layers to improve the response time. These devices work even better with reversed bias, therefore, they are designed to have low leakage current and high breakdown voltage.

3. The Schottky photodiodes (Fig. 13-6E) have a thin gold coating sputtered onto the *n* layer to form a Schottky *pn*-junction. Since the distance from the outer surface to the junction is small, UV sensitivity is high.

4. The *avalanche photodiodes* (Fig. 13-6F) are named so because if a reverse bias is applied to *pn*-junction and a high intensity field is formed with the depletion layer, photon carriers will be accelerated by the field and collide with the atoms producing the secondary carriers. In turn, the new carriers are accelerated again resulting in the extremely fast avalanche-type increase in current. Therefore, these diodes work as amplifiers making them useful for detecting extremely small levels of light.

There are two general operating modes for a photodiode: the *photoconductive* (PC) and the *photovoltaic* (PV). No bias voltage is applied for the photovoltaic mode. The result is that there is no dark current, so there is only thermal noise present. This allows much better sensitivities at low light levels. However, the speed response is worst due to an increase in C_j and responsively to longer wavelengths is also reduced.

Fig. 13-7A shows a photodiode connected in a PV mode. In this connection, the diode operates as a current generating device which is represented in the equivalent circuit by a current source i_p (Fig. 13-7B). The load resistor R_b determines the voltage developed at the input of the amplifier and the slope of the load characteristic is proportional to that resistor (Fig. 13-7C).

When using a photodiode in a photovoltaic mode, its large capacitance C_j may limit the speed response of the circuit. During the operation with a direct resistive load, as in Fig. 13-7A, a photodiode exhibits a bandwidth limited mainly by its internal capacitance C_j. Fig. 13-7B models such a bandwidth limit. The photodiode acts primarily as a current source. A large resistance R and the diode capacitance shunt the source. The capacitance ranges from 2 to 20,000 pF depending for the most part on the diode area. In parallel with the shunt is the amplifier's input capacitance (not shown) which results in a combined input capacitance C. The diode resistance usually can be ignored as it is much smaller than the load resistance R_b. The net input network determines the input circuit response rolloff. The resulting input circuit response has a break frequency $f_1=1/2\pi R_L C$, and the response is [4]

$$V_{out} = \frac{-R_L i_p}{1 + j\dfrac{f}{f_1}} \, .$$

(13.11)

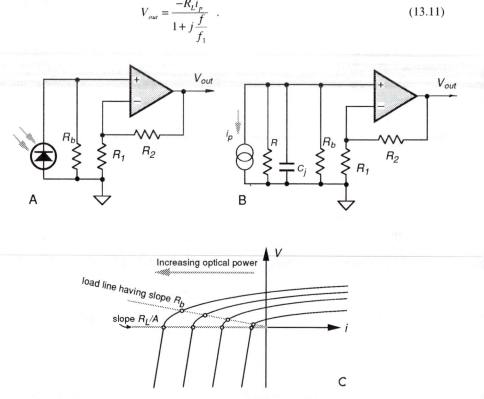

Fig. 13-7 Connection of a photodiode in a photovoltaic mode to a noninverting amplifier (A); the equivalent circuit (B); and a loading characteristic (C)

For a single-pole response, the circuit's 3-dB bandwidth equals the pole frequency. The expression reflects a typical gain-vs-bandwidth compromise. Increasing R_b gives a greater gain but reduces f_1. From a circuit perspective, this compromise results from impressing the signal voltage on the circuit capacitances. The signal voltage appears across the input capacitance $C=C_j + C_{OPAM}$. To avoid the compromise, it is desirable to develop input voltage across the resistor and prevent it from charging the capacitances. This can be achieved by employing a current-to-voltage amplifier (I/V) as shown in Fig. 13-8A. The amplifier and its feedback resistor R_L translate the diode current into a buffered output voltage with excellent linearity. Added to the figure is a feedback capacitor C_L that provides a phase compensation. An ideal amplifier holds its two inputs at the same voltage (ground in the figure), thus the inverting input is called a virtual ground. The photodiode operates at zero voltage across its terminals which improves the response linearity and prevents charging the diode capacitance. This is illustrated in Fig. 13-7C where the load line virtually coincides with the current axis, because the line's slope is inversely proportional to the amplifier's open loop gain A.

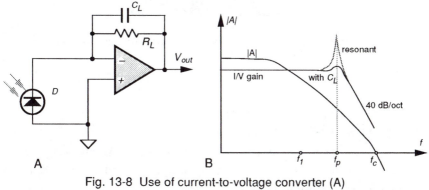

Fig. 13-8 Use of current-to-voltage converter (A)
and the frequency characteristics (B)

In practice, the amplifier's high, but finite open-loop gain limits the performance by developing small, albeit non-zero voltage across the diode. Then, the break frequency is defined as

$$f_p = \frac{A}{2\pi R_L C} \approx A f_1 \; , \tag{13.12}$$

where A is the open-loop gain of the amplifier. Therefore, the break frequency is increased by a factor A as compared with f_1. It should be noted that when frequency increases, gain, A, declines, and the virtual load attached to the photodiode appears to be inductive. This results from the phase shift of gain A. Over most of the amplifier's useful frequency range, A has a phase lag of 90°. The 180° phase inversion by the amplifier converts this to a 90° phase lead which is specific for the inductive impedance. This inductive load resonates with the capacitance of the input circuit at a frequency equal to f_p (Fig. 13.8B) and may result in an oscillating response (Fig. 13-9) or the circuit instability. To restore stability, a compensating capacitor C_L is placed across the feedback resistor. Value of the capacitor can be found from:

$$C_L = \frac{1}{2\pi R_L f_p} = \sqrt{C C_c} \; , \tag{13.13}$$

where $C_c = 1/(2\pi R_L f_c)$, and f_c is the unity-gain crossover frequency of the operational amplifier. The capacitor boosts the signal at the inverting input by shunting R_L at higher frequencies.

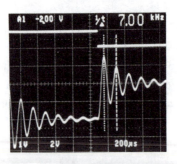

Fig. 13.9 Response of a photodiode with uncompensated circuit
(Courtesy of Hamamatsu Photonics K.K.)

When using photodiodes for the detection of low level light, noise floor should be seriously considered. There are two main components of noise in a photodiode: shot noise and Johnson noise (see Section 4.9). Besides the sensor, amplifier's and auxiliary component noise also should be accounted for [Eq. (4.9.3)].

For the photoconductive operating mode, a reverse bias voltage is applied to the photodiode. The result is a wider depletion region, lower junction capacitance C_j, lower series resistance, shorter rise time, and linear response in photocurrent over a wider range of light intensities. However, as the reverse bias is increased, shot noise increases as well due to increase in a dark current. The PC mode circuit diagram is shown in Fig. 13-10A and the diode's load characteristic is in Fig. 13-10B. The reverse bias moves the load line into the third quadrant where the response linearity is better than that for the PV mode (the second quadrant). The load lines crosses the voltage axis at the point corresponding to the bias voltage E, while the slope is inversely proportional to the amplifier's open-loop gain A. The PC mode offers bandwidths to hundreds of MHz, providing an accompanying increase in the signal-to noise ratio.

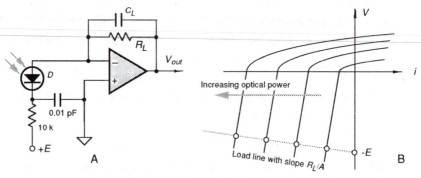

Fig. 13-10 Photoconductive operating mode
A: a circuit diagram; B: a load characteristic

13.3   PHOTOTRANSISTOR

A photodiode directly converts photons into charge carriers, specifically one electron and one hole (hole-electron pair) per a photon. The phototransistors can do the same, and in addition to provide current gain, resulting in a much higher sensitivity. The collector-base junction is a reverse biased diode which functions as described above. If the transistor is connected into a circuit containing a battery, a photo-induced current flows through the loop which includes the base-emitter region. This current is amplified by the transistor in the same manner as in a conventional transistor, resulting in a significant increase in the collector current.

The energy bands for the phototransistor are shown in Fig. 13-11. The photon-induced base current is returned to the collector through the emitter and the external circuitry. In so doing, electrons are supplied to the base region by the emitter where they are pulled into the collector by the electric field. The sensitivity of a phototransistor is a function of the collector-base diode quantum efficiency and also of the dc current gain of the transistor. Therefore, the overall sensitivity is a function of collector current.

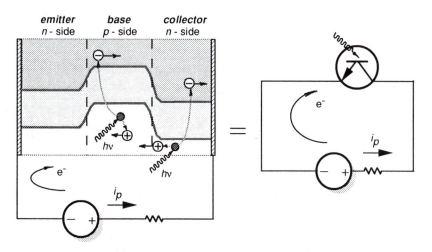

Fig. 13-11 Energy bands in a phototransistor

When subjected to varying ambient temperature, collector current changes linearly with a positive slope of about 0.00667 per °C. The magnitude of this temperature coefficient is primarily a result of the increase in current gain versus temperature, since the collector-base photo current temperature coefficient is only about 0.001 per °C. The family of collector current versus collector voltage characteristics is very much similar to that of a conventional transistor. This implies that circuits with phototransistors can be designed by using regular methods of transistor circuit techniques, except that its base should be used as an input of a photoinduced current which is supplied by its collector. Since the actual photogeneration of carriers occurs in the collector-base region, the larger the area of this region, the more carriers are generated, thus, the phototransistor is so designed to offer a large area to impinging light. A phototransistor can be either a two-lead or a three-lead device. In the latter case, the base lead is available and the transistor may be used as a standard bipolar transistor with or without the additional capability of sensing light, thus giving a designer greater flexibility in circuit development. However, a two-lead device is the most popular as a dedicated photosensor.

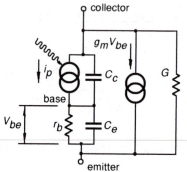

Fig. 13-12 An equivalent circuit of a phototransistor

When the base of the transistor is floating, it can be represented by an equivalent circuit shown in Fig. 13-12. Two capacitors C_c and C_e represent base-collector and base-emitter capacitances which are the speed limiting factors. Maximum frequency response of the phototransistor may be estimated from

$$f_1 \approx \frac{g_m}{2\pi C_e}, \qquad (13.14)$$

where f_1 is the current-gain-bandwidth product and g_m is the transistor's forward transconductance.

Whenever a higher sensitivity of a photodetector is required, especially if high response speed is not of a concern, an integrated Darlington detector is recommended. It is comprised of a phototransistor whose emitter is coupled to the base of a bipolar transistor. Since a Darlington connection gives current gain equal to a product of current gains of two transistors, the circuit proves to be an efficient way to make a sensitive detector. To illustrate typical specifications of a photosensor, Table 13-2 presents data for the MRD711 detector produced by Motorola.

Table 13-2 Typical characteristics of MRD711 Darlington photodetector
(adapted from [5]

Characteristic	Symbol	Value	Note
Maximum Collector-emitter voltage	V_{CEO}	60V	shows maximum voltage permissible across terminals
Collector dark current	I_D	100nA max.	useful for calculating dynamic range
Capacitance (@5V, 1 MHz)	C_{ce}	3.9pF	determines speed response
Collector light current (@5V H=500 µW/cm², l=940 nm)	I_L	25mA typ.	shows sensitivity
Turn-on time @5V, H=500 µW/cm², l=940 nm, R_L=100Ω)	t_{on}	125µs typ.	speed response on light-on
Turn-off time (@5V, H=500 µW/cm², l=940 nm, R_L=100 Ω)	t_{off}	125µs typ.	speed response on light-off
Saturation voltage (@5V, H=500 µW/cm², l=940 nm I_c=2 mA)	$V_{ce(sat)}$	150 µs	useful to calculate load for optimum speed response
Wavelength of maximum sensitivity	I_s	0.8 µm	peak spectral response (near infrared)

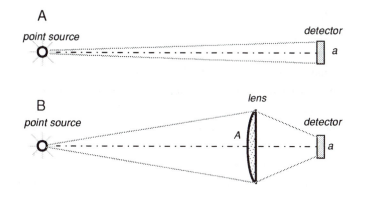

Fig. 13-13 Efficiency of a detector depends on its surface area (A)
or the area of the focusing system (B)

Spatial resolutions of both the light source and the detector must be seriously considered for many sensor applications. Whenever higher efficiency of sensing is required, optical components come in handy. Let's, for instance, take a point light source which should be detected by a photodetector (Fig. 13-13A). According to equation (13.10), the sensor's output is proportional to received photonic power, which, in turn, proportional to receiver's surface area. Fig. 13-13B shows that the use of a focusing lens can dramatically increase the area. Efficiency of a single lens depends on its refractive index n. The overall improvement in the sensitivity can be estimated by employing equations (3.15.9) and (3.15.12)

$$k \approx \frac{A}{a}\left[1 - 2\left(\frac{n-1}{n+1}\right)^2 \right] \ , \qquad\qquad (13.15)$$

where A and a are respective effective areas of the lens and the sensing area of a photodetector. For glasses and most plastics operating in the visible and near infrared spectral ranges, the equation can be simplified to

$$k \approx 0.92\frac{A}{a} \ . \qquad\qquad (13.16)$$

It should be pointed out that arbitrary placement of a lens may be more harmful than helpful. That is, a lens system must be carefully planed to be effective. For instance, many photodetectors have built-in lenses which are effective for parallel rays. If an additional lens is introduced in front of such a detector, it will create nonparallel rays at the input resulting in misalignment of the optical system and poor performance. Thus, whenever additional optical devices need to be employed, detector's own optical properties must be considered.

13.4   PHOTORESISTORS

As a photodiode, a photoresistor is a photoconductive device. The most common materials for its fabrication are cadmium sulfide (CdS)[1] and cadmium selenide $(CdSe)$ which are semiconductors whose resistances change upon light entering the surface. For its operation, a photoresistor requires a power source as it does not generate photocurrent - a photoeffect is manifested in change in the material's electrical resistance. Fig. 13-14A shows a schematic diagram of a photoresistive cell. An electrode is set at each end of the photoconductor. In darkness, the resistance of the material is high. Hence, applied voltage V results in small dark current which is attributed to temperature effect. When light is incident on the surface, current i_p flows.

The reason for the current increase is the following. Directly beneath the conduction band of the crystal is a donor level and there is an acceptor level above the valence band. In darkness, the electrons and holes in each level are almost crammed in place in the crystal, resulting in a high resistance of the semiconductor.

---

[1] Information on CdS photoresistors is courtesy of Hamamatsy Photonics K.K.

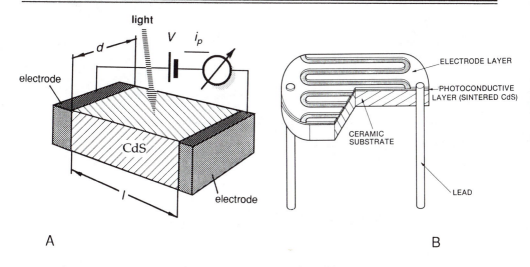

Fig. 13-14 Structure of a photoresistor (A)
and a plastic-coated photoresistor having a serpentine shape (B)

When light illuminates the photoconductive crystal, photons are absorbed which result in the added-up energy in the valence band electrons. This moves them into the conduction band, creating free holes in the valence band, increasing the conductivity of the material. Since near the valence band is a separate acceptor level that can capture free electrons not as easily as free holes, the recombination probability of the electrons and holes is reduced and the number of free electrons in the conduction band is high. Since CdS has a band gap of 2.41 eV, the absorption edge wavelength is $\lambda = c/\nu \approx 515$ nm, which is in the visible spectral range. Hence, the CdS detects light shorter than 515 nm wavelengths. Other photoconductors have different absorption edge wavelengths. For instance, CdS is most sensitive at shorter wavelengths range, while Si and Ge are most efficient in the near infrared.

The conductance of a semiconductor is given by

$$\Delta\sigma = ef(\mu_n\tau_n + \mu_p\tau_p) \quad , \tag{13.17}$$

where μ_n and μ_p are the free electron and hole movements (cm/V·sec), τ_n and τ_p are the free electron and hole lives (sec), e is the charge of an electron, and f is the number of generated carriers per second per unit of volume. For a CdS sell $\mu_n\tau_n \gg \mu_p\tau_p$, hence, conductance by free holes can be ignored. Then the sensor becomes an n-type semiconductor. Thus

$$\Delta\sigma = ef\mu_n\tau_n \quad , \tag{13.18}$$

We can define sensitivity b of the photoresistor through a number of electrons generated by one photon (until the carrier lifespan ends):

$$b = \frac{\tau_n}{t_t} , \tag{13.19}$$

where $t_t = l^2/V\mu_n$ is the transit time for the electron between the sensor's electrodes, l is distance between the electrodes and V is applied voltage. Then, we arrive to:

$$b = \frac{\mu_n \tau_n V}{l^2} . \tag{13.20}$$

For example, if $\mu_n = 300$ cm^2/V·s, $\tau_n = 10^{-3}$s, $l = 0.2$ mm, and $V = 1.2$V, then the sensitivity is 900, which means that a single photon releases for conduction 900 electrons, making a photoresistor work as a photomultiplier. Indeed, a photoresistor is a very sensitive device.

It can be shown that for better sensitivity and lower cell resistance, a distance l between the electrodes should be reduced, while width of the sensor d should be increased. This suggests that the sensor should be very short and very wide. For practical purposes, this is accomplished by fabricating a sensor in a serpentine shape (Fig. 13-14B) where the electrodes are connected to the leads.

Depending on the manufacturing process, the photoresistive cells can be divided into the sintered type, single crystal type, and evaporated type. Of these, the sintered type offers high sensitivity and easier fabrication of large sensitive areas, which eventually translated into lower cost devices. The sintering process of CdS cells consists of the following steps.

1. Highly pure CdS powder is mixed with appropriate impurities and a fusing agent;

2. The mixture is dissolved in water;

3. The solution in a form of paste is applied on the surface of a ceramic substrate and allowed to dry;

4. The ceramic subassemblies are sintered in a high-temperature oven to form a multicrystal structure. At this stage a photoconductive layer is formed;

5. Electrode layers and leads (terminals) are attached;

6. The sensor is packaged into a plastic or metal housing with or without a window.

To tailor a spectral response of a photoresistor, the powder of step 1 can contain some variations, for instance, the addition of selenide or even the replacement of CdS for CdSe shifts the spectral response toward longer wavelengths (orange and red).

To illustrate, how the photoresistors can be used, Fig. 13-15 shows two circuits. Circuit (A) shows an automatic light switch which turns lights on when illumination drops (a turn-off part of the circuit is not shown). Circuit (B) shows a beacon with a free running multivibrator, which is enabled at darkness, when the resistance of a photoresistor becomes high.

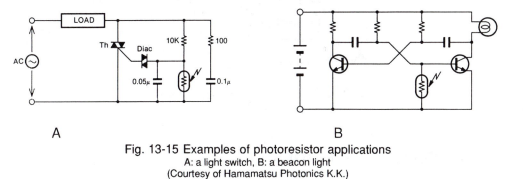

A   B

Fig. 13-15 Examples of photoresistor applications
A: a light switch, B: a beacon light
(Courtesy of Hamamatsu Photonics K.K.)

13.5 LIGHT-TO-LIGHT CONVERTERS

In many applications, light intensity is so small that a photodetector output has a very poor signal-to-noise ratio, and any further electronic amplification is useless. Thus, it would be highly desirable to amplify light optically before it strikes a photodetector. Vacuum tube photomultipliers are well known for this purpose. They can achieve a high speed of response (on the range of 100ps) and a very high gain: up to $2 \cdot 10^6$ from the ultraviolet to the near infrared spectral ranges. However, the photomultipliers require high voltage for the operation, ranging from 240 to 1,500 V.

Another solution is a solid state light-to-light converter [6]. It consists of an integrated photo-transistor (HPT) and a laser diode (LD), which utilize an internal positive optical feedback (Fig. 13-16). The transducer is comprised of several semiconductor layers of p-InP and n-InP, together with InGaAsP. The AuGe/Ni/Au mesh electrode is deposited on the input side of the device, and the AuZn/Au electrode is deposited on the substrate. An entire transducer is bonded on the silver heat sink. The transducer needs only 4 volts as a bias voltage and achieves a very high optical gain - up to $6 \cdot 10^5$. The device can emit the coherent laser beam with 4 mW power when it detects the input light with the power as small as 7nW.

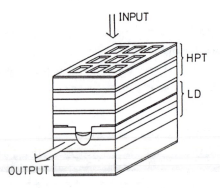

Fig. 13-16 Solid state light-to-light transducer
(from [6])

13.6 COOLED DETECTORS

For measurements of objects emanating photons on the range of 2 *eV* or higher, quantum detectors having room temperature are generally used. For the smaller energies (longer wavelengths) narrower band gap semiconductors are required. However, even if a quantum detector has a sufficiently small energy band gap, at room temperatures its own intrinsic noise is much higher than a photoconductive signal. Noise level is temperature dependent, therefore, when detecting long wavelength photons, a signal-to-noise ratio may become so small that accurate measurement becomes impossible. This is the reason, why for the operation in the near and far infrared spectral ranges a detector not only should have a sufficiently narrow energy gap, but its temperature has to be lowered to the level where intrinsic noise is reduced to an acceptable level. Fig. 13-17 shows typical spectral responses of some detectors with recommended operating temperatures. The operating principle of a cryogenically cooled detector is about the same as that of a photoresistor, except that it operates at far longer wavelengths at much lower temperatures. Thus, the sensor design becomes quite different. Depending on the required sensitivity and operating wavelength, the following crystals are typically used for this type of sensors: lead sulfide (PbS), indium arsenide (InAs), germanium (G), lead selenide (PbSe), and mercury-cadmium-telluride (HgCdTe).

Cooling shifts responses to longer wavelengths and increases sensitivity. However, response speeds of PbS and PbSe become slower with cooling. Methods of cooling include dewar cooling using dry ice, liquid nitrogen, liquid helium (Fig. 13-18), or thermoelectric coolers operating on the Peltier effect (see Section 3.9).

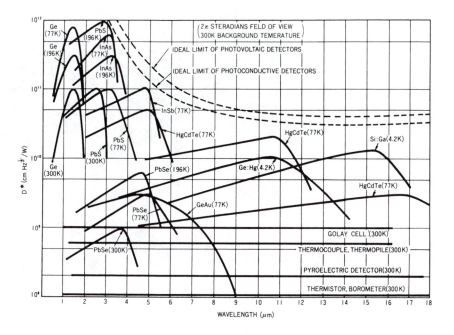

Fig. 13-17 Operating ranges for some infrared detectors

As an example, Table 13-3 lists typical specifications for an MCT photoconductive detector. MCT stands for the mercury-cadmium-telluride type of a sensitive element.

Table 13-3 Typical specifications for MCT far infrared detectors

Sensitive area (mm)	Temperature (°C)	I_p (μm)	I_c (μm)	FOV (°)	Dark resist	Rise time	Max current	D* @ I_p
1 x 1	- 3 0	3.6	3.7	6 0	1 kΩ	10 μs	3 mA	10^9
1 x 1	-196	1 5	1 6	6 0	20 kΩ	1 μs	40 mA	$3 \cdot 10^9$

Applications of the cryogenically cooled quantum detectors include measurements of optical power over a broad spectral range, thermal temperature measurement and thermal imaging, detection of water content and gas analysis.

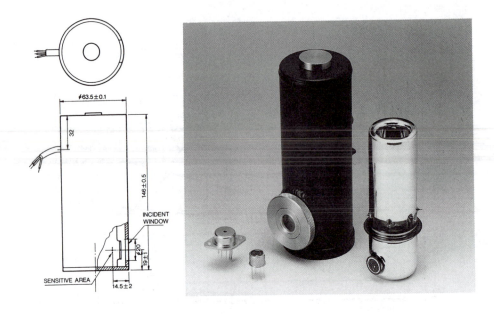

Fig. 13-18 Cryogenically cooled MCT quantum infrared detectors
A: Dimensional drawing of a Dewar type (in mm);
B: Outside appearances of canned and Dewar detectors
(Courtesy of Hamamatsu Photonics K.K.)

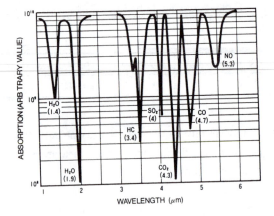

Fig. 13-19 Absorption spectra of gaseous molecules

Fig. 13-19 depicts gas absorption spectra for various molecules. Water strongly absorbs at 1.1, 1.4, 1.9, and 2.7μm. Thus, to determine the moisture content, for example, in coal, the monochromatic light is projected on the test and reference samples. The reflected light is detected and the ratio is calculated for the absorption bands. The gas analyzer makes use of absorption in the infrared region of the spectrum. This allows us to measure gas density. Thus, it is possible to measure automobile exhaust gases (CO, HC, CO_2), emission control (CO, SO, NO_2) fuel leakage (CH_4, C_3H_2), etc.

13.7   THERMAL DETECTORS

Thermal infrared detectors are primarily used for the noncontact temperature measurements which has been known in industry for about 60 years under the name of *pyrometry* from the Greek word *pur* (fire) and the respective instruments are called radiation pyrometers. Today, non-contact methods of measurement embrace a very broad range, including subzero temperatures, which are quite far away from that of flame. Therefore it appears that *radiation thermometry* is a more appropriate term for this technology.

A typical infrared noncontact temperature sensor consists of:

1. A sensing element - a component which is responsive to electromagnetic radiation in the infrared wavelength range. Main requirements to the element are fast, predictable and strong response to thermal radiation, and a good long term stability.

2. A supporting structure to hold the sensing element and to expose it to the radiation. The structure should have low thermal conductivity to minimize heat loss.

3. A housing, which protects the sensing element from the environment. It usually should be hermetically sealed and often filled either with dry air or inert gas, such as argon or nitrogen.

4. Contacts which are wires or conductive epoxy to provide an electrical connection between the sensing element and the terminals.

5. A protective window which is impermeable to environmental factors and transparent in the wavelength of detection. The window may have surface coatings to improve transparency, and to filter out undesirable portions of the spectrum.

Below mid infrared, thermal detectors are much less sensitive than quantum detectors. Their operating principle is based on a sequential conversion of thermal radiation into heat, and then, conversion of heat level or flow into an electrical signal by employing conventional methods of heat detection. In principle, any heat detector (Chapter 16) can be used for the detection of thermal radiation. However, according to Eq. (3.14.10) the infrared flux which is absorbed by a thermal detector is proportional to a geometry factor A which for a uniform spatial distribution of radiation is equal to the sensor's area. For instance, if a thermal radiation sensor at 25°C, having 5 mm² surface area and ideal absorptivity, is placed inside a radiative cavity whose temperature is 100°C, the sensor will receive an initial radiative power of 3.25 mW. Depending on the sensor's thermal capacity, its temperature will rise until thermal equilibrium between the sensor and the object occurs. It should be noted that the sensing element's temperature never reaches that of an object. The reason is that while the element receives heat by radiation, a portion of the heat is lost through a supporting structure and wires, as well as through gravitational convection. Thus, the equilibrium temperature is always somewhere in-between the object's temperature and the initial temperature of the thermal detector.

All thermal radiation detectors can be divided into two classes: *passive* infrared (PIR) and *active* far infrared (AFIR) detectors. Passive detectors absorb incoming radiation and convert it to heat, while active detectors emit thermal radiation toward the object (see below).

13.7.1 Thermopile sensors

Thermopiles belong to a class of PIR detectors. Their operating principle is the same as that of thermocouples. In effect, a thermopile are serially connected thermocouples. Originally, it was invented by Joule to increase the output signal of a thermoelectric sensor. He connected several thermocouples in a series and joined together their hot junctions. Nowadays, thermopiles have a different configuration. Their prime application is thermal detection of far infrared radiation.

An equivalent schematic of a thermopile sensor is shown in Fig. 13-19A. The sensor consists of a base having a relatively large thermal mass which is the place where the "cold" junctions are positioned. The base may be thermally coupled with a reference temperature sensor or attached to a thermostat having a known temperature. The base supports a thin membrane whose thermal capacity and thermal conductivity are small. The membrane is the surface where the "hot" junctions are positioned.

Methods of construction of thermopiles may differ to some extent, but all incorporate vacuum deposition techniques and evaporation masks to apply the thermoelectric materials, such as bismuth and antimony. The number of junctions varies from 20 to several hundreds. The "hot" junctions are often blackened (for instance, with goldblack or organic paint) to improve their absorptivity of the infrared radiation. A thermopile is a dc device whose output voltage follows its "hot" junction temperature quite well. A thermopile can be modeled as a thermal flux-controlled voltage source which is connected in series with a fixed resistor. The sensor is hermetically sealed in a metal can with a hard infrared transparent window, for instance, silicon, germanium or zinc selenide (Fig. 13-20B). The output voltage V_s is nearly proportional to the incident radiation. The thermopile operating frequency limit is mainly determined by thermal capacity and thermal conductivity of the membrane, which are manifested through a thermal time constant. The sensor exhibits quite a low noise which is equal to the thermal noise of the sensor's equivalent resistance, that is of 20 - 50 kΩ. Typical properties of thermopile sensors are given in Table 13-4.

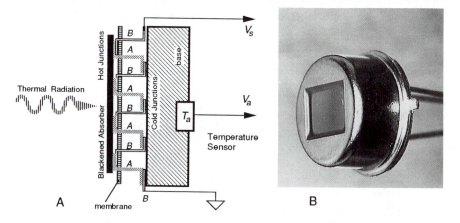

Fig. 13-20 Thermopile for detecting thermal radiation
A: an equivalent schematic with a reference temperature sensor attached, *A* and *B* are different metals;
B: sensor in a TO-5 packaging

Table 13-4 Typical specifications of thermopiles

Parameter	value	unit	conditions
Sensitive area	0.5 - 2	mm²	
Responsivity	50	V/W	6-14 µm, 500k
Noise	30	nV/√Hz	25°C, rms
Equivalent resistance	50	kΩ	25°C
Thermal time constant	60	ms	
TCR	0.15	%/K	
Temperature coefficient of responsivity	-0.2	%/K	
Operating temperature	-20 to +80	°C	
Storage temperature	-40 to 100	°C	
Price	8 - 50	US$	single quantity

13.7.2 Pyroelectric sensors

Pyroelectric sensors belong to a class of PIR detectors. A typical construction of a solid-state pyroelectric sensor is shown in Fig. 13-21A. It is housed in a metal TO-5 or TO-39 can for better shielding and is protected from the environment by a silicon or any other appropriate window. The inner space of the can is often filled with dry air or nitrogen. Usually, two sensing elements are oppositely, serially or in parallel, connected for better compensation of rapid thermal changes and mechanical stresses resulting from acoustical noise and vibrations. Sometimes, one of the elements is coated with heat absorbing paint or gold-black, while the other element is shielded from radiation and gold plated for better reflectivity. For applications in PIR motion detectors, both elements are exposed to the window.

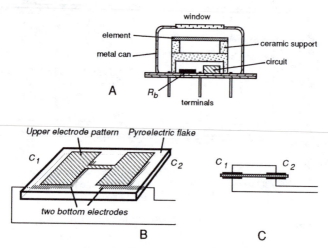

Fig. 13-21 A dual pyroelectric sensor
A: structure of a sensor in a metal can; B: metal electrodes are deposited on the opposite sides of a material; C: equivalent circuit of a dual element

A dual element is often fabricated from a single flake of a crystalline material (Fig. 13-21B). Metallized pattern on both sides of the flake form two serially connected capacitors C_1 and C_2. Fig. 13-21C shows an equivalent circuit of a dual pyroelectric element. This design has the benefit of a good balance of both elements, thus resulting in a better rejection of common-mode interferences. A major problem in design of pyroelectric detectors is in its sensitivity to mechanical stress and vibrations. All pyroelectrics are also piezoelectrics, therefore, while being sensitive to thermal radiation, the pyroelectric sensors are susceptible to interferences which sometimes are called "microphonics". For better noise rejection, the crystalline element must be mechanically decoupled from the outside, especially from connecting pins and the metal can.

A pyroelectric element can be modeled by a capacitor connected in parallel with a leakage resistor. The value of that resistor is on the orders of 10^{12}-10^{14} Ω. In practice, the sensor is connected to the circuit which contains a bias resistor R_b and an impedance converter ("circuit" in Fig. 13-21). The converter may be either a voltage follower (for instance, JFET transistor) or a current-to-voltage converter. The voltage follower (Fig. 13-22A) converts a high output impedance of the sensor (capacitance C in parallel with a bias resistance R_b into the output resistance of the follower which in this example is determined by the transistor's transconductance in parallel with 47kΩ. The advantage of this circuit is in its simplicity, low cost and low noise. A JFET follower is the most cost effective and simple, however it suffers from two major drawbacks. The first is the dependence of its speed response of the so-called *electrical time constant*, which is a product of the sensor's capacitance C and the bias resistor R_b

$$\tau_e = CR_b \quad . \tag{13.21}$$

For example, a typical dual sensor may have C=40pF and R_b=50GΩ, which yield τ_e=2s, corresponding to a first order frequency response with the upper cutoff frequency at 3 dB level equal to about 0.08 Hz - a very low frequency indeed. This makes the voltage follower suitable only for limited applications, where speed response in not too important. An example is the detection of movement of people (see Chapter 6). The second drawback of the circuit is a large offset voltage across the output resistor. This voltage depends on the type of the transistor and is temperature dependent. Thus, the output V_{out} is the sum of two voltages: the offset voltage which can be as large as several volts, and the alternate pyroelectric voltage which may be on the order of millivolts.

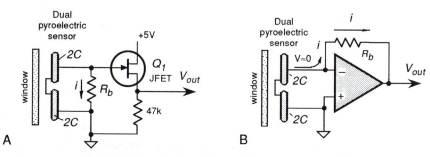

Fig. 13-22 Impedance converters for pyroelectric sensors
A: a voltage follower with JFET;
B: a current-to-voltage converter with operational amplifier

A more efficient, however, more expensive circuit for a pyroelectric sensor is an *I/V* (current-to-voltage) converter (Fig. 13-22B). Its advantage is in faster response and insensitivity to the capacitance of the sensor element. The sensor is connected to an inverting input of the operational amplifier which possesses properties of the so-called virtual ground (similar circuits are shown in Figs. 13-5, 13-8, and 13-10). That is, the voltage in the inverting input is constant and almost equal to that of a non-inverting input, which in this circuit is grounded. Thus, the voltage across the sensor is forced by the feedback to stay near zero. The output voltage follows the shape of the electric current (a flow of charges) generated by the sensor (Fig. 3-7.4). The circuit should employ an operational amplifier with a very low bias current (on the order of 1 pA). There are three major advantages in using this circuit: a fast response, insensitivity to the capacitance of the sensor and a low output offset voltage. However, being a broad bandwidth circuit, a current-to-voltage converter may suffer from higher noise.

At very low frequencies, both circuits, the JFET and I/V converter, transform pyroelectric current i_p into output voltage. According to Ohm's law

$$V_{out} = i_p R_b \quad .$$
(13.22)

For instance, for the pyroelectric current of 10pA (10^{-11} A) and the bias resistor of $5 \cdot 10^{10} \Omega$ (50 gigohm), the output voltage is 500 mV. Either the JFET transistor or operational amplifier must have low input bias currents (I_B) over an entire operating temperature range. A CMOS OPAMs are generally preferable as their bias currents are on the order of 1pA (see Table 4-1, amplifiers OP-41, AD546, and TLC271BCP).

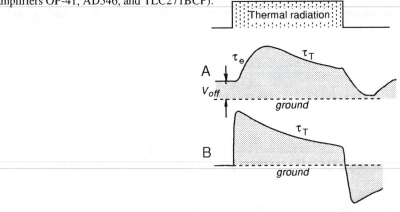

Fig. 13-23 Output signals of the voltage follower (A) and current-to-voltage converter (B) in response to a step function of a thermal radiation

It should be noted that the circuits described above (Fig. 13-22) produce output signals of quite different shapes. The voltage follower's output voltage is a repetition of voltage across the element and R_b (Fig. 13-23A). It is characterized by two slopes: the leading slope having an electrical time constant $\tau_e = C R_b$, and the decaying slope having thermal time constant τ_T. Voltage across the element in the current-to-voltage converter is essentially zero and, contrary to the follower, the input impedance of the converter is low. In other words, while the voltage follower acts as a voltmeter, the current-to-voltage converter acts as an ampermeter. The leading edge of its output voltage is fast (determined by a stray capacitance across R_b) and the decaying slope is characterized by τ_T. Thus, the converter's output voltage repeats the shape of the sensor's pyroelectric current (Fig. 13-23B).

A fabrication of gigohm-range resistors is not a trivial task. High quality bias resistors must have good environmental stability, low temperature coefficient of resistance (TCR), and low voltage coefficient of resistance (VCR). The VCR is defined as

$$\xi = \frac{R_1 - R_{0.1}}{R_{0.1}} \cdot 100\% \quad , \tag{13.23}$$

where R_1 and $R_{0.1}$ are the resistances measured, respectively, at 1 and 0.1 volts. Usually, VCR is negative, that is, the resistance value drops with an increase in voltage across the resistor (Fig. 13-24A). Since the pyroelectric sensor's output is proportional to the product of the pyroelectric current and the bias resistor, VCR results in nonlinearity of an overall transfer function of the sensor. A high impedance resistor may by fabricated by depositing a thin layer of a semiconductive ink on a ceramic (alumina) substrate, firing it in a furnace and subsequent covering the surface with a protective coating. A high quality, relatively thick (at least 50 µm thick) hydrophobic coating is very important for protection against moisture, since even a small amount of water molecules may cause oxidation of the semiconductive layer. This causes a substantial increase in the resistance and poor long term stability. A typical design of a high impedance resistor is shown in Fig. 13-24B.

In applications, where high accuracy is not required, such as thermal motion detection, the bias resistor can be replaced with one or two zero-biased parallel-opposite connected silicon diodes.

For the detection of thermal radiation, a distinction exists between two cases in which completely different demands has to be met with respect to the pyroelectric material and its thermal coupling to the environment [7]:

a) *Fast* sensors detect radiation of high intensity but very short duration (nanoseconds) of laser pulses, with a high repetition on the order of 1 MHz. The sensors are usually fabricated from single-crystal pyroelectrics, such as lithium tantalate ($LiTaO_3$) or triglicinesulfate (TGS). This assures a high linearity of response. Usually, the materials are bonded to a heat sink.

b) *Sensitive* sensors detect thermal radiation of low intensity, however, with a relatively low rate of change. Examples are infrared thermometry and motion detection [8, 9, 10]. These sensors are characterized by a sharp temperature rise in the field of radiation. This generally requires a good thermal coupling with a heat source. Optical devices, such as focusing lenses and waveguides are generally employed. A heat transfer to the environment (sensor's housing) must be minimized. If well designed, such a sensor can have a sensitivity approaching that of a cryogenically cooled quantum detector [7]. Commercial pyroelectric sensors are implemented on the basis of single crystals, such as TGS and LiTaO3, or lead zirconate titanate (PZT) ceramics. PVDF film is also occasionally used thanks to its high speed response and good lateral resolution.

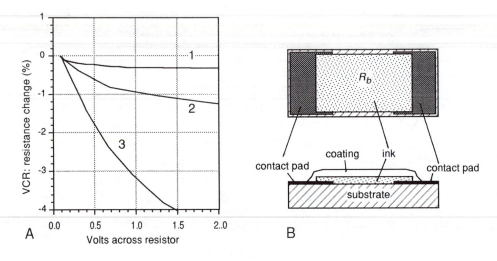

Fig. 13-24 High impedance resistor
A: VCR for three different types of the resistor
B: a structure of a resistor on an alumina substrate

13.7.3 Active far infrared sensors

In the active far infrared (AFIR) sensor, a process of measuring thermal radiation flux is different from previously described passive (PIR) detectors. Contrary to a PIR sensing element, whose temperature depends on both the ambient and object's temperatures, the AFIR sensor's surface is actively controlled to have a defined temperature T_s which in most applications is maintained constant during an entire measurement process. To control the sensor's surface temperature, electric power P is provided by a control (or excitation) circuit (Fig. 13-25A). To regulate T_s, the circuit measures element's surface temperature and compares it with an internal reference. We select a positive sign for power P when it is directed toward the element. Obviously, the incoming power maintains T_s higher than ambient. In some applications, T_s is selected higher than the highest temperature of the object. Hence, the element radiates far infrared flux Φ *toward* the object, rather than passively absorbs it, as in the PIR detector. Of course, the radiative net flux is governed by the fundamental Eq. (3.14.14) which is known as the Stefan-Boltzmann law. If the AFIR element is provided with a cooling element (for instance, a thermoelectric device operating on Peltier effect[1]), T_s may be maintained at or below ambient. However, from the practical standpoint, it is easier to warm the element up rather than to cool it down. Below, we discuss the AFIR sensors where the surface is warmed up either by an additional heating element or due to a self-heating effect in a temperature sensor [8, 11, 12, 13].

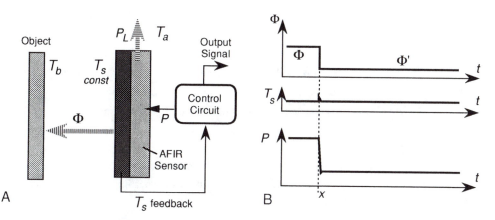

Fig. 13-25 AFIR element radiates thermal flux Φ toward the object (A); timing diagrams for radiative flux, surface temperature, and supplied power (B)

Dynamically, temperature T_s of any thermal element, either active or passive, in general terms may be described by the first order differential equation

$$cm\frac{dT_s}{dt} = P - P_L - \Phi \quad, \tag{13.24}$$

where P is the power supplied to the element from a power supply or an excitation circuit (if any), P_L is a nonradiative thermal loss which is attributed to thermal conduction and convection, m and c are the sensor's mass and specific heat, respectively.

In the PIR detector, no external power is supplied ($P = 0$), hence, the speed response depends only on the sensor's thermal capacity and heat loss, and is characterized by a thermal time

---

[1] see Section 3.9.

constant τ_T. In the AFIR element, after a warm-up period, the control circuit forces the element's surface temperature T_s to stay constant, which means

$$\frac{dT_s}{dt} = 0 \ , \tag{13.25}$$

and Eq. (13.24) becomes algebraic:

$$P = P_L + \Phi \ . \tag{13.26}$$

Contrary to PIR sensors, the AFIR detector acts as an "infinite" heat source. It follows from the above that under idealized conditions, its response does not depend on thermal mass and is not a function of time. If the control circuit is highly efficient, power P should track changes in the radiated flux Φ with high fidelity. A magnitude of that power may be used as the sensor's output signal. Eq. (13.26) predicts that an AFIR element, in theory, is a much faster device if compared with the PIR. The efficiency of the AFIR detector is a function of both: its design and the control circuit.

Nonradiative loss P_L is a function of ambient temperature T_a and a loss factor α_s:

$$P_L = \alpha_s(T_s - T_a) \ . \tag{13.27}$$

To generate heat in the AFIR sensor, it is provided with a heating element having electrical resistance R. During the operation, electric power dissipated by the heating element is a function of voltage V across that resistance

$$P = V^2/R \ . \tag{13.28}$$

Substituting Eqs. (3.14.19, 13.27, and 13.28) into (13.26), assuming that $T=T_b$ and $T_s>T_b$, after simple manipulations the object's temperature may be presented as a function of voltage V across the heating element:

$$T_b = \sqrt[4]{T_s^4 - \frac{1}{A\sigma\varepsilon_s\varepsilon_b}\left[\frac{V^2}{R} - \alpha_s(T_s - T_a)\right]} \ . \tag{13.29}$$

Coefficient α_s has a meaning of a thermal conductivity from the AFIR detector to the environment. Fig. 13-26 shows a family of curves calculated for different loss coefficients and the ideal case when $\alpha_s=0$.

One way to fabricate an AFIR element is to use a film thermistor having a relatively large surface area (4-20 mm^2). The resistance of a negative temperature coefficient (NTC) thermistor at temperature T_s may be modeled by Eq. (3.5.11). Electric current passing through the thermistor results in a self-heating effect which elevates the temperature of the thermistor above ambient. A control circuit regulates current in such a manner as to maintain the thermistor's temperature constant at a predetermined level T_s. A constant temperature yields a constant resistance R_s.

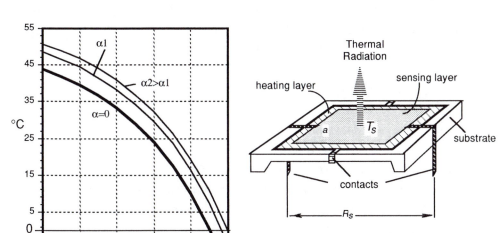

Fig. 13-26 Temperature of an object as represented by voltage across an AFIR heating element for three different α (A); structure of an AFIR sensor (B)

Alternatively, a film thermistor may be given a resistive underlayer which serves as a heating element (Fig. 13-26B) while the thermistor serves as a temperature sensitive layer to provide a feedback to the control circuit. Thermal radiation is emanated from the upper layer. Voltage V across the heating layer contacts may serve as the output signal. Therefore, contrary to a self-heating thermistor, a surface temperature measurement and a heating are separated. These functions are performed by electrically isolated layers. To fabricate a multilayer element, a thin substrate (for instance, a glass plate or a silicon membrane) is given a heating layer which is a thin metal film resistor of about 500 Ω. The heater is covered with a thin isolating coating (not shown). On the top of the coating, a temperature sensitive layer is vacuum deposited.

In many applications, it is highly desirable to compensate for variations in ambient conditions and to generate the output signal which is proportional to a difference in the IR fluxes emanated from two identical sensors. This can be achieved by use of a differential AFIR element which is depicted in Fig. 13-27A. In this example, two temperature sensitive layers (element 1 and element 2) and fabricated on a common substrate. Each element has its own heating underlayer (H_1 and

H_2) to compensate for emanated flux. Both parts share another heating element (common heater, H_c) which may provide heat for compensating nonradiative (conductive and convective) heat loss. Such a compensation is very important for better sensor's response. Eq. (13.26) may be modified to incorporate heat P_C from the compensating network: $P = P_L - P_C + \Phi$. Under the idealized conditions, compensating power must be equal to loss power: $P_c = P_L$, then Eq. (13.26) and subsequently (13.29) become independent of ambient conditions

$$T_b = \sqrt[4]{T_s{}^4 - \frac{V^2}{AR\sigma\varepsilon_s\varepsilon_b}} \ . \qquad (13.30)$$

Both parts of a differential element should be thermally decoupled, otherwise, a crosstalk may result in instability of the control circuits.

A control circuit must include the following essential components: a reference to preset temperature, an error amplifier, and a driver stage. In addition, it may include RC network for correcting a loop response function and for stabilizing its operation, otherwise an entire system may be prone to oscillations [14].

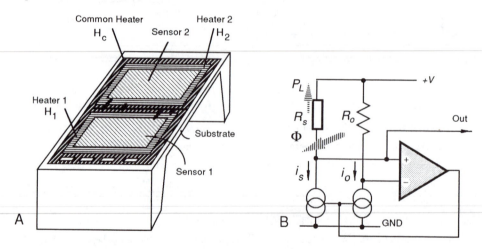

Fig. 13-27 Dual AFIR sensor is fabricated on a common substrate where the heating
and sensing layers are deposited in a sandwich-like structure (A);
a self-balancing bridge circuit with voltage-controlled current sinks (B)

A self-heating NTC thermistor may be controlled by a circuit which in a simplified form is shown in Fig. 13-26B. Two current generators regulate electric currents through a self-heating thermistor (the AFIR element) and the reference resistor R_o. These currents i_o and i_s are controlled by the amplifier's output voltage and are at a fixed ratio, for instance 1:1. The feedback from the amplifier forces the thermistor to stay at constant temperature T_s. A resistance of the thermistor will be maintained at a fixed level to assure the inverse proportion to currents:

$$\frac{R_s}{R_o} = \frac{i_o}{i_s} \ .$$ (13.31)

If the current generators are identical ($i_o = i_s$), the circuit will maintain $R_s = R_o$. Naturally, there is a great variety of other control circuits which may be useful to work with the AFIR detectors.

13.7.4 Coatings for thermal absorption

All thermal radiation sensors, either PIR or AFIR, rely on the absorption or emission of electromagnetic waves in the far infrared spectral range. According to Kirchhoff's discovery, absorptivity α and emissivity ε is the same thing (see Section 3.14.3). Their value for the efficient sensor's operation must be maximized, i.e., it should be made as close to unity as possible. This can be achieved by either a processing surface of a sensor to make it highly emissive, or by covering it with a coating having a high emissivity. Any such coating should have a good thermal conductivity and a very small thermal capacity, which means it must be very thin.

Several methods are know to give a surface the emissive properties. Among them a deposition of thin metal films (like nichrome) having reasonably good emissivity, a galvanic deposition of porous platinum black [15], and evaporation of metal in atmosphere of low pressure nitrogen [16]. The most effective way to create a highly absorptive (emissive) material is to form it with a porous surface. Particles with sizes much smaller than the wavelength generally absorb and diffract light. High emissivity of a porous surface covers a broad spectral range, however, it decreases with the increased wavelength. A film of goldblack with a thickness corresponding to $500\ \mu g/cm^2$ has an emissivity over 0.99 in the near, mid and far infrared spectral ranges.

To form porous platinum-black, the following electroplating recipe can be used [17]:

Platinum chloride	H_2PTCl_6 aq	2 g
Lead Acetate	$Pb(OOCCH_3)_2\ 3H_2O$	16 mg
Water		58 g

Out of this galvanic bath the films were grown at room temperature on silicon wafers with a gold underlayer film. A current density was $30 mA/cm^2$. To achieve an absorption better than 0.95, a film of $1.5 g/cm^2$ is needed.

To form a goldblack by evaporation, the process is conducted in a thermal evaporation reactor in a nitrogen atmosphere of 100Pa pressure. The gas is injected via a microvalve, and the gold source is evaporated from the electrically heated tungsten wire from a distance of about 6cm. Due to collisions of evaporated gold with nitrogen, the gold atoms loose their kinetic energy and are slowed down to thermal speed. When they reach the surface, their energy is too low to allow a surface mobility and they stick to the surface on the first touch event. Gold atoms form a surface structure in form of needles with linear dimensions of about 25nm. The structure resembles a surgical cotton wool. For the best results, goldblack should have a thickness in the range from 250 to 500µg/cm^2.

Another popular method to enhance emissivity is to oxidize a surface metal film to form metal oxide, which generally is highly emissive. This can be done by a metal deposition in a partial vacuum.

Another method of improving the surface emissivity is to coat a surface with an organic paint (visible color of the paint is not important). These paints have far infrared emissivity from 0.92 to 0.97, however, the organic materials have low thermal conductivity and can not be effectively deposited with thicknesses less than 10µm. This may significantly slow the sensor's speed response. In micromachined sensors, the top surface may be given a passivation glass layer, which not only provides an environmental protection, but has emissivity of about 0.95 in the far infrared spectral range.

13.8　GAS FLAME DETECTORS

Detection of a gas flame is very important for security and fire prevention systems. In many respects it is a more sensitive way to detect fire than a smoke detector, especially outdoors where smoke concentration may not reach a threshold level for the alarm triggering.

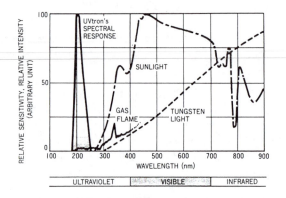

Fig. 13-28 Electromagnetic spectra of various sources
(Courtesy of Hamamatsu Photonics K.K.)

To detect burning gas, it is possible to use a unique feature of the flame - a noticeable portion of its optical spectrum is located in the ultraviolet (UV) spectral range (Fig. 13-28). Sunlight, after passing through the atmosphere loses a large portion of its UV spectrum located below 250nm, while a gas flame contains UV components down to 180nm. This makes it possible to design a narrow-bandwidth element for the UV spectral range which is selectively sensitive to flame and not sensitive to the sunlight or electric lights.

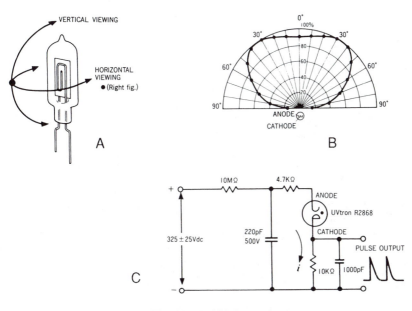

Fig. 13-29 UV flame detector
A: a glass-filled tube; B: angle of view in horizontal plane
C: recommended operating circuit
(Courtesy of Hamamatsu Photonics K.K.)

An example of such a device is shown in Fig. 13-29A. The element is a UV detector that makes use of a photoelectric effect in metals along with the gas multiplication effect. The detector is a rare-gas filled tube. The UV-transparent housing assures wide angles of view in both horizontal and vertical planes. The device needs high voltage for operation and under normal conditions is not electrically conductive. Upon being exposed to a flame, the high energy UV photons strike the cathode releasing free electrons to the gas-filled tube interior. Gas atoms receive an energy burst from the emitted electrons, which results in gas luminescence in the UV spectral range. This, in turn, cause more electrons to be emitted, which cause more UV luminescence. Thus, the element develops a fast avalanche-type electron multiplication making the anode-cathode region electrically conductive. Hence, upon being exposed to a gas flame, the element works as a current switch producing a strong positive voltage spike at its output (Fig. 13-29C). It follows from the above description that the element generates UV radiation in response to the flame detection. Albeit being of a low intensity, the UV does not present harm to people, however it may lead to crosstalk between the similar neighboring sensors.

REFERENCES

1. Chappell, A., ed. Optoelectronics: theory and practice. McGraw Hill, New York, 1978
2. Spillman, W. B., Jr. Optical Detectors. In: Fiber optic sensors, Eric Udd, ed., pp: 69-97 John Wiley & Sons, Inc, 1991
3. Verdeyen, J. T. Laser Electronics, Prentice-Hall, Englewood Clifs, N.J., 1981
4. Graeme, J. Phase compensation optimizes photodiode bandwidth. *EDN*, May 7, pp: 177-183, 1992.
5. Optoelectronic device data , Motorola,1988
6. Sasaki, A. and Noda, S. High gain and very sensitive light-to-light transducer. In: *Transducers'91. International conference on solid-state sensors and actuators. Digest of technical papers*, pp. 282-285, IEEE, 1991
7. Meixner, H., Mader, G., and Kleinschmidt, P. Infrared Sensors Based on the Pyroelectric Polymer Polyvinylidene Fluoride (PVDF). *Siemens Forsch.-u. Entwicl. Ber. Bd.* vol. 15, No. 3, p: 105-114, 1986
8. Fraden, J. Noncontact Temperature Measurements in Medicine. Chap. 17, In: Bioinstrumentation and Biosensors, D. Wise, ed., Marcel Dekker, Inc., pp: 511-549, 1991
9. Fraden, J. Infrared electronic thermometer and method for measuring temperature. *U.S. Patent* No. 4,797,840, Jan. 10, 1989
10. Fraden, J. Motion Detector, *U.S. Patent* No. 4,769,545, Sept. 6,1988
11. Fraden, J. Active far infrared detectors. In: *Temperature Its Measurement and Control in Science and Industry*, American Institute of Physics, vol. 6, part 2, pp: 831-836, 1992
12. Fraden, J. Radiation Thermometer and Method for Measuring Temperature, *U.S. Patent* No. 4,854,730, Aug. 8, 1989
13. Fraden, J. Active Infrared Motion Detector and Method for Detecting Movement, *U.S. Patent* No. 4,896,039, Jan. 23, 1990
14. Mastrangelo, C.H. and Muller, R.S. Design and performance of constant-temperature circuits for microbridge-sensor applications. In: *Transducers'91. International conference on solid-state sensors and actuators. Digest of technical papers*, pp: 471-474, IEEE, 1991
15. von Hevisy, G. and Somiya, T. Über platinschwarz. *Zeitschrift für phys. Chemie A.* vol 171, pp: 41, 1934
16. Harris, L., McGinnes, R., and Siegel, B. *J. of the Opt. Soc. of Am.*, vol. 38, pp: 7, 1948
17. Lang, W., Kühl, K., and Sandmaier, H. Absorption layers for thermal infrared detectors. In: *Transducers'91. International conference on solid-state sensors and actuators. Digest of technical papers*, pp: 635-638, IEEE, 1991

14

Radiation Detectors

Fig. 3-14.3 shows a spectrum of electromagnetic waves. On its left hand side, there is a region of the γ radiation. However, this is not the shortest possible length of electromagnetic waves. Besides, a spontaneous radiation from the matter not necessarily should be electromagnetic: there is the so-called nuclear radiation which is emission of particles from the atomic nuclei. It can be of two types: the charged particles (α and β particles, and protons), and uncharged particles which are the neutrons. Some particles are complex like the α-particles which are nuclei of helium atoms consisting of two neutrons, while other particles are generally simpler, like the β-particles which are either electrons or positrons. The γ and X rays belong to the nuclear type of electromagnetic radiation. In turn, X rays depending on the wavelengths are divided into hard, soft and ultrasoft rays.

Certain naturally occurring elements are not stable but slowly decompose by throwing away a portion of their nucleus. This is called *radioactivity*. It was discovered in 1896 by Henry Becquerel when he found that uranium atoms (Z=92)[1] give off radiation which fogs photographic plates. Besides the naturally occurring radioactivity, there are many manmade nuclei which are radioactive. These nuclei are produced in nuclear reactors, which may yield highly unstable elements. Regardless of the sources or ages of radioactive substances, they decay in accordance with the same mathematical law. The law is stated in terms of number N of nuclei still undecayed, and dN, the number of nuclei which decay in a small interval dt. It was proven experimentally, that

$$dN = -\lambda N dt \ , \tag{14.1}$$

where λ is a decay constant specific for a given substance. From (14.1) it can be defined as the fraction of nuclei which decay in unit time

$$\lambda = -\frac{1}{N}\frac{dN}{dt} \ . \tag{14.2}$$

---

[1] Z is the atomic number.

The SI unit of radioactivity is the *becquerel* (Bq) which is equal to the activity of radionuclide decaying at the rate of one spontaneous transition per second. Thus the becquerel is expressed in a unit of time: $Bq=s^{-1}$. To convert to the old historical unit which is the *curie*, the becquerel should be multiplied by $3.7 \cdot 10^{10}$ (Table 1-10).

The absorbed dose is measured in *grays* (Gy). A gray is the absorbed dose when the energy per unit mass imparted to matter by ionizing radiation is 1 joule per kg. That is, $Gy=J/kg$.

When it is required to measure exposure to X and γ rays, the dose of ionizing radiation is expressed in coulombs per kg, which is an exposure resulting in the production of 1 C of electric charge per 1 kg of dry air. In SI, a unit of C/kg replaces an older unit of *roentgen*.

The function of any radiation detector depends on the manner in which the radiation interacts with the material of the detector itself. There are many excellent texts available on the subject of detecting radioactivity, for instance [1, 2]. There are three general types of radiation detectors: the scintillation detectors, the gaseous detectors, and the semiconductor detectors. Further, all detectors can be divided into two groups according to their functionality: the collision detectors and the energy detectors. The former merely detect the presence of a radioactive particle, while the latter can measure the radiative energy. That is, all detectors can be either quantitative or qualitative.

14.1 SCINTILLATING DETECTORS

The operating principle of these detectors is based on the ability of certain materials to convert nuclear radiation into light. Thus, an optical photon detector in a combination with a scintillating material can form a radiation detector. It should be noted, however, that despite of a high efficiency of the conversion, the light intensity resulting from the radiation is extremely small. This demands photomultipliers to magnify signals to a detectable level.

The ideal scintillation material should posses the following properties:

1. It should convert the kinetic energy of charged particles into detectable light with a high efficiency;

2. The conversion should be linear. That is, the light produced should be proportional to the input energy over a wide dynamic range;

3. The post luminescence (the light decay time) should be short to allow fast detection;

4. The index of refraction of the material should be near that of glass to allow efficient optical coupling of the light to the photomultiplier tube.

The most widely used scintillators include the inorganic alkali halide crystals (of which sodium iodine is the favorite), and organic-based liquids and plastics. The inorganics are more sensitive, but generally slow, while organics are faster, but yield less light. A general simplified arrangement of a scintillating sensor is shown in Fig. 14-1 in conjunction with a photomultiplier.

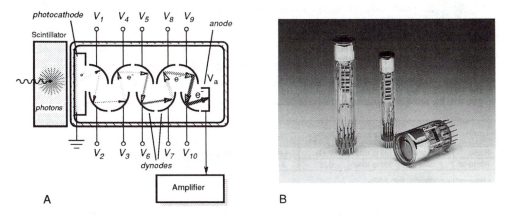

Fig. 14-1 A: A scintillation detector with a photomultiplier;
B: T-11 low-noise tubes with bialkali coated photocathodes
(Hamamatsu Photonics K.K.)

The scintillator is attached to the front end of the photomultiplier (PM). The front end contains a photocathode which is maintained at a ground potential. There is a large number of special plates called dynodes positioned inside the PM tube in an alternating pattern, reminding a shape of a "venetian blind". Each dynode is attached to a positive voltage source in a manner that the farther the dynode from the photocathode, the higher is its positive potential. The last component in the tube is an anode, which has the highest positive potential, sometimes on the order of several thousand volts. All components of the PM are enveloped into a glass vacuum tube, which may contain some additional elements, like focusing electrodes, shields, etc.

Although the PM is called a photomultiplier, in reality it is an electron multiplier, as there are no photons, only electrons inside the PM tube during its operation. For the illustration, let us assume that a gamma ray particle has a kinetic energy of 0.5MeV (megaelectron-volt). It is deposited on the scintillating crystal resulting in a number of liberated photons. In a thallium activated sodium iodine, the scintillating efficiency is about 13%, therefore total of 0.5x0.13 = 0.065MeV, or 65keV of energy is converted into visible light with an average energy of 4eV. Therefore, about 15,000 scintillating photons are produced per gamma pulse. This number is too small to be detected by an ordinary photodetector, hence a multiplication effect is required before the actual detection takes place. Of 15,000 photons probably about 10,000 reach the photocathode, whose quantum efficiency is about 20%. The photocathode serves to convert incident light photons into low-energy electrons. Therefore, the photocathode produces about 2,000 electrons per pulse. The PM tube is a linear device, that is, its gain is almost independent of the number of multiplied electrons.

Since all dynodes are at positive potentials (V_1 to V_{10}), an electron released from the photocathode is attracted to the first dynode, liberating at impact with its surface several very low energy electrons. Thus, a multiplication effect takes place at the dynode. These electrons will be easily guided by the electrostatic field from the first to the second dynode. They strike the second dynode and produce more electrons which travel to the third dynode, and so on. The process results in an increasing number of available electrons (avalanche effect). An overall multiplication ability of a PM tube is in the order of 10^6. As a result, about $2 \cdot 10^9$ electrons will be available at a high voltage anode (V_a) for the production of electric current. This is a pretty strong electric current which can be easily processed by an electronic circuit. A gain of a PM tube is defined as

$$G = \alpha \delta^N \quad , \tag{14.3}$$

where N is the number of dynodes, α is the fraction of electrons collected by the PM tube, δ is the efficiency of the dynode material, that is, the number of electrons liberated at impact. Its value ranges from 5 to 55 for a high yield dynode. The gain is sensitive to the applied high voltage, because δ is almost a linear function of the inter-dynode voltage.

One of the major limitations of scintillation counters is their relatively poor energy resolution. The sequence of events which leads to the detection involves many inefficient steps. Therefore, the energy required to produce one information carrier (a photoelectron) is in the order of $1,000 eV$ or more, and the number of carriers created in a typical radiation interaction is usually no more than a few thousand. For example, the energy resolution for sodium iodine scintillators is limited to about 6% when detecting $0.662 MeV$ γ rays, and is largely determined by the photoelectron statistical fluctuations. The only way to reduce the statistical limit on energy resolution is to increase the number of information carriers per pulse. This can be accomplished by the use of semiconductor detectors which are described below.

14.2   IONIZATION DETECTORS

These detectors rely on the ability of some gaseous and solid materials to produce ion pairs in response to the ionization radiation. Then, positive and negative ions can be separated in an electrostatic field and measured.

Ionization happens because charged particles upon passing at a high velocity through an atom can produce sufficient electromagnetic forces, resulting in the separation of electrons, thus creating ions. Remarkably, the same particle can produce multiple ion pairs before its energy is expended. Uncharged particles (like neutrons) can produce ion pairs at collision with the nuclei.

14.2.1   Ionization chambers

These radiation detectors are the oldest and most widely used. The ionizing particle causes ionization and excitation of gas molecules along its passing track. As a minimum, the particle must transfer an amount of energy equal to the ionization energy of the gas molecule to permit the ionization process to occur. In most gasses of interest for radiation detection, the ionization energy for the least tightly bound electron shells is between 10 and 20eV [2]. However, there are other mechanisms by which the incident particle may lose energy within gas that do not create ions, for instance, moving gas electrons to a higher energy level without removing it. Therefore, the average energy lost by a particle per ion pair formed (called *W-value*) is always greater than the ionizing energy. The W-value depends on gas (Table 14-1), the type of radiation, and its energy.

Table 14-1 W-values for different gases
[adapted from [2])

Gas	W-value (in eV/Ion Pair)	
	Fast electrons	Alphas
A	27.0	25.9
He	32.5	31.7
N_2	35.8	36.0
Air	35.0	35.2
CH_4	30.2	29.0

In the presence of an electric field, the drift of the positive and negative charges represented by the ions end electrons constitutes an electric current. In a given volume of gas, the rate of the formation of the ion pair is constant. For any small volume of gas, the rate of formation will be exactly balanced by the rate at which ion pairs are lost from the volume, either through recombination, or by diffusion or migration from the volume. If recombination is negligible and all charges are effectively collected, the steady state current produced is an accurate measure of the rate of ion pair formation. Fig. 14-2 illustrates a basic structure of an ionizing chamber and the current/voltage characteristic.

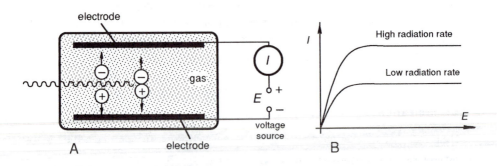

Fig. 14-2 Simplified schematic of an ionization chamber (A)
and a current vs. voltage characteristic (B)

A volume of gas is enclosed between the electrodes which produce an electric field. An electric current meter is attached in series with the voltage source E and the electrodes. There is no electrical conduction and no current under the no-ionization conditions. Incoming radiation produces, in the gas, positive and negative ions which are pulled by the electric field toward the corresponding electrodes forming an electric current. The current versus voltage characteristic of the chamber is shown in Fig. 14-2B. At relatively low voltages, the ion recombination rate is strong and the 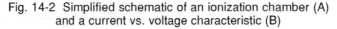 output current is proportional to the applied voltage, because higher voltage reduces the number of recombined ions. A sufficiently strong voltage completely suppress all recombinations by pulling all available ions toward the electrodes, and the current becomes voltage independent. However, it still depends on the intensity of irradiation. This is the region which is called *saturation* and where the ionization chamber normally operates.

4.2.2   Proportional chambers

The proportional chamber is a type of a gas-filled detector which almost always operates in a pulse mode and relies on the phenomenon of a gas multiplication. This is why these chambers are called the proportional counters. Due to gas multiplication, the output pulses are much stronger than in conventional ion chambers. These counters are generally employed in the detection and spectroscopy of low energy X radiation and for the detection of neutrons. Contrary to the ionization chambers, the proportional counters operate at higher electric fields which can greatly accelerate electrons liberated during the collision. If these electrons gain sufficient energy, they may ionize a neutral gas molecule, thus creating an additional ion pair. Hence, the process is of an avalanche type resulting in a substantial increase in the electrode current. The name for this process is

Townsend avalanche. In the proportional counter, the avalanche process ends when the electron collides with the anode. Since in the proportional counter the electron must reach the gas ionization level, there is a threshold voltage after which the avalanche process occurs. In typical gases at atmospheric pressure, the threshold field level is on the order of 10^6 V/m.

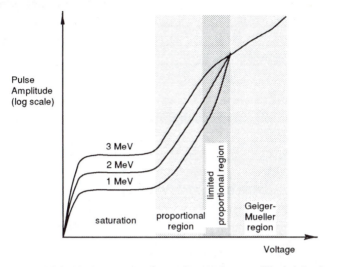

Fig. 14-3 Various operating voltages for gas-filled detectors
(adapted from [2])

Differences between various gas counters are illustrated in Fig. 14-3. At very low voltages, the field is insufficient to prevent the recombination of ion pairs. In the saturation level, all ions are drifted to the electrodes. A further increase in voltage results in gas multiplication. Over some region of the electric field, the gas multiplication will be linear, and the collected charge will be proportional to the number of original ion pairs created during the ionization collision. An even further increase in the applied voltage can introduce nonlinear effects, which are related to the positive ions, due to their slow velocity.

4.2.3 Geiger-Mueller Counters

The Geiger-Mueller (G-M) counter was invented in 1928 and is still in use thanks to its simplicity, low cost, and ease of operation. The G-M counter is different from other ion chambers by its much higher applied voltage (Fig. 14-3). In the region of the G-M operation, the output pulse amplitude doesn't depend on the energy of ionizing radiation and is strictly a function of the applied voltage. A G-M counter is usually fabricated in the form of a tube with an anode wire in the center (Fig. 14-4). The tube is filled with a noble gas, such as helium or argon. A secondary component is usually added to the gas for the purpose of *quenching*, which is preventing of a retriggering of the counter after the detection. The retriggering may cause multiple pulses instead of the desired one. The quenching can be accomplished by several methods, among which are a short-time reduction of the high voltage applied to the tube, use of a high impedance resistors in series with the anode, and adding the quench gas at concentrations of 5%-10%. Many organic molecules possess the proper characteristics to serve as a quench gas. Of these, ethyl alcohol and ethyl formate have proven to be the most popular.

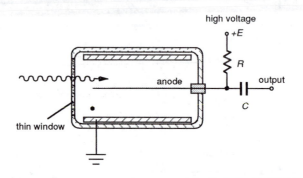

Fig. 14-4 Circuit of a Geiger-Mueller counter
Symbol • indicates gas

In a typical avalanche created by a single original electron, the secondary ions are created. In addition to them, many excited gas molecules are formed. Within a few nanoseconds, these excited molecules return to their original state through the emission of energy in the form of ultraviolet (UV) photons. These photons play an important role in the chain reaction occurring in the G-M counter. When one of the UV-photons interacts by photoelectric absorption in some other region of the gas, or at the cathode surface, a new electron is liberated which can subsequently migrate toward the anode, and will trigger another avalanche. In a Geiger discharge, the rapid propagation of the chain reaction leads to many avalanches which initiate at random radial and axial positions throughout the tube. Secondary ions are therefore formed throughout the cylindrical multiplying region which surrounds the anode wire. Hence, the discharge grows to envelope the entire anode wire, regardless of the position at which the primary initiating event occurred.

Once the Geiger discharge reaches a certain level, however, collective effects of all individual avalanches come into play and ultimately terminate the chain reaction. This point depends on the number of avalanches and not on the energy of the initiating particle. Thus, the G-M current pulse is always of the same amplitude, which makes the G-M counter just an indicator of irradiation, because all information on the ionizing energy is lost.

In the G-M counter, a single particle of a sufficient energy can create about 10^9 to 10^{10} ion pairs. Because a single ion pair formed within the gas of the G-M counter can trigger a full Geiger discharge, the counting efficiency for any charged particle that enters the tube is essentially 100%. However, the G-M counters are seldom used for counting neutrons because of a very low efficiency of counting. The efficiency of G-M counters for γ rays is higher for those tubes constructed with a cathode wall of high-Z material. For instance, bismuth (Z=83) cathodes have been widely used for the γ-detection in conjunction with gases of high atomic numbers, such as xenon and krypton, which yield a counting efficiency up to 100% for photon energies below about 10keV.

14.2.4 Semiconductor detectors

The best energy resolution in modern radiation detectors can be achieved in the semiconductor materials, where a comparatively large number of carriers for a given incident radiation event occurs. In these materials the basic information carriers are *electron-hole pairs* created along the path taken by the charged particle through the detector. The charged particle can be either primary radiation, or a secondary particle. The electron-hole pairs in some respects are analogous to the ion pairs produced in the gas-filled detectors. When an external electric field is applied to the semiconductive material, the created carriers form a measurable electric current. The detectors operating on this principle are called solid-state or semiconductor diode detectors. The operating principle of these radiation detectors is the same as that of the semiconductor light detectors. It is based on the transition of electrons from one energy level to another when they gain or lose energy. For the description of the energy band structure in solids the reader should refer to Section 13.1.

When a charged particle passes through a semiconductor with the band structure shown in Fig. 13-1, the overall significant effect is the production of many electron-hole pairs along the track of the particle. The production process may be either direct or indirect, in that the particle produces high-energy electrons (or Δ rays) which subsequently lose their energy in production more electron-hole pairs. Regardless of the actual mechanism involved, what is of interest to our subject is the average energy expended by the primary charged particle produces one electron-hole pair. This quantity is often called the "ionization energy". The major advantage of semiconductor detectors lies in the smallness of the ionization energy. The value of it for silicon or germanium is about 3eV, compared with 30eV required to create an ion pair in typical gas-filled detectors. Thus, the number of charge carriers is about ten times greater for the solid-state detectors for a given energy of a measured radiation.

To fabricate a solid-state detector, at least two contacts must be formed across a semiconductor material. For detection, the contacts are connected to the voltage source which enables carrier movement. The use of a homogeneous Ge or Si, however, would be totally impractical. The reason for that is in an excessively high leakage current caused by the material's relatively low resistivity (50 kΩ·cm for silicon). The external voltage, when applied to the terminals of such a detector may cause a current which is 3-5 orders of magnitude greater than a minute radiation-induced electric current. Thus, the detectors are fabricated with the blocking junctions, which are reverse biased to dramatically reduce leakage current. In effect, the detector is a semiconductor diode which readily conducts (has low resistivity) when its anode (p side of a junction) is connected to a positive terminal of a voltage source and the cathode (an n side of the junction) to the negative. The diode conducts very little (it has very high resistivity) when the connection is reversed, thus the name reverse biasing is implied. If the reverse bias is made very large, in excess of the manufacturer specified limit, the reverse leakage current abruptly increases (the breakdown effect) which often may lead to a catastrophic deterioration of detecting properties or to the device destruction.

Several configurations of silicon diodes are currently produced. Among them diffused junction diodes, surface barrier diodes, ion implanted detectors, epitaxial layer detectors and others. The diffused junction and surface barrier detectors find widespread applications for the detection of α particles and other short-range radiation. A good solid-state radiation detector should possess the following properties:

1. Excellent charge transport;

2. Linearity between the energy of the incident radiation and number of electron-hole pairs;

3. Absence of free charges (low leakage current);

4. Production of a maximum number of electron-hole pairs per unit of radiation;

5. High detection efficiency;

6. Fast response speed;

7. Large collection area;

8. Low cost.

When using semiconductor detectors, several factors should be seriously considered. Among them are the dead band layer of the detector and the possible radiation damage. If heavy charged particles or other weakly penetrating radiations enter the detector, there may be a significant energy loss before the particle reaches the active volume of the semiconductor. The energy can be lost in the metallic electrode and in a relatively thick silicon body immediately beneath the electrode. This thickness must be measured directly by the user if an accurate compensation is desirable. The simplest and most frequently used technique is to vary the angle of incidence of a monoenergetic charged particle radiation [2]. When the angle of incidence is zero (that is, perpendicular to the detector's surface) the energy loss in the dead layer is given by

$$\Delta E_o = \frac{dE_o}{dx}t \ , \tag{14.4}$$

where t is the thickness of the dead layer. The energy loss for an angle of incidence of Θ is

$$\Delta E(\theta) = \frac{\Delta E_o}{\cos\theta} \ . \tag{14.5}$$

Therefore, the difference between the measured pulse height for angles of incidence of zero and Θ is given by

$$E^{'} = \left[E_o - \Delta E_o\right] - \left[E_o - \Delta E(\theta)\right] = \Delta E_o \left(\frac{1}{\cos\theta} - 1\right) . \tag{14.6}$$

If a series of measurements is made as the angle of incidence is varied, a plot of $E^{'}$ as a function of ($\frac{1}{\cos\Theta}$ - 1) should be a straight line whose slope is equal to ΔE_o. Using tabular data for dE_o/dx for the incident radiation, the dead layer thickness can be calculated from Eq. (14.4).

Any excessive use of the detectors may lead to some damage to the lattice of the crystalline structure, due to disruptive effects of the radiation being measured as it passes through the crystal. These effects tend to be relatively minor for lightly ionizing radiation (β-particles or γ rays), but can become quite significant under typical conditions of use for heavy particles. For example, prolonged exposure of silicon surface barrier detectors to fusion fragments will lead to a measurable increase in leakage current and a significant loss in energy resolution of the detector. With extreme radiation damage, multiple peaks may appear in the pulse height spectrum recorded for monoenergetic particles.

Mentioned above diffused junction diodes and surface barrier diodes are not quite suitable for the detection of penetrating radiation. The major limitation is in the shallow active volume of these sensors, which rarely can exceed 2-3 mm. This is not nearly enough, for instance, for a γ-ray spectroscopy. A practical method to make detectors for a more penetrating radiation is the so-called ion-drifting process. The approach consists of creating a thick region with a balanced number of donor impurities, which add either p or n properties to the material. Under ideal conditions, when the balance is perfect, the bulk material would resemble the pure (intrinsic) semiconductor without either properties. However, in reality the perfect pn balance never can be achieved. In Si or Ge, the pure material with the highest possible purity tends to be of p type. To accomplish the desired compensation, the donor atoms must be added. The most practical compensation donor is lithium. The fabrication process involves a diffusing of lithium though the *p* crystal so that the lithium donors greatly outnumber the original acceptors, creating a *n* type region near the exposed surface. Then, temperature is elevated and the junction is reverse biased. This results in a slow drifting of lithium donors into the *p* type for the near perfect compensation of the original impurity. The process may take as long as several weeks. To preserve the achieved balance, the detector must be maintained at low temperature: 77K for the germanium detectors. Silicon has very low ion mobility, thus the detector can be stored and operated at room temperature. However, the lower atomic number for silicon (Z=14) as compared with germanium (Z=32) means that the efficiency of silicon for the detection of γ rays is very low and it is not widely used in the general γ-ray spectroscopy.

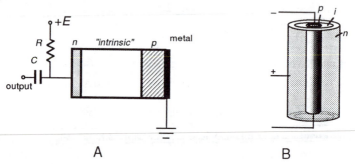

Fig. 14-5 Lithium-drifted *pin*-junction detector
A: structure of the detector; B: coaxial configuration of the detector

A simplified schematic of a lithium-drifted detector is shown in Fig. 14-5A. It consists of three regions where the "intrinsic" crystal is in the middle. In order to create detectors of a larger active volume, the shape can be formed as a cylinder (Fig. 14-5B), where the active volumes of Ge up to 150cm^3 can be realized. The germanium lithium-drifted detectors are designated as Ge(Li).

Table 14-2 Properties of some semiconductive materials
(adapted from [2])

Material (operating temperature in K)	Z	Band Gap, eV	Energy per electron-hole pair, eV
Si (300)	14	1.12	3.61
Ge (77)	32	0.74	2.98
CdTe (300)	48-52	1.47	4.43
HgI$_2$ (300)	80-53	2.13	6.5
GaAs (300)	31-33	1.43	4.2

Regardless of the widespread popularity of the silicon and germanium detectors, they are not the ideal from certain standpoints. For instance, germanium must always be operated at cryogenic temperatures to reduce thermally generated leakage current, while silicon is not efficient for the detection of γ rays. There are some other semiconductors which are quite useful for detection of radiation at room temperatures. Among them are cadmium telluride (CdTe), mercuric iodine (HgI_2), gallium arsenide (GaAs), bismuth trisulfide (Bi_2S_3), and gallium selenide (GaSe). Useful for radiation detectors properties of some semiconductive materials are given in Table 14-2.

Probably the most popular at the time of this writing is cadmium telluride which combines a relatively high Z-value (48 and 52) with a large enough band gap energy (1.47 eV) to permit room temperature operation. Crystals of high purity can be grown from CdTe to fabricate the intrinsic detector. Alternatively, chlorine doping is occasionally used to compensate for the excess of acceptors and to make the material of a near-intrinsic type. Commercially available CdTe detectors range in size from 1 to 50 mm in diameter and can be routinely operated at temperatures up to 50°C without an excessive increase in noise. Thus, there are two types of CdTe detectors available: the pure intrinsic type and the doped type. The former has high volume resistivity up to 10^{10} Ω·cm, however, its energy resolution is not that great. The doped type has significantly better energy resolution, however, its lower resistivity (10^8 Ω·cm) leads to a higher leakage current. Besides, these detectors are prone to polarization which may significantly degrade their performance.

In the solid-state detectors, it is also possible to achieve a multiplication effect as in the gas-filled detectors. An analog of a proportional detector is called an *avalanche detector* which is useful for the monitoring of low-energy radiation. The gain of such a detector is usually in the range of several hundreds. It is achieved by creating within a semiconductor high level electric fields. Also, the radiation PSDs are available whose operating principle is analogous to the similar sensors functioning in the near infrared region (see Section 5.8.4).

REFERENCES

1. Evans, R.D. The atomic nucleus. McGraw-Hill, New York, 1955
2. Knoll, G.F. Radiation detection and measurement. John Wiley and Sons, New York, 1979

15

Electromagnetic Field Detectors

15.1 MAGNETIC FIELD SENSORS

Magnetic field measurements are extensively used in geomagnetic studies, navigation, manufacturing processes, medicine, etc. The instruments which measure magnetic fields are called magnetometers. Most of them employ electronics and usually contain no moving parts.

Probably, the first sensor ever invented was a magnetic compass. In 2634 B.C. in China, a chariot was guided by a piece of magnetite suspended by a silk thread [1]. In 1070, the Chinese first used a magnetic compass for marine navigation [2]. In Europe, the marine compass is known since 1269 where it was brought from China by Marco Polo. To the aeronautic community it was introduced during the 1911 Paris-to-Madrid air race by Jean Canneau, a French Navy lieutenant [3].

The magnetic field is one of the most important characteristics of earth. Its strength is in the average of 0.5 gauss. The field consists of two parts. The inner part is 90% of total. It is attributed to a differential rotation between the earth's core and its mantle. The outer part is 10% of the total and attributed to ionospheric ring currents. The field can be represented by a dipole which fluctuates both in time and space. The best-fit dipole location is about 440 km off center and is inclined about 11° to the earth axis rotation.

In modern magnetic compasses, the magnetized needle is replace by more sensitive magnetic field detectors. The main reason for that is in several disadvantages of needle compasses. First, the motion of the needle assembly of a mechanical compass is particularly difficult to damp, since any damping medium used must be contained in a receptacle attached to the framework of the moving vehicle. Hence, motions of the vehicle are transmitted to the needle assembly through the damping medium. Second, any electromagnetic or ferromagnetic anomalies, occurring in the vehicle, influence the compass in the same way as does the earth's magnetic field. These local magnetic fields to some extent, can be canceled by the manipulation of small permanent magnets or electromagnets in the vicinity of the compass. Another problem, which especially is evident in aircraft, is the so-called "northerly turning error", which refers to the increase in error when the aircraft is headed north. Important work to develop a replacement for a mechanical compass has been done at the NASA Langley Research Center [4].

There are two classes of magnetometers or magnetic detectors. One is a nondirectional device which is called the "total-field magnetometer" which produces an output that is a function of the magnitude of the magnetic field passing though it, no matter what its vector relation to the instrument might be. This class includes the proton-precession magnetometers and optically pumped magnetometers [5]. The second class embraces vector magnetometers which register only the magnitude of that component of the magnetic field which lies parallel to their sensitive axes. This includes the saturable-core, the Hall-effect, and the magnetoresistive magnetometers.

The measurement of magnetic fields *per ce* is an important application for the magnetic sensors. However, a really broad area for their use in practical systems are position, force and pressure measurements, proximity detection, sensing of angular velocity, and strong electric currents.

15.1.1 Flux-gate sensors

Many sensors which are primarily intended for an accurate measurement of weak magnetic fields are based on the so called "flux-gate" principle. Fig. 15-1A shows the general shape of the *B-H* curve of a typical high-permeability core material such as mumetal, permalloy or super malloy. A magnetic field *H* applied to the core, induces a magnetic flux *B*

$$B = \mu H \ , \tag{15.1}$$

where μ is the permeability of the material. For higher *B*, the material saturates, that is, its magnetic flux can not be further increased. In other terms, its permeability at saturation becomes extremely small. There is some hysteresis in these materials which causes the curve to follow a somewhat different path for the increasing and decreasing values of *B*. When the core is not saturated (μ is high), the core material acts as a low impedance path to lines of magnetic flux in the surrounding space. The flux lines in the material immediate vicinity are drawn into the core as shown in Fig. 15-1B. When the core is saturated, the magnetic field lines direction are no more affected by the core (Fig. 15-1C). Each time the core passes from the unsaturated state to the saturated state and backward, the magnetic field lines undergo change. There is a pickup coil wound around the core. Whenever magnetic flux lines are drawn out of core, they induce a positive current spike in the coil. Whenever they are drawn inside the core, they generate a negative spike. The amplitude of the spikes is directly proportional to the intensity of the flux vector which lies parallel to the axis of the sense coil. The pulse polarity indicates direction of that vector.

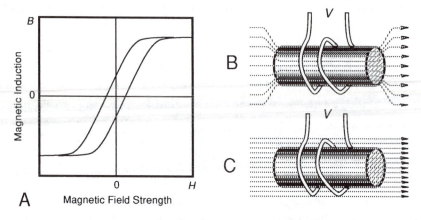

Fig. 15-1 Magnetization curve of a ferromagnetic core (A);
magnetic field around unsuturated (B), and saturated (C) core

To make this effect useful for magnetic field sensing, the core should be driven into and out of saturation. It can be accomplished by adding a second coil (the drive coil) which is excited by an alternate current (Fig. 15-2A). It is obvious, that the excitation current will induce a corresponding current in the sense coil which will be superimposed on the external flux induced pulses (Fig. 15-2B). These signals can be electrically separated, however, with a significant loss in accuracy.

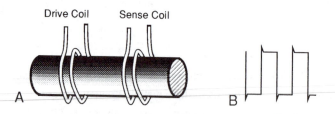

Fig. 15-2 Dual-coil sensing method (A) and output pulses (B)

A better approach is to position the excitation coil in such a manner as it would saturate the core without inducing an undesirable signal in the sensor coil. The method called ortogonal (or transflux) is based on directing the excitation flux at right angle to the axis of the sense coil, so it can not induce current in it. Fig. 15-3A shows a popular practical implementation of the method. It uses a toroidal core with the excitation (drive) winding and a cylindrical winding for the sense coil.

Fig. 15-3B shows pulses in the sense coil. In geomagnetic applications, they are induced by passage of the earth's magnetic field into and out of the core.

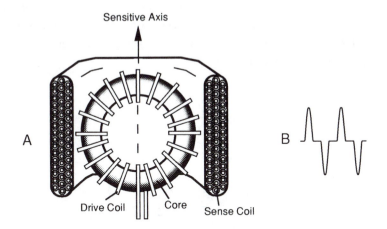

Fig. 15-3 A magnetometer with a toroidal core (A) and output pulses (B)

15.1.2   Hall effect sensors

The Hall effect sensors are responsive to both direction and intensity of magnetic fields. The sensor can be pictured as a resistive bridge where voltage sources are connected in series with the resistors.

One problem associated with a Hall sensor is its relatively large offset voltage. This voltage may be trimmed during the instrument calibration. However, a much more reliable way is to use a dynamic cancellation of the offset. Fig. 15-4A shows an equivalent circuit of a Hall sensor with no magnetic field present. Due to resistor mismatch, the sensor exhibits an offset voltage. The offset voltage is a result of many causes including processing technologies and the piezoresistive effect. When a magnetic field is applied perpendicular to and pointing out of the sensor plane, the output voltage increases by an amount proportional to the magnitude of the magnetic flux density at the sensor. This can be presented as [6]

$$\Delta V_a = V_2 - V_1 = (\mu_H \frac{W}{L} E)B + \frac{R' - R}{2(R + R')}E \quad , \qquad (15.1)$$

where μ_H is the electron Hall mobility, W/L is the effective width to length ratio for the Hall element, E is the excitation voltage, B is the magnetic flux density, R is a nominal bridge resistance, and R' is a mismatched bridge resistor. There are several methods to reduce an error defined by the second term in (15.1). One way is to trim the offset during a calibration process of the instrument. Another and more efficient method is a dynamic cancellation of the offset [7]. The method is based on the switching of the bridge terminals by using fast analog switches having low on-resistance and high off-resistance. Fig. 15-4B shows the same sensor with its terminal connected inversely, that is, its V_2 output and excitation terminals are switched over, as well as ground and V_1 terminals. In this instance, the new Hall output voltage is

$$\Delta V_b = V_2 - V_1 = (\mu_H \frac{W}{L} E)B - \frac{R' - R}{2(R + R')} E \quad . \tag{15.3}$$

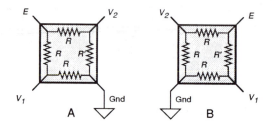

Fig. 15-4 Equivalent circuit of a Hall sensor
with two different methods of connection.

The magnetic field-dependent term is unaffected by the alteration, while the offset term changes sign. This suggests a way to cancel the Hall offset in the presence of an applied field. Adding voltages ΔV_a and ΔV_b we arrive at

$$\Delta V = 2(\mu_H \frac{W}{L} E)B \quad , \tag{15.4}$$

where it is assumed that B changes negligibly between the bridge samplings.

The operation of summation can be performed either by an analog summator or by a conventional switched capacitor circuit. The cancellation circuit consists of a fully differential switched-capacitor gain stage interfaced with a Hall sensor. In addition to providing gain, the stage incorporates a sample-and-hold function (φ_3 switches) that permits the addition of two successive samples from the Hall sensor which is useful for an A/D conversion [8]. The amplifier cancels its own d.c. offset and suppresses operational amplifier low-frequency noise. An additional differential stage provides a conversion from symmetrical to a non-symmetrical output and additional gain A. The output voltage of the offset cancellation circuit after the proper sequencing of the clock is

$$V_{out} = 2A \frac{C_o}{C_f} (\mu_H \frac{W}{L} E)B \quad . \tag{15.5}$$

A practical implementation of the circuit shown in Fig. 15-5 is a bit more complex, because of the need to cancel a differential charge injection from the feedback switches, φ_5. This can be accomplished with a more complex amplifier design [9]. It should be noted that output voltage is temperature dependent because μ_H is a function of temperature. To compensate for this dependence, standard techniques of temperature compensation can be used (see Section 4.7.3).

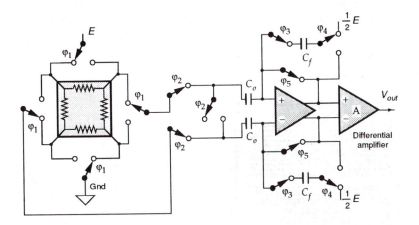

Fig. 15-5 Switched-capacitor circuit for cancellation of offset in a Hall sensor

An effective method of temperature compensation for a silicon n-doped Hall sensor is the use of a temperature stable current source I_o [6]. The Hall sensor is powered by I_o rather than by a hard voltage source. Then, excitation voltage is

$$E = I_o R_H = I_o \frac{L}{q\mu_N N_D W t} \; , \tag{15.6}$$

where μ_N is the electron drift mobility, t is a thickness of the Hall element, N_D is the n-well doping density, and q is the electronic charge. Substituting in Eq. (15.5) yields

$$V_{out} = 2A \frac{C_o}{C_f} \frac{\mu_H I_o}{q\mu_N N_D t} B \; . \tag{15.7}$$

Since μ_H is approximately equal to μ_N, the output voltage is substantially more temperature stable than that defined by Eq. (15.5).

15.1.3 Magnetoresistive sensors

The magnetoresistive effect is the well known property of a current-carrying magnetic material to change its resistivity in the presence of an external magnetic field. This change is brought about by the rotation of the magnetization relative to the current direction. For example, in the case of permalloy[1] a 90° rotation of the magnetization will produce a 2%-3% change in resistivity. At low temperatures, some materials, for instance bismuth, may change resistances by a factor as large as 10^6. Most conductors have a positive magnetoresistivity, i.e., their resistance increases in the presence of a magnetic field, while very few exhibit a negative magnetoresistivity, which seems to be related to an inhomogeneous structure of the conductor. The basic cause for the magnetoresistivity is the Lorentz force, which causes the electrons to move in curved paths between collisions (see Section 3.3.1). For small values of the magnetic field, the change in resistance is proportional to the square of the magnetic field strength, while for large magnetic fields, it may continue to follow the square law, to saturate or to follow a complex pattern.

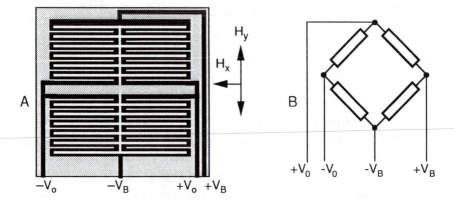

**Fig. 15-6 Four magnetoresistors form Wheatstone bridge (A)
and its equivalent circuit (B)**

The chip incorporates special resistors that are trimmed during the manufacture to give zero offset at 25°C

A magnetoresistive sensor is fabricated of permalloy strips positioned on a silicon substrate[2]. Each strip is arranged in a meander pattern (Fig. 15-6) and forms an arm of a Wheatstone bridge. The degree of bridge imbalance is then used to indicate the magnetic field strength, or more precisely, the variation in magnetic field in the plane of the permalloy strips normal to the direction of current. For trimming purposes, two additional fixed resistors may be formed on a substrate. These resistors may be laser trimmed to balance the bridge under zero-magnetic field conditions. During the manufacturing process, a strong magnetic field is applied parallel to the strip axis. This imparts a preferred magnetization direction to the permalloy strips. So, even in the absence of an

---

[1] A ferromagnetic alloy containing 20% iron and 80% nickel.
[2] Information on magnetoresistive KMZ10 sensors is courtesy of Philips Semiconductors BV, Eindhoven, The Netherlands.

external magnetic field, the magnetization will always tend to align with the strips. Therefore, the internal magnetization of the sensor strips has two stable positions, so that if for any reason, the sensor should come under the influence of a powerful magnetic field opposing the internal magnetic field, the magnetization may flip from one position to another, and the strips become magnetized in the opposite direction (for instance, from +x to –x direction).

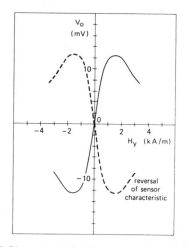

Fig. 15-7 Characteristic of a magnetoresistive sensor.
The solid line is for the "normal" sensor with magnetization oriented in +x direction, while the broken line shows the characteristic of a "flipped" sensor
(Courtesy of Philips Semiconductors BV)

As it follows from Fig. 15-7 this may lead to drastic changes in sensor characteristics. The solid line shows the characteristic of a normal sensor (i.e., with the magnetization in the +x direction), and the broken line is for the "flipped" sensor. The magnetic field, H_{-x}, needed to flip the sensor magnetization depends on the magnitude of the transverse field H_y. The greater the field H_y, the smaller the field H_{-x}. The reason for that is the following. For greater fields H_y, the magnetization closely approaches 90°, and hence the easier it will be to flip it into a corresponding stable position in the –x direction. This is illustrated in Fig. 15-8A which shows the sensor's output voltage V_0 as a function of H_x, for several values of H_y.

For example, low intensity transverse field $H_y = 0.5$ kA/m yields stable sensor characteristics for all possible values of H_x, and a reverse field of about 1 kA/m is required for the flipping. At $H_y = 4$ kA/m, on the other hand, the sensor will flip at even positive values of H_x (at about 1kA/m). The figure also illustrates that the flipping is not instantaneous. This is a result of the different flipping of different permalloy strips. Further, it can be seen that there is a substantial hysteresis effect. And, finally (Fig. 15-8) the sensitivity of the sensor falls with an increase in H_x.

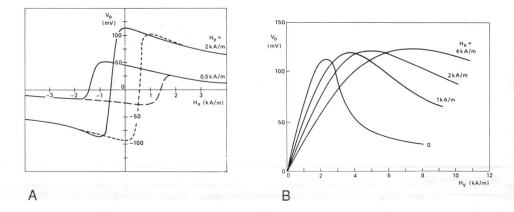

Fig. 15-8 Sensor's output voltage V_O as function of H_x for several values of H_y (A)
and sensor's output voltage V_O as function of H_y for several values of H_x (B)
(Courtesy of Philips Semiconductors BV)

The above discussion can be summarized in several practical recommendations.

1. Avoid operating the sensor in an environment where it's likely to be subjected to negative external fields H_{-x}. Preferably, apply a positive auxiliary field H_x of a sufficient magnitude to prevent any likelihood of flipping within the operating range of H_y.

2. Use the minimum auxiliary field that will assure stable operation. The larger the auxiliary field, the lower the sensitivity. For the KMZ10B sensor, maximum recommended auxiliary field is about 1 kA/m.

3. Before using the sensor for the first time, apply a positive auxiliary field of at least 3 kA/m. This will effectively erase the sensor's history and will ensure that no residual hysteresis remains.

When measuring weak magnetic fields, like that of the Earth, the sensor's internal magnetization may be in danger of being "flipped" by stray magnetic fields. A further problem of measuring weak magnetic fields is that the measuring accuracy is limited by the drift of both the sensor and the amplifier's offset. These problems can be solved by using the "flipping" characteristics of the KMZ10 sensors and supplying the sensor with a coil, the axial field of which is periodically reversed by successive positive and negative going pulses. As a consequence, the internal magnetization of the sensor, which is located inside the coil (Fig. 15-9a) is continuously flipped from its normal into its reverse polarity and back (Fig. 15-9b). Thus, the influence of a stray magnetic field is removed. Besides, an offset is also eliminated by using an ac amplifier. Then, the output can be rectified in a synchronous demodulator, which is controlled by the clock common for both the flipping source and the demodulator. The resulting dc output voltage is proportional to the magnetic field to be measured.

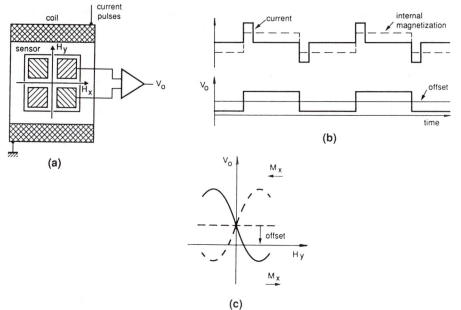

(a)

(b)

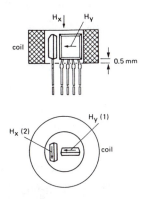

(c)

Fig. 15-9 Measuring weak magnetic fields with the magnetoresistive sensor
a: setup using a coil, the magnetic field of which is periodically reversed. Note that the sensor's leads must be parallel to the coil axis; b: pulse diagram; c: sensor's output characteristics
(Courtesy of Philips Semiconductors BV)

Fig. 15-10 A magnetic compass using two mutually ortogonal magnetoresistive sensors inside a coil
(Courtesy of Philips Semiconductors BV)

As the sensor has an internal magnetization that is parallel to the leads of the sensor, the sensor must be mounted inside the coil with leads parallel to the coil axis. The switching field applied by the coil should be no less than 3kA/m. Since there is no other auxiliary field in the arrangement, the sensitivity of the assembly is approximately 22 mV/V/(kA/m) and fields up to 50 A/m can be measured without stability problems.

As an example, Fig. 15-10 shows a magnetic compass which utilizes two perpendicularly mounted KMZ10A1 sensors. Again, the field inside the coil is periodically switched to obtain offset- and drift- independent output signals. This ratiometric compass doesn't need a temperature correction, as the field direction depends on the ratio of two output signals, but not on their absolute values. A suitable coil for the compass would be 100 turns of a 0.35 mm copper wire wound on a coil former, giving a resistance of 0.8 Ω, an inductance of 87 μH, and an axial magnetic field of 8.3 (kA/m)/A.

15.2 BOLOMETERS

Bolometers are miniature RTDs or thermistors (see Section 16.1.3) which are mainly used for measuring r.m.s. values of electromagnetic signals over a very broad spectral range from microwaves to far infrared. Applications include infrared temperature detection and imaging, measurements of local fields of high power, the testing of microwave devices, RF antenna beam profiling, testing of high power microwave weapons, monitoring of medical microwave heating, and others. The operating principle is based on a fundamental relationship between the absorbed electromagnetic signal and dissipated power [10]. The conversion steps in a bolometer are as follows:

1. An ohmic resistor is exposed to electromagnetic radiation. The radiation is absorbed by the resistor and converted into heat;

2. The heat elevates resistor's temperature above the ambient;

3. A temperature increase is a representation of the electromagnetic power. Naturally, this temperature differential can be measured by any suitable method. The methods of temperature measurements are covered in Chapter 16. Here, we just briefly outline the most common methods of bolometer fabrications.

A basic circuit for the voltage biased bolometer application is shown in Fig. 15-11A. It consists of a bolometer having resistance R, a stable reference resistor R_o, and a bias voltage source E. The voltage V across R_o is the output signal of the circuit. It has the highest value when both resistors are equal. Sensitivity of the bolometer to the incoming electromagnetic (EM) radiation can be defined as [11]

$$\beta_v = \frac{\alpha \varepsilon Z_T E}{4\sqrt{1 + (\omega\tau)^2}} \quad , \tag{15.8}$$

where $\alpha = (dR/dT)/R$ is the TCR (temperature coefficient of resistance) of the bolometer, ε is the surface emissivity, Z_T is the bolometer thermal resistance, which depends on its design and the supporting structure, τ is the thermal time constant, which depends on Z_T and the bolometer's thermal capacity, and ω is the frequency.

Since bolometer's temperature increase, ΔT is

$$\Delta T = T - T_o \approx P_E Z_T = \frac{E^2}{4R} Z_T \quad , \tag{15.9}$$

and the resistance of RTD bolometer can be represented by a simplified Eq. (16.7)

$$R = R_0(1 + \alpha_o \Delta T) \quad , \tag{15.10}$$

then Eq. (15.8) can be rewritten as:

$$\beta_v = \frac{1}{2} \varepsilon \alpha_o \sqrt{\frac{R_o Z_T \Delta T}{(1 + \alpha_o \Delta T)\left[1 + (\omega\tau)^2\right]}} \tag{15.11}$$

Therefore, to improve the bolometer's responsivity, its electrical resistance and thermal impedance should be increased.

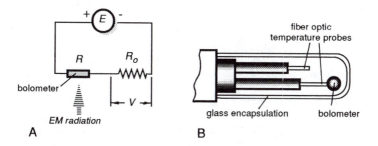

Fig. 15-11   Equivalent circuit of electrically biased bolometer (A)
and a design of an optical bolometer (B)

The bolometers are often fabricated as miniature thermistors, suspended by tiny wires. Another popular method of bolometer fabrication is the use of metal film depositions [11, 12], usually of nichrome. In many modern bolometers, a thermoresistive thin film material is deposited on the surface of a micromachined silicon, or a glass membrane which is supported by a silicon frame.

Fig. 15-12A shows a metal film bolometer which is formed on a substrate with a cavity [13]. The substrate can be prepared, depending on the application, of aluminum oxide ceramic, a glass, an insulated metal, or a silicon slice. A small cavity is formed in the substrate. Silicon slices can be etched, while alumina ceramics are either drilled with high speed diamond drills, or preformed with appropriate cavities. Then, the bonding pads are prepared on the substrate and its surface is given a thin organic or inorganic film on the top of which a thin metal film is deposited. The metal film (nickel, gold, or bismuth) forms a bolometer. Alternatively, the thin membrane may be back-etched in a silicon wafer with the subsequent metal vacuum deposition.

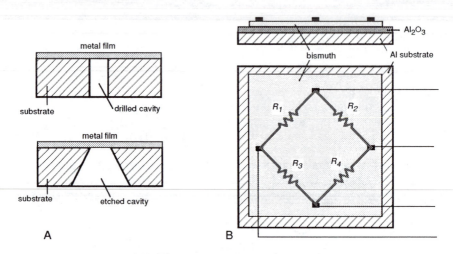

Fig. 15-12 Thin film bolometers
A: Two types of cavities: drilled and etched; B: Two dimensional bolometer

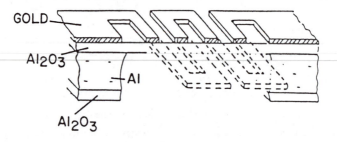

Fig. 15-13 Gold film bolometer formed on the aluminum oxide layer

Fig. 15-13 shows a meander-shaped measuring layer made of gold [14]. The bolometer is fabricated by first producing a carrier made of aluminum foil, which is polished and oxidized. The conductive gold layer is evaporated through a mask in a high vacuum on the oxide layer. The thickness of the conductive layer is on the order of 20 nm. The back side of the carrier is selectively etched out to form a very thin layer of aluminum oxide supporting the gold bolometric resistor.

When it is desirable to detect the position of the incoming focused electromagnetic radiation, a two dimensional bolometer may be formed as it is shown in Fig. 15-12B [15]. The bolometer is formed of a thin layer of bismuth with four contacts. The bismuth layer forms distributed temperature sensitive resistors between the contacts. The resistors R_1 - R_4 are separately measured. Whenever radiation energy is focused on the sensor's surface, the values of resistors change in proportion with the proximity of the radiation spot to the corresponding contact.

As it follows from Eq. (15.11) one of the critical issues which always must be resolved when designing a bolometer (or any other accurate temperature sensor, for that matter) is to assure good thermal insulation of the sensing element from a supporting structure, connecting wires, and interface electronics. Otherwise, heat loss from the element may result in large errors and reduced sensitivity. One method to achieve this is to completely eliminate any metal conductors and to measure the temperature of the bolometer by using a fiber optic technique, as its has been implemented in the E-field probe fabricated by Luxtron, Mountain View, CA (US Patent No. 4,816,634). In the design (Fig. 15-11B), a miniature bolometer is suspended in the end of an optical probe, and its temperature is measured by a fluoroptic® temperature sensor (see Section 16.4.1), while another similar optical sensor measures ambient temperature to calculate ΔT.

REFERENCES

1. Carter, E. F. ed., Dictionary of inventions and discoveries. Crane, Russak and Co., N.Y., © 1966 by F. Muller.
2. Andrew and McMeel. The universal almanac. A University Press Syndicate Company, Kansas City, New York, 1990
3. Prendergast, C. The first aviators. Life-Time Books, Alexandria, VA, 1980
4. Garner, H. D. Practical design of magnetic heading sensors. *Sensors Expo Proceedings.* North American Technology, Inc., 1986
5. Primdahl, F. The fluxgate magnetometer. *J. Phys. E. Sci.* vol. 12, 1979
6. Sensor signal conditioning: an IC designer's perspective. *Sensors*, pp: 23-30, Nov. 1991
7. Gregorian, R. and Temes, G. Analog MOS integrated circuits for signal processing. Chap. 6.1, Wiley and Sons, New York, 1986
8. Li, P. W., Chin, M. J., Gray, P. R. and Castello, R. A ratio independent algorithmic analog-to-digital conversion technique. *IEEE J. Solid-state circuits*, Vol. sc-19, No. 6, pp. 828-36, Dec. 1984

9. Armstrong, M., Ohara, H., Ngo, H., Rahim, C., Grossman, A. and Gray, P. A. CMOS programmable self-calibrating 13b eight-channel analog interface processor. *1987 ISSCC Digest of technical papers*

10. Astheimer, R.W. Thermistor infrared detectors. *SPIE*, No. 443, pp: 95-109, 1984

11. Shie, J-S and Weng, P.K. Fabrication of micro-bolometer on silicon substrate by anizotropic etching technique. In: *Transducers'91. International conference on solid-state sensors and actuators. Digest of technical papers*, pp: 627-630, ©IEEE, 1991

12. Vogl, T.P., Shifrin, G.A., and Leon, B.J. Generalized theory of metal-film bolometers. *J. Opt. Soc. Am.*, vol. 52, pp: 957-964, 1962

13. Liddiard, K. C. Infrared radiation detector. *U.S. Patent* No. 4,574,263. March 4, 1986

14. Hartmann, R., Krah, A., Egle, H., Ziegler, R. Method of making a bolometric radiation detector. *U.S. Patent* No. 4,544,441. Oct. 1, 1985

15. Goranson, R.W. and Wick, R.V. Bolometer. *U.S. Patent* No. 4,061,917. Dec. 6, 1977

16

Temperature Sensors

From prehistoric times people were aware of heat and trying to assess its intensity by measuring temperature. Perhaps the simplest, and certainly the most widely used phenomenon for temperature sensing is thermal expansion. This forms the basis of the liquid-in-glass thermometers. For electrical transduction, different methods of sensing are employed. Among them are: resistive, thermoelectric, semiconductive, optical, and piezoelectric detectors. Taking a temperature essentially requires the transmission of a small portion of the object's thermal energy to the sensor, whose function is to convert that energy into an electrical signal. When a contact sensor is placed inside or on the object, heat conduction takes place through the interface between the object and the probe. The probe warms up or cools down, i.e., it exchanges heat with the object. Any probe, no matter how small, will disturb the measurement site. This applies to any method of sensing: conductive, convective, and radiative. Thus, it is an engineering task to minimize the error by an appropriate sensor design and a correct measurement technique.

A contact temperature measurement is complete when there is no thermal gradient between the contact surface and the interior of the probe[1]. This process may take significant time, because, after the probe placement, reaching thermal equilibrium between the object and the sensor may be a slow process, especially if the contact area is dry.

In a contact sensing, amount of transferred heat is proportional to a temperature gradient between the thermometer's sensing element of instantaneous temperature T and that of the object T_1:

$$dQ = aA(T_1 - T)dt \ , \tag{16.1}$$

where a is the thermal conductivity of the sensor-object interface and A is the heat transmitting surface. If the sensor has specific heat c and mass m the absorbed heat is

---

[1] An equilibrium is not required when the so called *predictive* method of temperature calculation is employed, where the equilibrium point is determined frough the rate of heat transfer.

$$dQ = mcdT .$$ (16.2)

If we ignore the heat lost from the sensor to the environment through the connecting and supporting structure, Eqs. (16.1) and (16.2) yield a first order differential equation

$$aA(T_1 - T)dt = mcdT .$$ (16.3)

We denote thermal time constant τ_T as

$$\tau_T = \frac{mc}{\alpha A} ,$$ (16.4)

then differential equation takes form

$$\frac{dT}{T_1 - T} = \frac{dt}{\tau_T} .$$ (16.5)

This equation has a solution

$$T = T_1 - Ke^{-t/\tau_T} ,$$ (16.6)

where K is constant[1]. The time transient of temperature T which corresponds to the above solution is shown in Fig. 16-1A. One time constant is equal to the time required for temperature T to reach 63.2% of the initial gradient between T and T_1. The smaller the time constant the faster the sensor responds to a change in temperature.

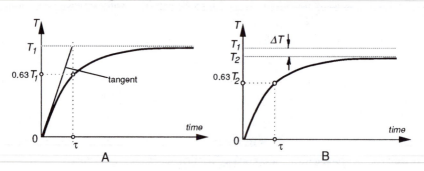

Fig. 16-1 Temperature changes of a sensor
A: The sensor is ideally coupled with the object;
B: The sensor has heat loss to its surroundings

---

[1] In this analysis we consider T_1 independent of the sensor's initial temperature. This corresponds to the case when the object has a thermal mass several orders of magnitude larger than that of the sensor and thermal conductivity of the object is high. Such an object may be called *infinite heat source (sink)*.

Theoretically, it takes infinite time to reach a perfect equilibrium between T_1 and T. However, since only finite accuracy is usually required, for the most practical cases, a quasi-equilibrium state may be considered after 5 to 10 time constants. If a sensor is coupled not only to the object whose temperature it transduces, but to some other stray objects as well, an additional error may be introduced. One example of a stray object is a connecting cable. One part of the cable is connected to the sensor while the other part is subjected to ambient temperature which may be quite different from that of the object. The cable conducts both an electric signal and some portion of heat from the sensor. Fig. 16-1B shows that in that case the sensor never reaches the actual temperature of the object T_1. It settles on a lower level T_2 which is smaller by a difference ΔT corresponding to heat loss.

A typical *contact* temperature sensor consists of the following components (Fig. 16-2A):

1. A sensing element - a material which is responsive to a change in its own temperature. A good element should have low specific heat, high thermal conductivity, strong and predictable temperature sensitivity.

2. Contacts are conductive pads or wires which interface between the sensing element and the external electronic circuit. The contacts should have the lowest possible thermal conductivity and electrical resistance. Also, they are often used to support the sensor.

3. A protective envelope is either a sheath or coating which physically separates a sensing element from the environment. A good envelope must have low thermal resistance and high electrical isolation properties. It must be impermeable to moisture and other factors which may spuriously affect the sensing element.

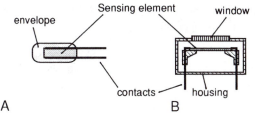

Fig. 16-2 General structure of temperature sensors
A: a contact sensor and B: is a thermal radiation sensor

A noncontact temperature sensor (Fig. 16-2B) is a thermal radiation sensor whose designs are covered in Chapter 13. Here, we just want to mention that as a contact sensor, it also contains a sensing element which is responsive to temperature. In addition, it may have an optical window and a built-in interface circuit.

16.1   THERMORESISTIVE SENSORS[1]

Sir Humphry Davy had noted as early as 1821 that electrical resistances of various metals depend on temperature [1]. Sir William Siemens, in 1871, first outlined the use of a platinum resistance thermometer. In 1887 Hugh Callendar published a paper [2] where he described how to practically use platinum temperature sensors. The advantages of thermoresistive sensors are in the simplicity of interface circuits, sensitivity, and long term stability. All such sensors can be divided into three groups: RTDs, *pn*-junction detectors and thermistors.

16.1.1   Resistance temperature detectors (RTD)

This term is usually pertinent to metal sensors, fabricated either in a form of a wire or a thin film. Temperature dependence of resistivities of all metals and most alloys gives an opportunity to use them for temperature sensing (Table 3-3). While virtually all metals can be employed for sensing, platinum is used almost exclusively because of its predictable response, long terms stability and durability. Tungsten RTDs are usually applicable for temperatures over 600°C. All RTDs have positive temperature coefficients. Several types of them are available from various manufacturers:

1. Thin film RTDs are often fabricated of a thin platinum or its alloys and deposited on a suitable substrate, such as a micromachined silicon membrane. The RTD is often made in a serpentine shape to ensure a sufficiently large length/width ratio.

2. Wire-wound RTDs, where the platinum winding is partially supported by a high temperature glass adhesive inside a ceramic tube. This construction provides a detector with the most stability for industrial and scientific applications.

According to the International Practical Temperature Scale (IPTS-68), precision temperature instruments should be calibrated at reproducible equilibrium states of some materials. This scale designates kelvin temperatures by symbol T_{68} and the Celsius scale by t_{68}. Equation (3.5.10) gives a best fit second-order approximation for platinum. In industry, it is customary to use different approximations for the cold and hot temperatures. Callendar-van Dusen approximations represent platinum transfer function:

For the range from -200°C to 0°C

$$R_t = R_o \left[1 + At + Bt^2 + Ct^3(t-100°) \right] \; . \tag{16.7a}$$

---

[1] Also, see section 3.5.

For the range from 0°C to 630°C is the same as (3.5.10)

$$R_t = R_o [1 + At + Bt^2] \quad . \tag{16.7b}$$

The constants A, B, and C are determined by the properties of platinum used in the construction of the sensor. Alternatively, the Callendar-van Dusen approximation can be written as

$$R_t = R_o \left\{ 1 + \alpha \left[t - \delta(\frac{t}{100})(\frac{t}{100} - 1) - \beta(\frac{t}{100})^3(\frac{t}{100} - 1) \right] \right\} \quad , \tag{16.8}$$

where t is the temperature in °C and the coefficients are related to A, B, and C as

$$A = \alpha(1 + \frac{\delta}{100}), \qquad B = -\alpha \cdot \delta \cdot 10^{-4}, \qquad C = -\alpha \cdot \beta \cdot 10^{-8} \quad . \tag{16.9}$$

The value of δ is obtained by the calibration at a high temperature, for example, at the freezing point of zinc (419.58°C) and β is obtained at the calibration at a negative temperature.

To conform with the IPTS-68, the Callendar-van Dusen approximation must be corrected. The correction is rather complex and the user should refer for details to the International Practical Temperature Scale of 1968. In different countries, some national specifications are applicable to RTDs. For instance, in Europe these are BS 1904: 1984; DIN 43760 - 1980; IEC 751: 1983. In Japan it is JIS C1604-1981. In the U.S.A. different companies have developed their own standards for α-values. For example, SAMA Standard RC21-4-1966 specifies $\alpha=0.003923$°C^{-1}, while in Europe DIN standard specifies $\alpha=0.003850$°C^{-1}, and the British Aircraft industry standard is $\alpha=0.003900$°C^{-1}.

Usually, RTDs are calibrated at standard points which can be reproduced in a laboratory with high accuracy (Table 16-1). Calibrating at these points allows for precise determination of approximation constants α and δ.

Typical tolerances for the wire-wound RTDs is ± 10mΩ which corresponds to about ± 0.025°C. Giving high requirements to accuracy, packaging isolation of the device should be seriously considered. This is especially true at higher temperatures where the resistance of isolators may drop significantly. For instance, a 10MΩ shunt resistor at 550°C results in resistive error of about 3 mΩ which corresponds to temperature error of -0.0075°C.

16.1.2 Silicon resistive sensors

Conductive properties of bulk silicon have been successfully implemented for the fabrication of temperature sensors with PTC characteristics. The so-called KTY temperature detectors manufactured by Philips have reasonably good linearity (which can be improved by use of simple compensating circuits), and high long term stability (typically ±0.05K per year)[1]. The positive temperature coefficient makes them inherently safe for operation in heating systems - a moderate overheating (below 200°C) results in RTD's resistance increase and a self-protection.

Table 16-1 Temperature reference points

point description	°C
Triple point[2] of hydrogen	- 259.34
Boiling point of normal hydrogen	- 252.753
Triple point[1] of oxygen	- 218.789
Boiling point of nitrogen	- 195.806
Triple point[1] of argon	- 189.352
Boiling point of oxygen	- 182.962
Sublimation point of carbon dioxide	- 78.476
Freezing point of mercury	- 38.836
Triple point[1] of water	0.01
Freezing point of water (water-ice mixture)	0.00
Boiling point of water	100.00
Triple point of benzoic acid	122.37
Freezing point of indium	156.634
Freezing point of tin	231.968
Freezing point of bismuth	271.442
Freezing point of cadmium	321.108
Freezing point of lead	327.502
Freezing point of zinc	419.58
Freezing point of antimony	630.755
Freezing point of aluminum	660.46
Freezing point of silver	961.93
Freezing point of gold	1064.43
Freezing point of copper	1084.88
Freezing point of nickel	1455
Freezing point of palladium	1554
Freezing point of platinum	1769

---

[1] Information on KTY sensors is courtesy of Philips Semiconductors BV, Einthover, The Netherlands.
[2] Tripple point is equilibrium between the solid, liquid and vapor phases

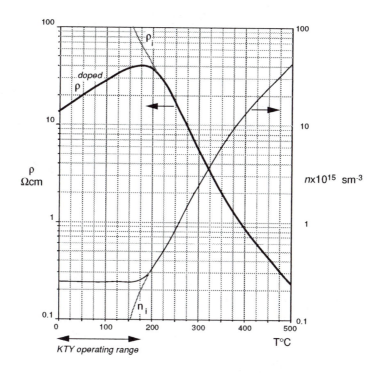

Fig. 16-3 Resistivity and number of free charge carriers for *n*-doped silicon

Pure silicon, either polysilicon or single crystal silicon, intrinsically has a negative temperature coefficient of resistance (Fig. 3-17.2A). However, when it is doped with an *n* type impurity, in a certain temperature range, its temperature coefficient becomes positive (Fig. 16-3). This is a result of the fall in the charge carrier mobility at lower temperatures. At higher temperatures, the number *n* of free charge carriers increases due to the number n_i of spontaneously generated charge carriers, and the intrinsic semiconductor properties of silicon predominate. Thus, at temperatures below 200°C resistivity ρ has a positive temperature coefficient while over 200°C it becomes negative. The basic KTY sensor consists of an *n* type silicon cell having approximate dimensions of 500x500x240 μm, metallized on one side and having contact areas on the other side. This produces an effect of resistance "spreading" which causes a conical current distribution through the crystal, significantly reducing the sensor's dependence on manufacturing tolerances. A KTY sensor may be somewhat sensitive to a current direction, especially, at larger currents and higher temperatures. To alleviate this problem, a serially-opposite design is employed where two of the sensors are connected with opposite polarities to form a dual sensor.

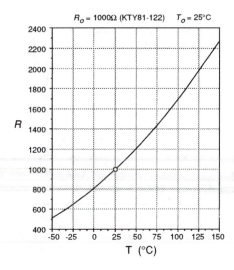

Fig. 16-4   Transfer function of a KTY silicon temperature sensor

A typical sensitivity of a PTC silicon sensor is on the order of 0.7%/°C, that is its resistance changes by 0.7% per every degree C. As any other sensor with a mild nonlinearity, the KTY sensor transfer function may be approximated by a second-order polynomial

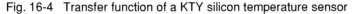

$$R_T = R_o[1 + A(T - T_o) + B(T - T_o)^2] \ , \qquad (16.10)$$

where R_o and T_o are the resistance (Ω) and temperature (K) at a reference point. For instance, for the KTY-81 sensors operating in the range from -55 to +150°C, the coefficients are: A=0.007874 K^{-1} and B=1.874·$10^{-5}K^{-2}$. A typical transfer function of the sensor is shown in Fig. 16-4.

16.1.3   Thermistors

The term thermistor is a contraction of words *temperature* and *resistor*. The name is usually applied to metal-oxide sensors fabricated in a form of droplets, bars, cylinders, and thick films. The thermistors are divided into two groups: NTC (negative temperature coefficient) and PTC (positive temperature coefficient).

NTC Thermistors

A conventional metal-oxide thermistor has a negative temperature coefficient (NTC), that is, its resistance decreases with the increase in temperature. The NTC thermistor element resistance, as of any resistor, is determined by it physical dimensions and material resistivity. The relationship between the resistance and temperature is highly nonlinear (Figs. 3-5.4). In practice, it can be approximated either by a polynomial or, what is most popular for the moderate accuracy applications, by an exponential function. Normally, for the exponential approximation, a thermistor is specified by two numbers: its nominal resistance R_{T_0} at a reference temperature (T_0=25°C) and a value of β (beta) which is a number representative of the sensitivity

$$R_T = R_{T_0} e^{\beta(\frac{1}{T} - \frac{1}{T_o})} \quad , \qquad (16.11)$$

where R_T is the resistance at temperature T measured in K. Beta (β) is called a *characteristic temperature* of a thermistor. Its value (in degrees kelvin) is determined by measuring resistances R_1 and R_2 at two different test temperatures T_1 and T_2

$$\beta = \frac{1}{\frac{1}{T_1} - \frac{1}{T_2}} \ln\frac{R_1}{R_2} \quad . \qquad (16.12)$$

This formula may be sufficiently accurate for a narrow temperature range application. Beta specifies a thermistor, but it does not directly describe its sensitivity, which is a negative temperature coefficient, α (NTC). It can be found by differentiating Eq. (16.11)

$$\alpha = \frac{1}{R_T}\frac{dR_T}{R_T} = -\frac{\beta}{T^2} \quad . \qquad (16.13)$$

It is seen that NTC depends on both: beta and a temperature. A thermistor is much more sensitive at lower temperatures and its sensitivity (NTC) drops fast with a temperature increase. In the reality, β is not constant, and depends on temperature quite noticeably. Therefore, the exponential expression (16.11) is good enough only for nondemanding applications. For the precision measurements, the so-called Steinhart-Hart relationship has a wide industry acceptance for computation of temperature in kelvin T through the thermistor's resistance. The equation is an empirical third order polynomial

$$\frac{1}{T} = A + B\ln R_T + C\ln^3 R_T \quad , \qquad (16.14)$$

where A, B, and C are coefficients derived experimentally. To find these coefficients, a system of three equations should be solved for three different temperatures. The Steinhart-Hart equation explicit in resistance takes the form

$$R_T = \exp\left\{\left[-\frac{\alpha}{2} + \sqrt{\frac{\alpha^2}{4} + \frac{\beta^3}{27}}\right]^{1/3} + \left[-\frac{\alpha}{2} - \sqrt{\frac{\alpha^2}{4} + \frac{\beta^3}{27}}\right]^{1/3}\right\}, \qquad (16.15)$$

where

$$\alpha = \frac{A - \frac{1}{T}}{C}, \quad \beta = \frac{B}{C}. \qquad (16.16)$$

Eq. (16.15) yields an accuracy of about ±0.02°C. For more precision measurements, higher order polynomials may be employed.

In the NTC thermistors, the sensitivity α varies over the temperature range from -2 (at the warmer side of the scale) to -8%/°C (at the cooler side of the scale), which implies that this is a very sensitive device, roughly an order of magnitude more temperature sensitive than RTD. This is especially important in applications where a high output signal over a relatively narrow temperature range is desirable.

Generally, thermistors can be classified into three major groups depending upon the method by which they are fabricated. The first group consists of bead type thermistors. The beads may be bare, or coated with glass (Fig. 16-5), epoxy, or encapsulated into a metal jacket. All these beads have platinum alloy leadwires which are sintered into the ceramic body. When fabricated, a small portion of mixed metal oxide with a suitable binder is placed onto a parallel leadwires, which are under slight tension. After the mixture has been allowed to dry, or has been partly sintered, the strand of beads is removed from the supporting fixture and placed for the final sintering into a tubular furnace. The metal oxide shrinks onto the leadwires during this firing process and forms an intimate electrical bond. Then, the beads are individually cut from the strand, and are given an appropriate coating.

Fig. 16-5   Glass coated bead thermistors
(Courtesy of Thermometrics)

Another type of thermistor is a chip thermistor with surface contacts for the leadwires. Usually, the chips are fabricated by a tape casting process, with subsequent screen printing, spraying, painting or vacuum metallization of the surface electrodes. The chips are either bladed or cut into desired geometry. If desirable, the chips can be ground to meet the required tolerances.

The third type of thermistors are fabricated by the depositing of semiconductive materials on a suitable substrate, such as glass, alumina, silicon, etc. These thermistors are preferable for integrated sensors and for a special class of thermal infrared detectors.

Among the metallized surface contact thermistors, flakes and uncoated chips are the least stable. A moderate stability may be obtained by epoxy coating. The bead type with leadwires sintered into the ceramic body permit operation at higher temperatures – up to 550°C. The metallized surface contact thermistors usually are rated up to 150°C.

Whenever a fast response time is required, bead thermistors are preferable, however, they are more expensive than the chip type. Besides, the bead thermistors are more difficult to trim to a desired nominal value.

While using the NTC thermistors, one must not overlook possible sources of error. One of them is aging, which for the low quality sensors may be as large as +1%/year. Fig. 16-6 shows typical percentage changes in resistance values for the epoxy encapsulated chip thermistors as compared with the sintered glass encapsulated glass thermistors. A good environmental protection and pre-aging, which is a powerful method of sensor characteristic stabilizing. During pre-aging, the thermistor is maintained at +300°C for at least 700 hours. For the better protection, it may be further encapsulated into a stainless steel jacket and potted with epoxy.

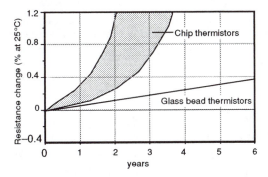

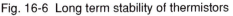

Fig. 16-6 Long term stability of thermistors

Another issue which is important for the thermistor performance is a self-heating effect. A thermistor is an active type of a sensor, that is, it does require an excitation signal for its operation. The signal is usually either a dc or ac current passing through the thermistor. The current causes a

Joule heating and a subsequent increase in temperature. In many applications, this is a source of error which may result in the wrong determination of the temperature of a measured object. In many other applications, the self-heating is successfully employed for sensing fluid flow, thermal radiation and other stimuli.

Let us analyze the thermal events in a thermistor, when electric power is applied. Fig. 16-7A shows a voltage source E connected to a thermistor R_T thorough a current limiting resistor R.

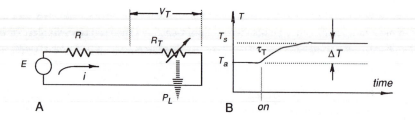

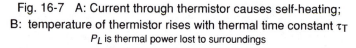

Fig. 16-7   A: Current through thermistor causes self-heating;
B: temperature of thermistor rises with thermal time constant τ_T
P_L is thermal power lost to surroundings

When electric power P is applied to the network (moment *on* in Fig. 16-7B), the rate at which energy is supplied to the thermistor must be equal the rate at which energy H_L is lost plus the rate at which energy H_s is absorbed by the thermistor body. The absorbed energy is stored in the thermistor's thermal capacity C. The power balance equation is

$$\frac{dH}{dt} = \frac{dH_L}{dt} + \frac{dH_s}{dt} \quad . \tag{16.17}$$

According to the law of conservation of energy, the rate at which thermal energy is supplied to the thermistor is equal to electric power delivered by voltage source E

$$\frac{dH}{dt} = P = \frac{V_T^2}{R} = V_T i \quad , \tag{16.18}$$

where V_T is the voltage drop across the thermistor.

The rate at which thermal energy is lost from the thermistor to its surroundings is proportional to temperature gradient ΔT between the thermistor and surrounding temperature T_a

$$P_L = \frac{dH_L}{dt} = \delta \Delta T = \delta(T_s - T_a) \quad , \tag{16.19}$$

where d is the so-called *dissipation factor* which is equivalent to thermal conductivity from the thermistor to its surroundings. It is defined as a ratio of dissipated power and temperature gradient (at a given surrounding temperature). The factor depends upon the sensor design, length and thickness of leadwires, thermistor material, supporting components, thermal radiation from the thermistor surface, and relative motion of medium in which the thermistor is located.

The rate of heat absorption is proportional to thermal capacity of the sensor assembly

$$\frac{dH_s}{dt} = C\frac{dT_s}{dt} \quad . \tag{16.20}$$

This rate produces the thermistor's temperature T_s to rise above its surroundings. Substituting Eqs. (16.19) and (16.20) into (16.18) we arrive at

$$\frac{dH}{dt} = P = Ei = d(T_s - T_a) + C\frac{dT_s}{dt} \quad . \tag{16.21}$$

The above is a differential equation describing the thermal behavior of the thermistor. Let us now solve it for two conditions. The first condition is the constant electric power supplied to the sensor: P=const. Then, solution of Eq. (16.21) is

$$\Delta T = (T_s - T_a) = \frac{P}{\delta}\left[1 - e^{-\frac{\delta}{C}t}\right] , \tag{16.22}$$

where e is the base of natural logarithms. The above solution indicates that upon applying electric power, the temperature of the sensor will exponentially rise above ambient. This specifies a transient condition which is characterized by a thermal time constant $\tau_T = C\frac{1}{\delta}$. Here, the value of $\frac{1}{\delta} = r_T$ has a meaning of thermal resistance between the sensor and its surroundings. The exponential transient is shown in Fig. 16-7B.

Upon waiting sufficiently long to reach a steady-state level T_s the rate of change in Eq. (16.21) becomes equal to zero ($\frac{dT_s}{dt} = 0$), then the rate of heat loss is equal to supplied power

$$d(T_s - T_a) = \delta\Delta T = V_Ti \quad . \tag{16.23}$$

If by selecting low supply voltage and high resistances, the current i is made very low, temperature rise, ΔT can be made negligibly small and self-heating is virtually eliminated. Then, from Eq. (16.21)

$$\frac{dT_s}{dt} = -\frac{\delta}{C}(T_s - T_a) \quad . \tag{16.24}$$

The solution of this differential equation yields an exponential function (16.6), which means that the sensor responds to change in environmental temperature with time constant τ_T. Since the time constant depends on the sensor's coupling to the surroundings, it is usually specified for certain conditions, for instance, $\tau_T = 1$ s @25°C, in still air, or 0.1 s @25°C, in stirred water. It should be kept in mind, that the above analysis represents a simplified model of the heat flows. In the reality, a thermistor response has a somewhat nonexponential shape.

All thermistor applications require the use of one of three basic characteristics:

1. The resistance vs. temperature characteristic as it is shown in Fig. 3-5.4. In most of the applications based on this characteristic, the self-heating effect is undesirable. Thus, the nominal resistance R_{T_0} of the thermistor should be selected high and its coupling to the object should be maximized (increase in δ). The characteristic is primarily used for sensing and measuring tempera-ture. Typical applications are contact electronic thermometers, thermostats and thermal breakers.

2. The current versus time (or resistance versus time) as shown in Fig. 16-7B

3. The voltage versus current characteristic (Fig. 16-8) is important for applications where the self-heating effect is employed, or otherwise can not be neglected. The power supply-loss bal-ance is governed by Eq. (16.23). If variations in δ are small (which is often the case) and the re-sistance versus temperature characteristic is known, then Eq. (16.23) can be solved for the static voltage versus current characteristic. That characteristic is usually plotted on log-log coordinates, where lines of constant resistance have a slope of +1 and lines of constant power have slope of -1 (Fig. 16-8).

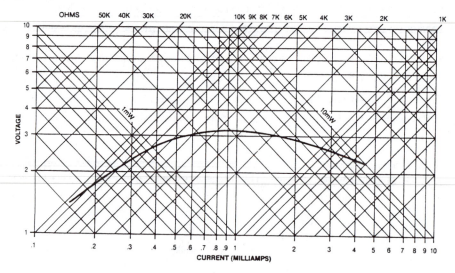

Fig. 16-8 Voltage-current characteristic
of an NTC thermistor in still air at 25°C
(Courtesy of Thermometrics, Inc.)

At very low currents (left side of the Fig. 16-8), the power dissipated by the thermistor is negligibly small, and the characteristic is tangential to a line of constant resistance of the thermistor at a specified temperature (25°C in Fig. 16-8). Thus, the thermistor behaves as a simple resistor. That is, voltage drop V_T is proportional to current i.

As the current increases, the self-heating increases as well. This results in a decrease in the resistance of the thermistor. Since the resistance of the thermistor is no longer constant, the characteristics start to depart from the straight line. The slope of the characteristic (dV_T/di), which is the resistance, drops with increase in current. The current increase leads to further resistance drop which, in turn, increases the current. Eventually, current will reach its maximum value i_p at a voltage maximum value V_p. It should be noted that, at this point, a resistance of the thermistor is zero. Further increase in current i_p will result in continuing decrease in the slope, which means that the resistance has a negative value (right side of Fig. 16-8). An even further increase in current will produce another reduction of resistance, where leadwire resistance becomes a contributing factor. A thermistor should never be operated under such conditions. A thermistor manufacturer usually specifies the maximum power rating for thermistors.

According to Eq. (16.23), self-heating thermistors can be used to measure variations in δ, ΔT, or V_T. The applications where δ varies, include vacuum manometers, anemometers, flow meters, fluid level sensors, etc. Applications where ΔT is the stimulus include microwave power meters, AFIR detectors (see below), etc. The applications where V_T varies are in some electronic circuits: automatic gain control, voltage regulation, volume limiting, etc.

PTC Thermistors

All metals may be called PTC materials, however, their temperature coefficients of resistivity (TCR) are quite low (Table 3-3) and change very little over the temperature range. In contrast, ceramic PTC materials in a certain temperature range are characterized by a very large temperature dependence. They are fabricated of polycrystalline ceramic substances, where the base compounds, usually barium titanate or solid solutions of barium and strontium titanate (highly resistive materials) made semiconductive by the addition of dopants [3]. Above the Curie temperature of a composite material, the ferroelectric properties change rapidly resulting in a rise in resistance, often several orders of magnitude. A typical transfer function curve for the PTC thermistor is shown in Fig. 16-9 in a comparison with the NTC and RTD responses. The shape of the curve does not lend itself to easy mathematical approximation, therefore, manufacturers usually specify PTC thermistors by a set of numbers:

1. Zero power resistance, R_{25}, at 25°C, where self-heating is negligibly small;

2. Minimum resistance R_m is the value on the curve where thermistor changes its TCR from positive to negative value (point m);

3. Transition temperature T_τ is the temperature where resistance begins to change rapidly. It coincides approximately with the Curie point of the material. A typical range for the transition temperatures is from -30 to +160°C (Keystone Carbon Co.);

4. TCR is defined in a standard form

$$\alpha = \frac{1}{R} \cdot \frac{\Delta R}{\Delta T} .$$ (16.25)

The coefficient changes very significantly with temperature and often is specified at point x, that is, at its highest value, which may be as large as 2/°C (meaning the change in resistance is 200% per °C);

5. Maximum voltage E_{max} is the highest value which the thermistor can withstand at any temperature;

6. Thermal characteristics are specified by a thermal capacity, a dissipation constant δ (specified under given conditions of coupling to the environment) and a thermal time constant (defines speed response under specified conditions).

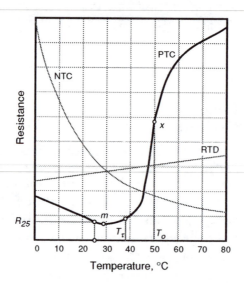

Fig. 16-9 Transfer functions of PTC and NTC thermistors as compared with RTD

It is important to understand that for the PTC thermistors two factors play a key role: environmental temperature and a self-heating effect. Either one of these two factors shifts the thermistor's operating point.

The temperature sensitivity of the PTC thermistor is reflected in a volt-ampere characteristic of Fig. 16-10. A regular resistor with the near zero TCR, according to Ohm's law has a linear characteristic. A NTC thermistor has a positive curvature of the volt-ampere dependence. An implication of the negative TCR is that if such a thermistor is connected to a hard voltage source[1], a self-heating due to Joule heat dissipation will result in resistance reduction. In turn, that will lead to further increase in current and more heating. If the heat outflow from the NTC thermistor is restricted, a self-heating may eventually cause overheating and a catastrophic destruction of the device.

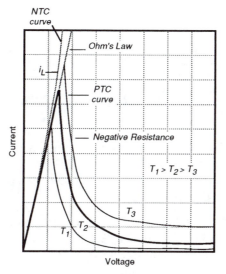

Fig. 16-10 Volt-ampere characteristic of a PTC thermistor

Thanks to positive TCRs, metals do not overheat when connected to hard voltage sources and behave as self-limiting devices. For instance, a filament in an incandescent lamp doesn't burn out because the increase in its temperature results in an increase in resistance which limits current. This self-limiting (self-regulating) effect is substantially enhanced in the PTC thermistors. The

---

[1] A hard voltage source means any voltage source having a near zero output resistance and capable of delivering unlimited current without a change in voltage.

shape of the volt-ampere characteristic indicates that in a relatively narrow temperature range the PTC thermistor possesses a negative resistance, that is

$$R_x = -\frac{V_x}{i} \ .$$

(16.26)

This results in the creation of an internal negative feedback which makes this device a self-regulating thermostat. In the region of negative resistance, any increase in voltage across the thermistor results in heat production which, in turn, increases the resistance and reduces heat production. As a result, the self-heating effect in a PTC thermistor produces enough heat to balance the heat loss on such a level that it maintains the device's temperature on a constant level T_o (Fig. 16-9). That temperature corresponds to point x where tangent to the curve has the highest value.

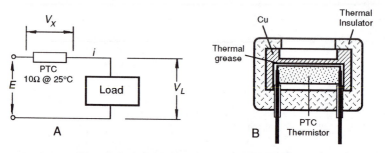

Fig. 16-11 Applications of PTC thermistors
A: a current limiting circuit; B: a micro thermostat

There are several applications where the self-regulating effect of a PTC thermistor may be quite useful. We briefly mention three of them.

1. Circuit protection. A PTC thermistor may operate as a nondestructible (resettable) fuse in electric circuits. Fig. 16-11A shows a thermistor connected in series with a power supply voltage E feeding the load with current i. The resistance of the thermistor at room temperature is quite low (typically from 10 to 140Ω). Current i develops voltage V_L across the load and voltage V_x across the thermistor. It is assumed that $V_L \gg V_x$. Power dissipated by the thermistor $P = V_x i$, is lost to the surroundings and the thermistor's temperature is raised above ambient by a relatively small value. Whenever either ambient temperature becomes too hot, or load current increases dramatically (for instance, due to internal failure in the load), the heat dissipated by the thermistor elevates its temperature to a T_τ region where its resistance starts increasing. This limits further current increase. Under the shorted-load conditions, $V_x = E$ and current i drops to its minimal level. This will be maintained until normal resistance of the load is restored and, it is said, that the fuse resets itself. It is important to assure that $E < 0.9 E_{max}$, otherwise a catastrophic destruction of the thermistor may occur.

2. A miniature self-heating thermostat (Fig. 16-11B) for the microelectronic, biomedical, chemical, and other suitable applications can be designed with a single PTC thermistor. Its transition temperature must be appropriately selected. A thermostat consists of a dish, which is thermally insulated from the environment and thermally coupled to the thermistor. Thermal grease is recommended to eliminate a dry contact. The terminals of the thermistor are connected to a voltage source whose value may be estimated from the following formula

$$E \geq 2\sqrt{\delta(T_\tau - T_a)R_{25}} \ ,$$ (16.27)

where δ is the heat dissipation constant which depends on thermal coupling to the environment and T_a is ambient temperature. The thermostat's set point is determined by the physical properties of the ceramic material (Curie temperature) and due to internal thermal feedback, the device reliably operates within relatively large range of power supply voltages and ambient temperatures. Naturally, ambient temperature must be always less than T_τ.

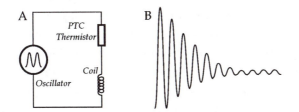

Fig. 16-12 A demagnetization device with a PTC thermistor attenuator
A: a circuit diagram; B: current through the coil

3. Time delay circuits can be created with the PTC thermistors thanks to a relatively long transition time between the application of electric power in its heating to a low resistance point. Fig. 16-12 shows a simple demagnetization device where electric current in a coil and the corresponding magnetic field decline in magnitude as the PTC thermistor warms up. When the oscillator is turned off, the thermistor is cold and its resistance is low. Upon turning the oscillator on, current through the coil warms up the thermistor resulting in a gradual increase in its temperature. This, in turn, increases its resistance thus reducing the current.

4. Flowmeter and liquid level detectors which operate on principle of heat dissipation can be made very simple with the PTC thermistors (see Section 10.3)

16.2　THERMOELECTRIC CONTACT SENSORS

Thermoelectric contact sensors are called *thermocouples* because at least two dissimilar conductors are required to make a sensor. Section 3.9 provides a physical background for a better understanding of their operation and Table 3-8 lists some popular thermocouples which are designated by letters originally assigned by the Instrument Society of America (ISA) and adopted by an American Standard in ANSI MC 96.1. A detailed description of various thermocouples and their applications can be found in many excellent texts, for instance in [1, 4, 5]. Below, we summarize the most important recommendations for the use if these sensors.

Type T: Cu (+) versus constantan (–) are resistant to corrosion in moist atmosphere and are suitable for subzero temperature measurements. Their use in air in oxidizing environment is restricted to 370°C (700°F) due to the oxidation of the copper thermoelement. They may be used to higher temperatures in some other atmospheres.

Type J: Fe (+) versus constantan (-) are suitable in vacuum and in oxidizing, reducing, or inert atmospheres, over the temperature range of 0 to 760°C (32 to 1400°F). The rate of oxidation in the iron thermoelement is rapid above 540°C (1000°F), and the use of heavy-gage wires is recommended when long life is required at the higher temperatures. This thermocouple is not recommended for use below the ice point because rusting and embrittlement of the iron thermoelement make its use less desirable than Type T.

Type E: 10% Ni/Cr (+) versus constantan (–) are recommended for use over the temperature range of –200 to 900°C (–330 to 1600°F) in oxidizing or inert atmospheres. In reducing atmospheres, alternately oxidizing or reducing atmospheres, marginally oxidizing atmospheres, and in vacuum, they are subject to the same limitations as type K. These thermocouples are suitable to subzero measurements since they are not subject to corrosion in atmospheres with a high moisture content. They develop the highest e.m.f. per degree of all the commonly used types and are often used primarily because of this feature (see Fig. 3-9.3).

Type K: 10% Ni/Cr (+) versus 5%Ni/Al/Si (–) are recommended for use in an oxidizing or completely inert atmosphere over a temperature range of -200 to 1260°C (–330 to 2300°F). Due to their resistance to oxidation, they are often used at temperatures above 540°C. However, type K should not be used in reducing atmospheres, in sulfurous atmospheres, and in a vacuum.

Types R and S: Pt/Rh (+) versus Pt (–) are recommended for continuous use in oxidizing or inert atmospheres over a temperature range of 0 to 1480°C (32 to 2700°F).

Type B: 30% Pt/Rh (+) versus 6%Pt/Rh (–) are recommended for continuous use in oxidizing or inert atmospheres over the range of 870 to 1700°C (1000 to 3100°F). They are also suitable for short term use in a vacuum. They should not be used in reducing atmospheres, nor those containing metallic or nonmetallic vapors. They should never be directly inserted into a metallic primary protecting tube or well.

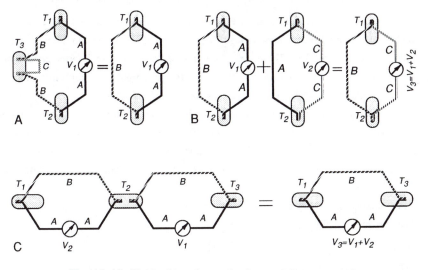

Fig. 16-13 Illustrations for or the laws of thermocouples
(see text)

For practical purposes, an application engineer must be concerned with three basic laws which establish the fundamental rules for proper connection of the thermocouples. It should be stressed, however, that an electronic interface circuit must always be connected to two identical conductors. These conductors may be formed from one of the thermocouple loop arms. That arm is broken to connect the metering device to the circuit. The broken arm is indicated as material *A* in Fig. 16-13A.

Law No. 1

—*A thermoelectric current can not be established in a homogeneous circuit by heat alone.*

This law provides that a nonhomogeneous material is required for the generation of the Seebeck potential. If a conductor is homogeneous, regardless of the temperature distribution along its length, the resulting voltage is zero. The junction of two dissimilar conductors provide a condition for voltage generation.

Law No. 2

—The algebraic sum of the thermoelectric forces in a circuit composed of any number and combination of dissimilar materials is zero if all junctions are at a uniform temperature.

The law provides that an additional material C can be inserted into any arm of the thermoelectric loop without affecting the resulting voltage V_1 as long as both additional joints are at the same temperature (T_3 in Fig. 16-13A). There is no limitation on the number of inserted conductors, as long as both contacts for each insertion are at the same temperature. This implies that an interface circuit must be attached in such a manner as to assure a uniform temperature for both contacts. Another consequence of the law is that thermoelectric joints may be formed by any technique, even if an additional intermediate material is involved (like, solder, for instance). The joints may be formed by welding, soldering, twisting, fusion, and so on without affecting the accuracy of the Seebeck voltage. The law also provides a rule of *additive materials* (Fig. 16-13B): if thermoelectric voltages (V_1 and V_2) of two conductors (B and C) with respect to a reference conductor (A) are known, the voltage of a combination of these two conductors is the algebraic sum of their voltages against the reference conductor.

Law No. 3

— If two junctions at temperatures T_1 and T_2 produce Seebeck voltage V_2, and temperatures T_2 and T_3 produce voltage V_1, then temperatures T_1 and T_3 will produce $V_3 = V_1 + V_2$ (Fig. 16-13C). This is sometimes called the law of intermediate temperatures. The law allows us to calibrate a thermocouple at one temperature interval and then to use it at another interval. It also provides that extension wires of the same combination may be inserted into the loop without affecting the accuracy.

The above laws provide for numerous practical circuits where thermocouples can be used in a great variety of combinations. They can be arranged to measure the average temperature of an object, to measure the differential temperature between two objects, and to use other than thermocouple sensors for the reference junctions, etc.

It should be noted that thermoelectric voltage is quite small and the sensors, especially with long connecting wires are susceptible to various transmitted interferences. A general guideline for the noise reduction may be found in Section 4.9.

Fig. 16-14A shows an equivalent circuit for a thermocouple and a thermopile. It consists of a voltage source and a serial resistor. The voltage source represents the Seebeck voltage whose magnitude is a function of a temperature differential. The terminals of the circuit are assumed being fabricated of the same material, copper in this example.

Traditionally, thermocouples were used with a cold junction immersed into a reference ice bath to maintain its temperature at 0°C. This presents serious limitations for many practical uses. The 2nd and 3rd thermoelectric laws provided above allow for a simplified solution. A "cold" junction can be maintained at any temperature, including ambient, as long as that temperature is precisely known. Therefore, a "cold" junction is thermally coupled to an additional temperature sensor which does not require a reference compensation. Usually, such a sensor is either thermoresistive or semiconductor.

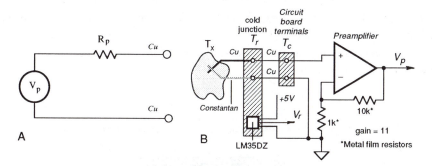

Fig. 16-13   Use of a thermocouple
A: an equivalent circuit of a thermocouple;
B: a front end of a thermometer with a semiconductor reference sensor (LM35DZ)

Fig. 16-13B shows the correct connection of a thermocouple to an electronic circuit and a "cold" junction reference sensor. Both the "cold" junction and the reference sensor must be positioned in intimate thermal coupling. Usually, they are imbedded into a chunk of copper. To avoid dry contact, thermally conductive grease or epoxy should be applied for better thermal tracking. A reference temperature sensor in this example is a semiconductor circuit LM35DZ manufactured by National Semiconductor, Inc. The circuit has two outputs - one for the signal representing the Seebeck voltage V_p and the other for the reference signal V_r. The schematic illustrates that connections to the circuit board input terminals and then to the amplifier's noninverting input and to the ground bus is made by the same type of wires (Cu). However, input terminals of the board not necessarily have to be at the "cold" junction temperature. It is especially important for the remote measurements, where circuit board temperature T_c may be different from the "cold" junction temperature T_r.

A complete thermocouple sensing assembly generally consists of one or more of the following: a sensing element assembly (the junction), a protective tube (ceramic or metal jackets), a thermowell (for some critical applications these are drilled solid bar stocks which are made to precise tolerances and are highly polished to inhibit corrosion), terminations (contacts which may be in the form of a screw type, open type, plug and jack-disconnect, military standard type connectors, etc.). Some typical thermocouple assemblies are shown in Fig. 16-15. The wires may be left bare, or given electrical isolators. For the high temperature applications, the isolators may be of a fish-spine or ball ceramic type, which provide sufficient flexibility. If thermocouple wires are not electrically isolated, a measurement error may occur. Insulation is affected adversely by moisture, abrasion, flexing, temperature extremes, chemical attack, and nuclear radiation. A good knowledge of particular limitations of insulating materials is essential for accurate and reliable measurement. Some insulations have a natural moisture resistance. Teflon, polyvinyl chloride (PVC), and some forms of polyimides are examples of this group. With the fiber type insulations, moisture protection results from impregnating with substances such as wax, resins, or silicone compounds. It should be noted, that only one-time exposure to over-extreme temperatures cause evaporation of the impregnating materials and loss of protection.

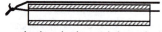

A. Bare thermocouple element, twisted and welded

B. Thermocouple wires glued on a tubular carrier (disposable type)

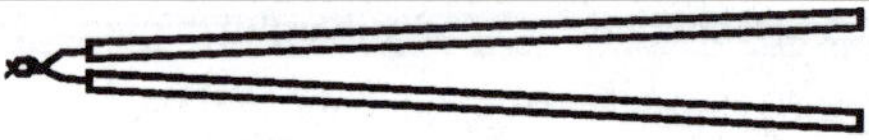

C. Insulated thermocouple, twisted and welded

D. Butt-welded thermocouple with fish-spine insulators

Fig. 16-15 Some thermocouple assemblies

The moisture penetration is not confined to the sensing end of the assembly. For example, if a thermocouple passes through hot or cold zones, condensation may produce errors in the measurement, unless adequate moisture protection is provided.

The basic types of flexible insulations for elevated temperature usage are fiber glass, fibrous silica, and asbestos (which should be used with proper precaution due to health hazard). In addition, thermocouples must be protected from atmospheres that are not compatible with the alloys. Protecting tubes serve the double purpose of guarding the thermocouple against mechanical damage and interposing a shield between the wires and the environment. The protecting tubes can be made of carbon steels (up to 540°C in oxidizing atmospheres), stainless steel (up to 870°C), ferric stainless steel (AISI 400 series), high-nickel alloys, Nichrome[1], Inconel[2], etc. (up to 1150°C in oxidizing atmospheres).

Practically all base-metal thermocouple wires are annealed or given a "stabilizing heat treatment" by the manufacturer. Such treatment generally is considered sufficient, and seldom it is found advisable to further anneal the wire before testing or using. Although a new platinum and platinum-rhodium thermocouple wire as sold by some manufacturers already is annealed, it has become a regular practice in many laboratories to anneal all types R, S, and B thermocouples, whether new or previously used, before attempting an accurate calibration. This is accomplished usually by heating the thermocouple electrically in air. The entire thermocouple is supported between two binding posts, which should be close together, so that the tension in the wires and stretching while hot are kept at a minimum. The temperature of the wire is conveniently determined with an optical pyrometer. Most of the mechanical strains are relieved during the first few minutes of heating at 1400 to 1500°C.

Thin film thermocouples are formed by bonding junctions of foil metals. They are available in a free filament style with a removable carrier and in a matrix style with a sensor embedded in a thin laminated material. The foil having a thickness in the order of 5 μm (0.0002") gives an extremely low mass and thermal capacity. Thin flat junctions may provide intimate thermal coupling with the measured surface. Foil thermocouples are very fast (a typical thermal time constant is 10 ms), and can be used with any standard interface electronic apparatuses. While measuring temperature with sensors having small mass, thermal conduction through the connecting wires always must be accounted for. Thanks to a very large length to thickness ratio of the film thermocouples (on the order of 1,000) heat loss via wires usually is negligibly small.

To attach a film thermocouple to an object, several methods are generally used. Among them are various cements and flame or plasma sprayed ceramic coatings. For ease of handling, the sensors often are supplied on a temporary carrier of polyimide film which is tough, flexible, and dimensionally stable. It is exceptionally heat resistant and inert. During the installation, the carrier can be easily pilled off or released by application of heat. The free foil sensors can be easily brushed into a thin layer, to produce an ungrounded junction. While selecting cements, care must be taken to avoid corrosive compounds. For instance, cements containing phosphoric acid are not recommended for use with thermocouples having copper in one arm.

---

[1] Trademark of the Driver-Harris Company.
[2] Trademark of the International Nickel Company.

16.3 SEMICONDUCTOR *PN*-JUNCTION SENSORS

A semiconductor *pn*-junction in a diode and a bipolar transistor exhibits quite a strong thermal dependence. If the junction is connected to a constant current generator (Section 4.3.1), the resulting voltage becomes a measure of the junction temperature (Fig. 16-17A). A very attractive feature of such a sensor is its high degree of linearity (Fig. 16-16). This allows a simple method of calibration using just two points to define a slope (sensitivity) and an intercept.

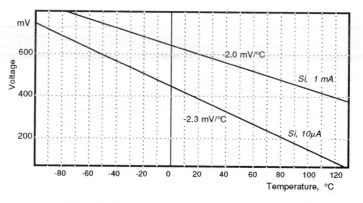

Fig. 16-16 Voltage-to-temperature dependence
of a forward biased semiconductor junction under constant current conditions

The current-to-voltage V equation of a *pn*-junction diode can be expressed as

$$I = I_o \exp(qV/2kT) \ , \tag{16.28}$$

where I_o is the saturation current, which itself is a strong function of temperature. It can be shown that the temperature-dependent voltage across the junction can be expressed as

$$V = \frac{E_g}{q} - \frac{2kT}{q} \ (\ln K - \ln I) \ , \tag{16.29}$$

where E_g is the energy band gap for silicon at 0K (absolute zero), q is the charge of an electron and K is a temperature independent constant. It follows from the above equation that when the junction is operated under constant current conditions, the voltage is linearly related to the temperature, and the slope is given by

$$b = \frac{dV}{dT} = -\frac{2k}{q} \ (\ln K - \ln I) \ . \tag{16.30}$$

Typically, for a silicon junction operating at 10μA, the slope (sensitivity) is approximately -2.3mV/°C and it drops to about -2.0mV/°C for a 1mA current. While any diode or transistor can be used as a temperature sensor, special devices are available for that particular purpose. An example is a sensor MTS102 from Motorola Semiconductor Products, Inc. A practical circuit for the transistor used as a temperature sensor is shown in Fig. 16-17B. A voltage source E and a stable resistor R is used instead of a current source. Current though the transistor is determined as

$$I = \frac{E\text{-}V}{R} \ .$$

(16.31)

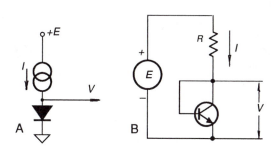

Fig. 16-17 Forward biased *pn*-junction temperature sensors
A: a diode; B: a diode-connected transistor

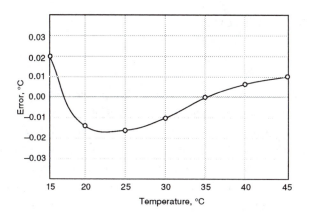

Fig. 16-18 An error curve for a silicon transistor (PN100) as a temperature sensor

It is recommended to use current on the order of $I=100\mu A$, therefore for $E=5V$ and $V\approx0.6V$, the resistor $R=(E\text{-}V)/I=44$ kΩ. When the temperature increases, voltage V drops which results in a minute increase in current I. According to Eq. (16.30), this causes some reduction in sensitivity which, in turn, is manifested as nonlinearity. However, the nonlinearity may be either small enough for a particular application, or it can be taken care of during signal processing. This makes a transistor (a diode) temperature sensor a very attractive device for many applications, due to its simplicity and low cost. Fig. 16-18 shows an error curve for the temperature sensors made with the PN100 transistor operating at 100μA. It is seen that the error is quite small and for many practical purposes no linearity correction is required.

A diode sensor can be formed in a silicon substrate in many monolithic sensors which require temperature compensation. For instance, it can be diffused into a micromachined membrane of a silicon pressure sensors to compensate for temperature dependence of piezoresistive elements.

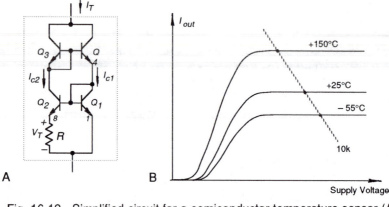

Fig. 16-19   Simplified circuit for a semiconductor temperature sensor (A) and current-to-voltage curves (B)

An inexpensive, yet precision semiconductor temperature sensor may be fabricated by using fundamental properties of transistors to produce voltage which is proportional to absolute temperature (in K). That voltage can be used directly or it can be converted into current [6]. The relationship between base-emitter voltage (V_{be}) and collector current of a bipolar transistor is the key property to produce a linear semiconductor temperature sensor. Fig. 16-19A shows a simplified circuit where Q_3 and Q_4 form the so-called current mirror. It forces two equal currents $I_{C1}=I$ and $I_{C2}=I$ into transistors Q_1 and Q_2. The collector currents are determined by resistor R. In a monolithic circuit, transistor Q_2 is actually made of several identical transistors connected in parallel, for example, 8. Therefore, the current density in Q_1 is 8 times higher than that of each of transistors Q_2. The difference between base emitter voltages of Q_1 and Q_2 is

$$\Delta V_{be} = V_{be1} - V_{be2} = \frac{kT}{q}\ln(\frac{k}{I_{ceo}}) - \frac{kT}{q}\ln(\frac{I}{I_{ceo}}) = \frac{kT}{q}\ln r \ , \qquad (16.32)$$

where r is a current ratio (equal to 8 in our example), k is the Boltzmann constant, q is the charge of an electron and T is the temperature in K. Currents I_{ceo} are the same for both transistors. As a result, a current across resistor R produces voltage $V_T = 179\mu V \cdot T$ which is independent on the collector currents. Therefore, the total current through the sensor is

$$I_T = 2\frac{V_T}{R} = \left(2\frac{k}{qR}\ln r\right)T \quad , \tag{16.33}$$

which for currents ratio $r = 8$ and resistor $R = 358\Omega$ produces a linear transfer function $I_T/T = 1\mu A/°K$.

Fig. 16-19B shows current-to-voltage curves for different temperatures. Note that the value in parenthesis of Eq. (16.33) is constant for a particular sensor design and may be precisely trimmed during the manufacturing process for a desired slope I_T/T. Current I_T may be easily converted into voltage. If, for example, a 10kΩ resistor is connected in series with the sensor, the voltage across that resistor will be a linear function of absolute temperature.

The simplified circuit of Fig. 16-19A will work according to the above equations only with perfect transistors ($\beta = \infty$). Practical monolithic sensors contain many additional components to overcome limitations of the real transistors. Several companies produce temperature sensors based on this principle. Examples are LM35 from National Semiconductors (voltage output circuit) and AD590 from Analog Devices (current output circuit).

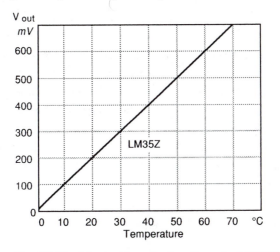

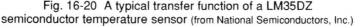

Fig. 16-20 A typical transfer function of a LM35DZ semiconductor temperature sensor (from National Semiconductors, Inc.)

Fig. 16-20shows a transfer function of a LM35Z temperature sensor which has a linear output internally trimmed for the Celsius scale with a sensitivity of 10 mV per °C. The function is quite linear where the nonlinearity error is confined within ±0.1°. The function can be modeled by

$$V_{out} = V_o + aT \quad , \tag{16.34}$$

where T is the temperature in degrees C. Ideally, V_o should be equal to zero, however part-to-part variations of its value may be as large as ±10mV which correspond to an error of 1°C. Slope a may vary between 9.9 and 10.1 mV/°C.

16.4   OPTICAL TEMPERATURE SENSORS

Sometimes, temperatures have to be measured at tough hostile environments when very strong electrical, magnetic or electromagnetic fields, or very high voltages make measurements either too susceptible to interferences, or too dangerous for the operator. One way to solve the problem is to use noncontact methods of temperature measurements, such as described in Section 13.7.

However, there are also contact optical sensors which can sense temperature and transmit information without a need for any electronic devices at the measurement site.

16.4.1 Fluoroptic sensors

These sensors rely on the ability of a special phosphor compound to give away a fluorescent signal in response to light excitation. The shape of the response pulse is a function of temperature. The decay of the response pulse is highly reproducible over a wide temperature range [7]. As a sensing material, magnesium fluoromagnetite activated with tetravalent manganese is used. This is phosphor, long known in the lighting industry as a color corrector for mercury vapor street lamps, is prepared as a powder by a solid-state reaction at approximately 1200°C. It is thermally stable, relatively inert and benign from a biological standpoint, and insensitive to damage by most chemicals or by prolonged exposure to ultraviolet (UV) radiation. It can be excited to fluoresce by either UV or blue radiation. Its fluorescent emission is in the deep red region, and the fluorescent decay is essentially exponential.

To minimize crosstalk between the excitation and emission signals, they are passed through the bandpass filters which reliably separate the related spectra (Fig. 16-21A). The pulsed excitation source, a Xenon flash lamp, can be shared among a number of optical channels in a multisensor system. The temperature measurement is made by measuring the rate of decay of the fluorescence, as shown in Fig. 16-21B. That is, a temperature is represented by a time constant τ which drops fivefold over the temperature range from -200 to +400°C. The measurement of time is usually the simplest and most precise operation that can be performed by an electronic circuit, thus, temperature can be measured with a good resolution and accuracy: about ±2°C over the range without calibration.

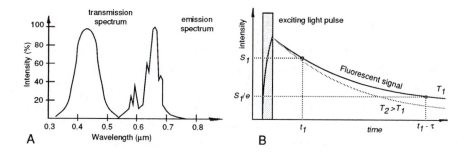

Fig. 16-21 Fluoroptic method of temperature measurement
A: spectral responses of the excitation and emission signals
B: exponential decay of the emission signal for two temperatures (T_1 and T_2);
e is the base of natural logarithms, and τ is a decay time constant
(adapted from [7])

Since the time constant is independent of excitation intensity, a variety of designs is possible. For instance, the phosphor compound can be directly coated onto the surface of interest and the optic system can take measurement without a physical contact (Fig. 16-22A). This makes possible continuous temperature monitoring without disturbing a measured site. In another design, a phosphor is coated on the tip of a pliable probe which can form a good contact area when brought in contact with the object (Fig. 16-22B and C).

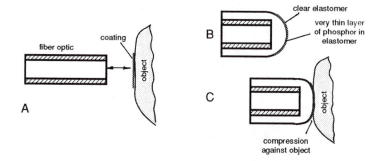

Fig. 16-22 Placement of a phosphor compound in fluoroptic method
A: on the surface of an object;
B and C: on the tip of the probe
(adapted from [7])

16.4.2 Interferometric sensors

Another method of temperature measurement is based on the modulation of light intensity by interfering two light beams. One beam is a reference, while the other's travel through a temperature sensitive medium is somewhat delayed depending on temperature. This results in a phase shift and a subsequent extinction of the interference signal. A similar principle was described in more detail in Section 9.8. For temperature measurement, a thin layer of silicon [8, 9] can be used because its refractive index changes with temperature, thus modulating a light travel distance.

Fig. 16-23 shows a schematic of a thin film optical sensor. The sensor was fabricated by sputtering of three layers onto the ends of the step-index multimode fibers with 100 μm core diameters and 140μm cladding diameters [10]. The first layer is silicone, then silicon dioxide. The FeCrAl layer on the end of the probe prevents oxidation of the underlying silicon. The fibers can be used up to 350°C, however much more expensive fibers with gold buffered coatings can be used up to 650°C. The sensor is used with LED light source operating in the range of 860 nm and a micro-optic spectrometer. A useful instrument for measurements of fiber optic interference signals is a fiber-optic refractometer. One such instrument (Model 1430) is produced by MetriCor Inc., Woodinville, WA.

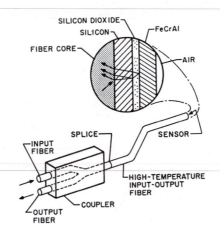

Fig. 16-23 A schematic of a thin film optical temperature sensor
(Reprinted with permission, Sensors Magazine, © 1990)

16.4.3 Thermochromic solution sensor

For biomedical applications, where electromagnetic interferences may present a problem, a temperature sensor can be fabricated with use of a thermochromic solution [11], such as cobalt chloride ($CoCl_2 \cdot 6H_2O$).

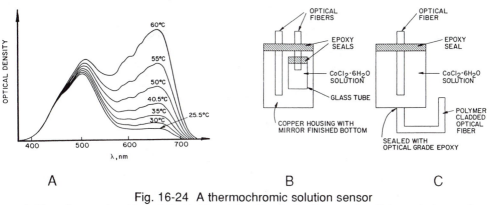

Fig. 16-24 A thermochromic solution sensor
A: Absorption spectra of the cobalt chloride solution; B: reflective fiber coupling; C: transmissive coupling
(From [11])

The operation of this sensor is based on the effect of a temperature dependence of a spectral absorption in the visible range of 400-800 nm by the thermochromic solution (Fig. 16-24A). This implies, that the sensor should consist of a light source, a detector and a cobalt chloride solution, which is thermally coupled with the object. Two possible designs are shown in Figs. 16-24B and 16-24C, where transmitting and receiving optical fibers are coupled through a cobalt chloride solution.

6.5 ACOUSTIC TEMPERATURE SENSOR

Under extreme conditions, temperature measurement may become a difficult task. These conditions include a cryogenic temperature range, high radiation levels inside nuclear reactors, etc. Another unusual condition is the temperature measurement inside a sealed enclosure with a known medium, in which no contact sensors can be inserted and the enclosure in not transmissive for the infrared radiation. Under such unusual conditions, acoustic temperature sensors may come in quite handy. An operating principle of such a sensor is based on a relationship between temperature of the medium and speed of sound. For instance, in dry air at a normal atmospheric pressure the relationship is

$$c \approx 331.5\sqrt{\frac{T}{273.15}} \text{ m/sec} ,\qquad\qquad (16.35)$$

where c is the speed of sound and T is the absolute temperature.

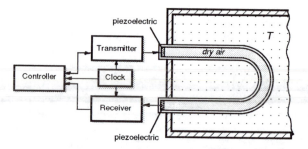

Fig. 16-25 An acoustic thermometer with an ultrasonic detection system

An acoustic temperature sensor (Fig. 16-25) is composed of three components: an ultrasonic transmitter, an ultrasonic receiver, and a gas filled hermetically sealed tube. The transmitter and receiver are ceramic piezoelectric plates which are acoustically decoupled from the tube to assure sound propagation primarily through the enclosed gas, which in most practical cases is dry air. Alternatively, the transmitting and receiving crystals may be incorporated into a sealed enclosure with a known content whose temperature has to be measured. That is, an intermediate tube in not necessarily required in cases where the internal medium, its volume and mass are held constant. When a tube is used, care should be taken to prevent its mechanical deformation and loss of hermeticity under the extreme temperature conditions. A suitable material for the tube is invar.

The clock of a low frequency (near 100 Hz) triggers the transmitter and disables the receiver. The piezoelectric crystal flexes transmitting an ultrasonic wave along the tube. The receiving crystal is enabled before the wave arrives to its surface and converts it into an electrical transient, which is amplified and sent to the control circuit. The control circuit calculates the speed of sound by determining propagation time along the tube. Then, the corresponding temperature is determined from the calibration numbers stored in a look-up table. In another design, the thermometer may contain only one ultrasonic crystal which alternatively acts either as a transmitter, or as a receiver. In that case, the tube has a sealed empty end. The ultrasonic waves are reflected from the end surface and propagate back to the crystal, which before the moment of the wave arrival is turned into a reception mode. An electronic circuit [12] converts the received pulses into a signal which corresponds to the tube temperature.

REFERENCES

1. Benedict, R. P. Fundamentals of temperature, pressure, and flow measurements. 3rd ed., John Wiley & Sons, New York, 1984

2. Callendar, H. L. On the practical measurement of temperature. *Phil. Trans. R. Soc.*, London, vol. 178, p:160, 1887

3. Keystone NTC and PTC thermistors. Catalogue © Keystone Carbon Company, St. Marys, PA, 1984

4. Caldwell, F.R. Thermocouple materials. NBS monograph 40. National Bureau of Standards. March 1, 1962

5. Manual on the use of thermocouples in temperature measurement. *ASTM Publication code number 04-470020-40*, ©ASTM, Philadelphia, 1981

6. Timko, M. P. A two terminal IC temperature transducer. IEEE Journal of Solid-State Circuits. *vol. SC-11*, pp: 784-788, Dec. 1976

7. Wickersheim, K.A. and Sun, M.H. Fluoroptic thermometry. Med. Electronics, pp: 84-91, Febr. 1987

8. Schultheis, L., Amstutz, H., and Kaufmann, M. Fiber-optic temperature sensing with ultra-thin silicon etalons. In: *Opt. Lett.* vol. 13, No. 9, pp: 782-784, 1988

9. Wolthuis, R., A., Mitchell, G.L., Saaski, E., Hartl, J.C., and Afromowitz, M.A. Development of medical pressure and temperature sensors employing optical spectral modulation. In: *IEEE Trans. on Biomed. Eng.* vol. 38, No. 10, pp: 974-981, Oct. 1991

10. Beheim, G., Fritsch, K., Azar, M.T. A sputtered thin film fiber optic temperature sensor. In: *Sensors*, pp: 37-43, Jan. 1990

11. Hao, T. and Lui, C.C. An optical fiber temperature sensor using a thermochromic solution. In: *Sensors and Actuators A*, vol. 24, pp: 213-216, 1990

12. Williams, J. Some techniques for direct digitization of transducer outputs, AN7, *Linear Technology Application Handbook*, 1990

17

Chemical Sensors

Chemical sensors are sensitive to stimuli produced by various chemical compounds or elements. The most important property of these sensors is selectivity. Another generally important property of a chemical sensor, is its very small output electrical signal. Any electrical loading on such a sensor would distort information. This commonly requires the use of high quality interface electronic devices.

17. 1 SELECTIVITY

Selectivity can be defined as the ability of a sensor to respond primarily to only one chemical element or compound (species) in the presence of other species. In most biological systems, specificity is achieved by shape recognition, which involves a comparison with some kind of a stereotype. High selectivity means that the contribution from the *primary* species dominates, and that the contribution from the *interfering* species is minimal. Therefore, one of the most important functions in the evaluation a chemical sensor performance is the qualification of its selectivity. It is common practice to evaluate the response of a sensor only for increasing the values of activity (concentration) of the primary species. This is mainly due to the fact that it is more convenient to prepare a continuously broad range of the test concentrations by adding increasing amounts of a concentrated (pure) primary species to the background sample than *vice versa*. An absolutely selective sensor really does not exist and there is always some interference present.

17.1.1 Enzyme sensors

There are several ways to achieve selectivity. One of them is by using sensors with enzymatic layers. Enzymes are a special kind of catalyst, proteins of molecular weight 6-4000 kdaltons found in living organisms. They have two remarkable properties: (1) they are extremely selective to a given substrate and (2) they are extraordinarily effective in increasing the rate of reactions. Therefore, they favorably contribute to both: the selectivity and the magnitude of the output signal. The maximum velocity of the reaction is proportional to the concentration of the enzyme. A general diagram of an enzymatic sensor is shown in Fig. 17-1 [1].

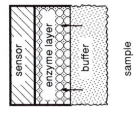

Fig. 17-1 Schematic diagram of an enzyme sensor

The sensing element can be a heated probe, an electrochemical sensor, or an optical sensor. Enzymes operate only in an aqueous environment, so they are incorporated into immobilization matrices which are gels, specifically hydrogels. The basic operating principle is as follows. An enzyme (a catalyst) is immobilized inside a layer into which the substrate diffuses. Hence, it reacts with the substrate and the product is diffused out of the layer into the sample solution. Any other species which participates in the reaction must also diffuse in and out of the layer.

17.1.2 Catalytic sensors

These sensors operate on the principle similar to thermal enzymatic sensors. Heat is liberated as a result of a catalytic reaction taking place at the surface of the sensor and the related temperature change inside the device is measured. On the other hand, the chemistry is similar to that of high temperature conductometric oxide sensors. Catalytic gas sensors have been designed specifically to detect low concentration of flammable gases in ambient air inside mines. These sensors often are called pellistors [2]. These sensors are quite simple. The platinum coil is imbedded in a pellet of ThO_2/Al_2O_3 coated with a porous catalytic metal: palladium or platinum. The coil acts as both the heater and the temperature sensor (RTD). Naturally, any other type of heating element and temperature sensor can be successfully employed. When the combustible gas reacts at the catalytic surface, the heat evolved from the reaction increases the temperature of the pellet and of the platinum coil, thus increasing its resistance. There are two possible operating modes of the sensor. One is isothermal, where an electronic circuit controls the current through the coil to maintain its temperature constant. In the nonisothermal mode the sensor is connected as a part of a Wheatstone bridge whose output voltage is a measure of the gas concentration.

17.2   THERMAL SENSORS

When the internal energy of a system changes, it is accompanied by an absorption or evolution of heat. This is called the first law of thermodynamics. Therefore, a chemical reaction which is associated with heat can be detected by an appropriate thermal sensor, such as those described in Chap. 16. These sensors operate on the basic principles which form the foundation of a microcalorimetry. An operating principle of a thermal sensor is simple: a temperature probe is coated with a chemically selective layer. Upon the introduction of a sample, the probe measures release of heat during the reaction between the sample and the coating.

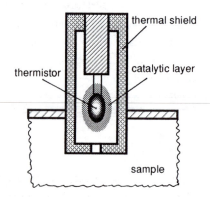

Fig. 17-2   Schematic diagram of a chemical thermal sensor

A simplified drawing of such a sensor is shown in Fig. 17-2. It contains a thermal shield to reduce heat loss to the environment and a catalytic layer coated thermistor. The layer may be an enzyme immobilized into a matrix. An example of such a sensor is the enzyme thermistor using an immobilized oxidize (GOD). The enzymes are immobilized on the tip of the thermistor, which is then enclosed in a glass jacket in order to reduce heat loss to the surrounding solution. Another similar sensor with similarly immobilized bovine serum albumin is used as a reference. Both thermistors are connected as the arms of a Wheatstone bridge [1]. Temperature increase as a result of a chemical reaction is proportional to the incremental change in the enthalpy dH

$$dT = -dH/C_p \quad ,$$ (17.1)

where C_p is the heat capacity.

The chemical reaction in the coating is

$$\beta\text{-D-glucose} + H_2O + O_2 \xrightarrow{\text{GOD}} H_2O + \text{D-gluconic acid}, \quad \Delta H_1 , \qquad (17.2)$$

and

$$H_2O_2 \xrightarrow{\text{catalase}} 1/2O_2 + H_2O, \quad \Delta H_2 , \qquad (17.3)$$

where ΔH_1 and ΔH_2 are partial enthalpies, the sum of which for the above reaction is approximately -80 kJ/mol. The sensor responds linearly with the dynamic range depending on the concentration of hydrogen peroxide (H_2O_2).

17.3 ELECTROCHEMICAL SENSORS

The electrochemical sensors are the most versatile and better developed than any other chemical sensors. Depending on the operating mode, they are divided into sensors which measure voltage (potentiometric), those which measure electric current (amperometric), and those which rely on the measurement of conductivity or resistivity (conductometric). In all these methods, special electrodes are used, where either a chemical reaction takes place, or the charge transport is modulated by the reaction. A fundamental rule of an electrochemical sensor is that it always requires a closed circuit, that is, an electric current (either dc, or ac) must be able to flow in order to make a measurement. Since electric current flow essentially requires a closed loop, the sensor needs at least two electrodes, one of which often is called a *return electrode*. It should be noted, that even if in the potentiometric sensors no flow of current is required for measurement, the loop still must be closed for the measurement of voltage.

In the electrochemical sensors, voltage, current, resistance, or capacitance are generally measured. Often, the sensor is called an electrochemical cell, and how the cell is used, mainly depends on the sensitivity, selectivity, and accuracy.

17.3.1 Potentiometric sensors

These sensors use the effect of the concentration on the equilibrium of the redox reactions occurring at the electrode-electrolyte interface in a electrochemical cell. An electrical potential may develop at this interface due to the redox reaction which takes place at the electrode surface, where Ox denotes the oxidant, and Red the reduced product [3]

$$Ox + Ze = Red \ . \tag{17.4}$$

This reaction occurs at one of the electrodes (cathodic reaction in this case) and is called a half-cell reaction. Under thermodynamical quasi-equilibrium conditions, the Nernst equation is applicable and can be expressed as

$$E = E_0 + \frac{RT}{nF} \ln \frac{C_0^*}{C_R^*} \ , \tag{17.5}$$

where C_o^* and C_R^* are concentrations of Ox and Red, respectively, n is the number of electrons transferred, F is the Faraday constant, R is the gas constant, T is the absolute temperature, and E_o is the electrode potential at a standard state. In a potentiometric sensor, two half-cell reactions will take place simultaneously at each electrode. However, only one of the reactions should involve the sensing species of interest, while the other half-cell reaction is preferably reversible, noninterfering, and known.

The measurement of the cell potential of a potentiometric sensor should be made under zero-current or quasi-equilibrium conditions, thus a very high input impedance amplifier (which is called an electrometer) is generally required. There are two types of electrochemical interface from the viewpoint of the charge transfer: ideally polarized (purely capacitive) and nonpolarized. Some metals (e.g., Hg, Au, Pt) in contact with solutions containing only inert electrolyte (e.g., H_2SO_4) approach the behavior of the ideally polarized interface. Nevertheless, even in those cases a finite charge-transfer resistance exists at such an interface and excess charge leaks across with the time constant given by the product of the double-layer capacitance and the charge-transfer resistance ($\tau = R_{ct}C_{dl}$).

An ion-selective membrane is the key component of all potentiometric ion sensors. It establishes the reference with which the sensor responds to the ion of interest in the presence of various other ionic components in the sample. An ion-selective membrane forms a nonpolarized interface with the solution. A well-behaved membrane, i.e., one which is stable, reproducible, immune to adsorption and stirring effects, and also selective, has both high absolute and relative exchange-current density.

17.3.2 CHEMFET Sensors

Chemical potentiometric sensors employing the field-effect transistors (FET) are known as CHEMFETs. They become more and more popular in the areas where small size and low power consumption is essential. Examples are biological and medical monitoring. The CHEMFETs are solid-state sensors suitable for batch fabrication. The surface field effect is a desirable mechanism for a generating potential that provides high chemical selectivity and sensitivity. The CHEMFET is essentially an extended gate field-effect transistor with the electrochemical potential inserted between the non-metal gate surface of the transistor, and the reference electrode. Generally, there are four types of CHEMFET sensors: the ion selective, the gas-selective, the enzyme-selective, and the immuno-selective sensors. Fig. 17-3A shows an ion selective CHEMFET with a silicon nitrate gate for measuring pH. This sensor is essentially a metaloinsulator semiconductor FET that is given a pH sensitivity by exposing the bare silicon nitrate gate insulator to the sample solution.

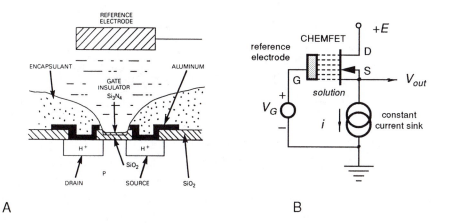

A   B

Fig. 17-3 Structure of an ion selective CHEMFET for monitoring pH (A) and an electrical connection in a source-follower mode (B)

The operating principle of the inorganic oxide CHEMFET is based on the site binding model with surface association and dissociation of charge species as shown. As the ionic concentration varies, the surface charge density at the CHEMFET gate sensing area changes as well. The ionic selectivity is determined by the surface complexation of the gate insulator. The selectivity of the sensor can be obtained by varying the composition of the gate insulator. In addition, conventional ion-selective membranes can be deposited on the top of the gate insulator to provide a large selection of different chemical sensors.

A change in the surface charge density affects the CHEMFET channel conductance which can be measured as a variation in the drain current (Fig. 17-3B). Thus, if a bias voltage E is applied to the drain and the source of the CHEMFET, the resulting current i controlled by the electrochemical potential is, in turn, a function of the concentration of activities of various electrolytes in the solution. Drain current variations as a response to the concentration of electrolytes can be treated as a conventional output voltage V_{out}. It is seen, that the CHEMFET sensor becomes essentially a three-terminal device. It consist of a drain (D), source (S), and the reference electrode (G).

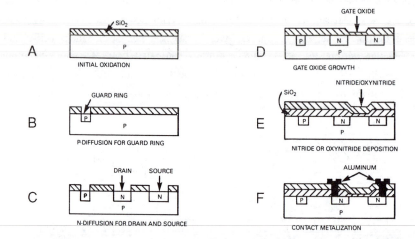

Fig. 17-4 Major steps in the fabrication of a CHEMFET sensor

The manufacturing process of the CHEMFET is similar to that of a standard MOSFET technology. Fig. 17-4 shows the major steps of the fabrication process of a pH CHEMFET. The process begins with a p-type <100> oriented silicon wafer. After the initial oxidation (A), a p^+ guard ring is formed to enclose the drain region and to minimize the leakage between the drain and the source electrodes due to surface inversion of the p-type substrate (B). In step C the drain and the source electrodes are formed by a phosphorus diffusion process. An approximately 600 Å thick dioxide layer is then formed by the thermal oxidation of the silicon substrate (D). A thin layer of silicon nitrate is deposited using chemical vapor deposition technique (E), and the final step (F) involves contact metallization on the drain and source electrodes.

Since the gate area of the CHEMFET sensor is in contact with an electrolyte, the packaging of the device must utilize an encapsulation where the sensitive region is exposed to the solution. Great care must be taken to insulate all lead wires and any conductive components from being exposed to the solution. This is one of the difficulties for implementing this technology for long term exposure to electrolytes. To illustrate how the sensor can be used, Fig. 17-5 shows a rendering of a medical catheter tip which is fabricated of silicone rubber (Dow Corning 3410 RTV). The sensing opening exposes the silver/silver-chloride reference electrode and the gate window.

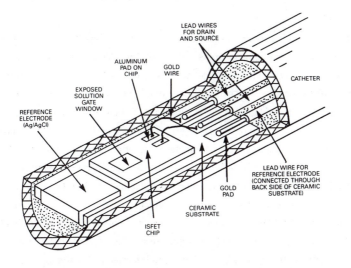

Fig. 17-5 Catheter tip with a CHEMFET pH sensor for medical applications

17.3.3 Conductometric sensors

An electrochemical conductivity sensor measures the change in conductivity of the electrolyte in an electrochemical cell. An electrochemical sensor may involve a capacitive impedance resulting from the polarization of the electrodes and faradic or charge transfer process.

In a homogeneous electrolytic solution, the conductance of the electrolyte $G(\Omega^{-1})$, is inversely proportional to L, which is the segment of the solution along the electrical field and directly proportional to A, which is the cross-sectional area perpendicular to the electric field [3]

$$G = \frac{\rho A}{L} , \tag{17.6}$$

where ρ ($\Omega^{-1}cm^{-1}$) is the specific conductivity of the electrolyte and is related quantitatively to the concentration and the magnitude of the charges of the ionic species. According to Kohlarausch [6], the equivalent conductance of the solution at any concentration, C in mol/l or any convenient units, is given by

$$\Lambda = \Lambda_0 - \beta C^{0.5} , \qquad (17.7)$$

where β is a characteristic of the electrolyte and Λ_0 is the equivalent conductance of the electrolyte at an infinite dilution.

Measurement techniques of electrolytic conductance by an electrochemical conductivity sensor has remained basically the same over the years. Usually, a Wheatstone bridge is used with the electrochemical cell (the sensor) forming one of the resistance arms of the bridge. However, unlike the measurement of the conductivity of a solid, the conductivity measurement of an electrolyte is often complicated by the polarization of the electrodes at the operating voltage. A faradic or charge transfer process occurs at the electrode surfaces. Therefore, a conductivity sensor should be operated at a voltage where no faradic process could occur. Another important consideration is the formation of a double layer adjacent to each of the electrodes when a potential is imposed on the cell. This is described by the so called Warburg impedance. Hence, even in the absence of the faradic process, it is essential to take into consideration the effect of the double layers during measurement of the conductance. The effect of the faradic process can be minimized by maintaining the high cell constant L/A of the sensor so that the cell resistance lies in the region between 1 and 50 kΩ. This implies using a small electrode surface area and large interelectrode distance. This, however, reduces the sensitivity of the Wheatstone bridge. Often the solution is in use of a multiple electrode configuration. Both effects of the double layers and the faradic process can be minimized by using a high frequency low amplitude alternating current. Another good technique would be to balance both the capacitance and the resistance of the cell by connecting a variable capacitor in parallel to the resistance of the bridge area adjacent to the cell.

17.3.4 Amperometric sensors

An example of an amperometric chemical sensor is a Clark oxygen sensor which was proposed in 1956 [4]. The operating principle of the electrode is based on the use of electrolyte solution contained within the electrode assembly to transport oxygen from an oxygen-permeable membrane to the metal cathode. The cathode current arises from a two-step oxygen-reduction process that may be represented as

$$O_2 + 2H_2O + 2e^- \rightarrow H_2O_2 + 2OH^-$$

$$H_2O_2 + 2e^- \rightarrow 2OH^- . \qquad (17.8)$$

Fig. 17-6A shows the membrane which is stretched across the electrode tip, allowing oxygen to diffuse through a thin electrolyte layer to the cathode. Both anode and cathode are contained within the sensor assembly, and no electrical contact is made with the outside sample. A first-order diffusion model of the Clark electrode is illustrated in Fig. 17-6B [5]. The membrane-electrolyte-electrode system is considered to act as a one dimensional diffusion system with the partial pressure at the membrane surface equal to the equilibrium partial pressure p_o and that at the cathode equal to zero. It can be shown that the steady steady-state electrode current is given by

$$I \approx \frac{4FA\alpha_m D_m p_o}{x_m} \; , \qquad (17.9)$$

where A is the electrode area, α_m is the solubility of oxygen in the membrane, F is the Faraday's constant, D_m is the diffusion constant, and x_m is the thickness of the membrane. It should be noted that the current is independent on the electrolyte thickness and diffusion properties. A Teflon® membrane is used as an oxygen permeable film. We may define the sensor's sensitivity as a ratio of the current to the oxygen partial pressure

$$S = \frac{I}{p_o} \; . \qquad (17.10)$$

For example, if the membrane is 25μm thick, and the cathode area is $2 \cdot 10^{-6}$ cm^2, then the sensitivity is approximately 10^{-12}A/mmHg.

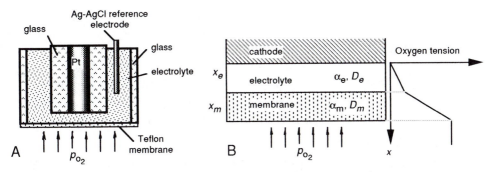

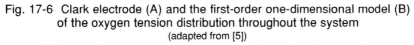

Fig. 17-6 Clark electrode (A) and the first-order one-dimensional model (B)
of the oxygen tension distribution throughout the system
(adapted from [5])

An enzymatic type amperometric sensor can be built with a sensor capable of measuring the relative oxygen deficiency caused by the enzymatic reaction by using two Clark oxygen electrodes. The operating principle of the sensor is shown in Fig. 17-7. The sensor consists of two identical oxygen electrodes, where one (A) is coated with an active oxidize layer, while the other (B) with an inactive enzyme layer. An example of the application is a glucose sensor, where inactivation can be carried out either chemically, or by radiation, or thermally. The sensor is encapsulated into a plastic carrier with glass coaxial tubes supporting two Pt cathodes and one Ag anode. In the absence of the enzyme reaction, the flux of oxygen to these electrodes, and therefore the diffusion limiting currents, are approximately equal to one another. When glucose is present in the solution and the enzymatic reaction takes place, the amount of oxygen reaching the surface of the active electrode is reduced by the amount consumed by the enzymatic reaction, which results in a current imbalance.

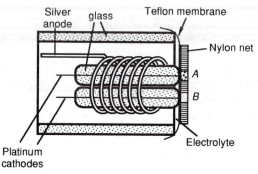

Fig. 17-7   Simplified schematic of an amperometric Clark oxygen sensor
adapted for detecting glucose

17.4   CONCENTRATION SENSORS

These sensors react to the concentration of a specific chemical material. The concentration of the material may alter (modulate) some physical property of the sensor, for instance, its resistance or capacitance. Generally speaking, no chemical reaction takes place in the sensor, thus these sensors are called *physical* sensors.

To detect the presence of a liquid, a sensor usually must be specific to that particular agent at a certain concentration. That is, it should be selective to the liquid's physical and/or chemical properties. An example of such a sensor is a resistive detector of hydrocarbon fuel leaks (originally devised in Bell Communication Research to protect buried telephone cables). A detector is made of silicone and carbon black composite. The polymer matrix serves as the sensing element and the conductive filler is used to achieve a relatively low volume resistivity, on the order of $10\Omega\cdot$cm in the initial stand-by state. The composition is selectively sensitive to the presence of a solvent with a large solvent-polymer interaction coefficient [7]. Since the sensor is not susceptible to polar solvents such as water or alcohol, it is compatible with the underground environment. The sensor is fabricated in the form of a thin film with a very large surface/thickness ratio. Whenever the solvent is applied to the film sensor, the polymer matrix swells resulting in the separation between conductive particles. This causes a conversion of the composite film from being a conductor to becoming an isolator with a resistivity on the order $10^9 \ \Omega \cdot$cm, or even higher. The response time for a film sensor is less than 1 second. The sensor returns to its normally conductive state when it is no longer in contact with the hydrocarbon fuel, making the device reusable.

Another example of a concentration sensor is the air cleanliness sensor, where a semiconductive material is deposited on a tubular ceramic substrate. The sensor needs to be heated above ambient temperature. A semiconductive material is a purely resistive compound, whose resistance drops in the presence of indoor odor gases such as methyl mercaption (CH_3SH), ethyl alcohol (C_2H_5OH), cigarette smoke, etc. Fig. 17-8A shows a sensor's circuit and Fig. 17-8B is its sensitivity characteristics with respect to several pollutants. The output of the sensor is voltage V developed across the ballast resistor R.

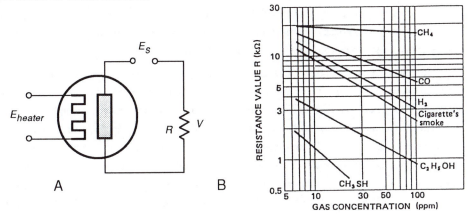

Fig. 17-8 Air cleanliness sensor
A: schematic and B: sensitivity characteristics
(Courtesy of Nippon Ceramic Co., Ltd.)

17.5   OSCILLATING (MASS) SENSORS

An idea behind the oscillating sensor is the shift in the resonant frequency of a piezoelectric crystal when an additional mass is deposited on its surface. A piezoelectric quartz oscillator resonates with a frequency which, depending on the circuit is called either a series (f_r) or a parallel

(f_{ar}) resonant (see Fig. 5-9.1B). Either frequency is a function of the crystal mass and shape. Molecules of a chemical compound deposit on the surface of the crystal increasing its mass, and subsequently lowering its resonant frequency. An electronic circuit measures the frequency shift, which is almost a linear measure of the chemical concentration in the sampled gas. Thus, this method is called a *microgravimetric* technique, as added mass is extremely small. The absolute accuracy of the method depends on such factors as the mechanical clamping of the crystal, temperature, etc., therefore, the over-the-range calibration is usually required. The oscillating sensors are extremely sensitive. For instance, a typical sensitivity is on the range of 5MHz·cm^2/kg, which means that 1Hz in frequency shift corresponds to about 17 ng/cm^2 added weight. The dynamic range is quite broad: up to 20μg/cm^2. To assure a selectivity, a crystal is coated with a chemical layer specific for the material of interest.

One of the possible applications for the technique is the monitoring of heterogeneous samples, such as aerosols and suspensions. The mass increase due to impacting and sticking particles (liquid-aerosol or solid-suspension) produce a strong frequency shift, however, it is also sensitive to a particle size, which means that it can be used either to detect the sizes of the particles, or to monitor samples with constant particle dimensions. To improve the "stickiness" of the crystal, it can be either treated chemically, or an electrostatic effect can be used.

Another type of oscillating detector is a surface acoustic wave (SAW) sensor. The SAW is a phenomenon of propagating mechanical waves along a solid surface which is in a contact with a medium of lower density, such as air [8]. These waves are sometimes called Reyleigh waves after the man who predicted them in 1885. A surface acoustic wave sensor is a transmission line with three essential components: the piezoelectric transmitter, the transmission line with a chemically selective layer, and the piezoelectric receiver. An electrical oscillator causes the electrodes of the transmitter to flex the substrate, thus producing a mechanical wave. The wave propagates along the transmission surface toward the receiver. The substrate is usually fabricated of $LiNbO_3$ with a high piezoelectric coefficient. The transmission line doesn't have to be piezoelectric, which opens several possibilities of designing the sensor of different materials. The transmission surface interacts with the sample according to the selectivity of the coating, thus modulating the propagating waves. The waves are received at the other end, and converted back to an electric form. Usually, there is another reference sensor whose signal is subtracted from the test sensor's output.

17.6   OPTICAL SENSORS

These sensors are based on the interaction of electromagnetic radiation with matter, which results in altering some properties of the radiation. Examples of such modulations are variations in intensity, polarization, and velocity of light in the medium. Optical modulation is studied by spectroscopy which provides information on various microscopic structures from atoms to the dynamics in polymers. In a general arrangement, the monochromatic radiation passes through a sample (which may be gas, liquid, or solid), and its properties are examined at the output. Alternatively, the sample may respond with a secondary radiation which is also measured.

Optical chemical sensors can be and are designed and built in a great variety of ways, which are limited only by the designer's imagination. Here, we will describe only one device just to illustrate how an optical sensor works. For the illustration, Fig. 17-9 shows a simplified configuration of a CO_2 sensor [9]. It consists of two chambers which are illuminated by a common light emitting diode (LED). Each chamber has metallized surfaces for better internal reflectivity. The left chamber has slots covered with a gas permeable membrane. The slots allow CO_2 to diffuse into the chamber. The bottom parts of the chambers are made of glass. Both wafers, A and B, form optical waveguides. The test chamber is filled with a reagent, while the reference chamber is not. The sample part of the sensor monitors the optical absorbency of a pH indicator in a dilute solution, where the optical absorbency changes in accordance with the Beer-Lambert law

$$I = I_o \exp\left[-a(\lambda,\text{pH})Cd\right] , \tag{17.11}$$

where I is the transmitted intensity, I_o is the source intensity, a is the molar absorptivity, λ is the wavelength, C is the concentration, and d is the optical path length.

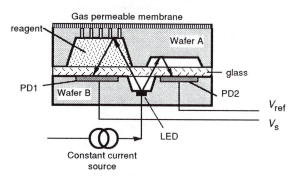

Fig. 17-9 Simplified configuration of an optical CO_2 sensor

Ambient CO_2 equilibrates with the bicarbonate ion buffer system in the reagent, as it is done in the traditional Severinghaus-Stow CO_2 electrode. Equilibrium between CO_2, H_2CO_3, and

HCO_3 produces a change in the pH of the solution. The solution contains a pH-indicator *Chlorophenol Red*, which exhibits a sharp, nearly linear change in the optical absorbency at 560nm from pH5 to pH7. The buffer concentration can be selected to exhibit pH changes in the range for partial CO_2 pressures from 0 to 140Torr. Since buffer pH varies linearly with the log of the partial pressure of carbon dioxide (pCO_2), changes in optical absorbency can also be expected to vary linearly with the log of pCO_2.

The LED common for both halves of the sensor transmits light through the pH sensitive sample to a test photodiode (PD1). The second photodiode (PD2) is for the reference purposes to negate variations in the light intensity of the LED. For temperature stability, the sensor should operate in a thermally stable environment.

REFERENCES

1. Janata, J. Principles of chemical sensors. Plenum Press, New Yourk, 1989
2. Gentry, S. J. Catalytic devices. In: *Chemical sensors*. Edmonds, T. E., ed. Chapman and Hall, New York, 1988
3. Tan, T.C. and Liu, C.C. Principles and fabrication materials of electrochemical sensors. In: *Chemical sensor technology*. vol. 3. © by Kodansha Ltd., 1991
4. Clark, L.C. Monitor and control of blood and tissue oxygen tension. In: *Trans. Am. Soc. Artif. Internal Organs*, vol. 2, p: 41-46, 1956
5. Cobbold, R.S.C. Transducers for biomedical measurements. John Wiley & Sons, 1974
6. Browning, D.R., ed. Electrometric methods. McGraw Hill, New York, 1969
7. Hydrocarbon fuel, HCl sensor look for trouble. *Sensors*, pp: 11-12, Febr. 1991
8. Ristic, V.M. Principles of acoustic devices. Wiley and Sons, New York, 1983
9. Morgan, C.H. and Cheung, P.W. An integrated optoelectronic CO_2 gas sensor. In: *Transducers'91. International conference on solid-state sensors and actuators. Digest of technical papers*, pp: 343-346, ©IEEE, 1991

INDEX

4.